THE INTERNATIONAL SERIE
MONOGRAPHS ON CHEMIS

THE INTERNATIONAL SERIES OF MONOGRAPHS ON CHEMISTRY

Chemical Oscillations and Instabilities

Non-linear Chemical Kinetics

Peter Gray

and

Stephen K. Scott

CLARENDON PRESS · OXFORD

1994

Oxford University Press, Walton Street, Oxford OX2 6DP
Oxford   New York   Toronto
Delhi   Bombay   Calcutta   Madras   Karachi
Kuala Lumpur   Singapore   Hong Kong   Tokyo
Nairobi   Dar es Salaam   Cape Town
Melbourne   Auckland   Madrid
and associated companies in
Berlin   Ibadan

Oxford is a trade mark of Oxford University Press

Published in the United States
by Oxford University Press Inc., New York

British Library Cataloguing in Publication Data
Gray, Peter
Chemical oscillations and instabilities.
1. Chemical reactions. Oscillations
I. Title   II. Scott, Stephen   III. Series
541.3'9

Library of Congress Cataloging in Publication Data
Gray, Peter, 1933–
Chemical oscillations and instabilities: non-linear chemical
kinetics/Peter Gray and Stephen K. Scott.
p. cm.—(The International series of monographs on
chemistry; 21)
Includes bibliographical references.
1. Chemical reaction, Rate of.   2. Nonlinear theories.   I. Scott,
Stephen K.   II. Title.   III. Series.
QD502.G73   1990   541.3'94—dc20   89-23870
ISBN 0 19 855864 3

Printed in Great Britain
by Bookcraft Ltd., Midsomer Norton, Bath

To
Hilary and Barbara

for their patience and understanding

PREFACE

In the spirit of Molière's *bourgeois gentilhomme*, chemical kineticists have been studying non-linear reactions almost all of their lives, often without realizing it. The only systems which are not 'non-linear' are the linear ones. In the present context, that means only first-order processes escape the non-linear net, and even these get caught if there is the slightest departure from isothermal operation.

Even oscillations, the best known of the recognizably non-linear phenomena which we discuss here, have had isolated reports in the chemical literature since at least 1834: cool flames, which appear in chapter 15, were known to and studied by Sir Humphry Davy. The 'modern era', however, can be dated from the mid 1960s. Belousov's first report of his oscillating reaction in 1958 was in such an obscure journal that it failed to catch attention, until Zhabotinskii rediscovered it. The latter's studies were reported in 1964, at almost exactly the same time as a series of important theoretical advances were being made. The combined experimental and theoretical onslaught on the disbelievers, hiding behind a misappreciation of the second law of thermodynamics, has now triumphed and seen an explosion of interest in all countries. In a wider context, chemical engineers made an earlier start, particularly through their attention to non-isothermal reactor engineering. Fortunately, the regular series of Gordon conferences and other workshops attracts non-linear scientists from a wide variety of disciplines, so we progress together rather than in isolation and by reinvention.

The present text is aimed partly at chemists, to encourage more of us to be less afraid of the mathematics, but also with the hope of catching the attention of mathematicians, engineers, etc., who should find that the chemical world offers exemplary non-linearities which can be realized in practice as well as *in numero*. The book is loosely divided into two unequal parts. The main thrust is to introduce a variety of the techniques used by the non-linear kineticist, with special attention to particularly simple model schemes. This part encompasses chapters 2 to 13. Other readers may prefer to begin with 'real' experimental results, and these are presented in chapters 14 and 15.

It is a pleasure to have an opportunity to thank the many colleagues who have helped us in our struggles with this subject. The following list is almost certainly incomplete, and like all other omissions and inaccuracies in the text we must accept full responsibility for this. There is also no particular significance to the order of names, with one exception. Both of us wish to give special mention to Dr John F. Griffiths as friend as well as coworker. Our appreciation is also due to: Professors Rutherford Aris, Brian Gray, Ken Showalter, Drs John Merkin, Terry Boddington, Dave Needham, David

Knapp, John Brindley, Steven Kay, Christian Kaas-Peterson, Malcolm Roberts, Professors Richard Noyes, Richard Field, Jim Murray, John Tyson, Dan Luss, Jack Hudson, Gregoire Nicolis, John Ross, Lanny Schmidt, Kedma Bar-Eli, Nils Jaeger, Fred Schneider, Uhlrich Franck, Pier–Giorgio Lignola, Graeme Wake, Christian Vidal, Chang-Gen Feng, and Irv Epstein, Drs Patrick De Kepper, Jacques Boissonade, and Peter Plath.

We have both been rather taken aback by the sheer magnitude of work involved in bringing this project to completion. It has required an obsessiveness which could seriously have damaged home lives. The fact it has not done so bears witness to the understanding of those to whom we gratefully dedicate this book.

Leeds and Cambridge   P.G.
March 1989   S.K.S.

CONTENTS

Part 2 Experiments

INTRODUCTION

This is a book about chemical kinetics—not necessarily the most familiar aspects of that subject, but nevertheless the various phenomena to be described arise primarily because reactions occur at finite rates, and different reactions may occur at different rates. Before proceeding along our kinetics course, however, it is worth while examining what information we can gain from thermodynamics. For most of us, the familiar aspects of thermodynamics are those dealing with systems at chemical equilibrium. Then we can 'use' concepts such as enthalpy and entropy to place strong restrictions on the final equilibrium composition attained from a given set of initial reactant concentrations.

Because of our familiarity with the concept of equilibrium, there is a tendency to regard this as the 'normal' situation and somewhat unremarkable. In fact, the equilibrium state is really a very special one. It is a dynamic state; individual molecules and atoms are in continual processes of bond breaking and bond formation, and so are redistributed between reactants, products and intermediates. At the equilibrium state, the net rates at which each of the different participating species are being formed all become zero together. Moreover, we can look not just at the overall processes, but also at the various elementary steps within the whole kinetic mechanism. At equilibrium, all of these must be perfectly balanced, with the forward rate exactly equal to the reverse rate, simultaneously. It is because of these stringent requirements that chemical reactions at equilibrium are very much constrained to show relatively simple behaviour. Chemical systems are not as free as mechanical ones; unlike a pendulum or a mass on a spring, they cannot approach their resting state in a damped oscillatory manner—isolated chemical systems finally attain their equilibrium composition (maximum entropy) through a monotonic approach. There are no oscillations, damped or undamped, about chemical equilibrium.

These rigorous constraints apply to all reactions in closed vessels (those which do not exchange material with the surroundings). When we recognize the special nature of the chemical equilibrium state, it is perhaps not so surprising that if a reaction can be studied under other circumstances, away from these conditions, a much wider range of behaviour becomes possible.

The restrictions which apply at equilibrium can also be extended a small way to cover the behaviour 'near to' equilibrium, i.e. at long but not infinitely long times in closed reactors. Even so these rules do not mean that closed systems are incapable at all of showing interesting time-dependent behaviour. The requirements of equilibrium cannot give any guide to the possible behaviour during early stages of reaction if the system starts from a state 'far' from its equilibrium composition, save that the total free energy must decrease monotonically. It is these early stages, particularly of oscillatory reactions, in which we are most interested here. When we turn to open systems, where there may be a continual inflow of fresh reactants, classical thermodynamics makes little restriction and gives little guide to the range of responses which may be shown.

One particular pattern of behaviour which can be shown by systems far from equilibrium and with which we will be much concerned is that of oscillations. Some preliminary comments about the thermodynamics of oscillatory processes can be made and are particularly important. In closed systems, the only concentrations which vary in an oscillatory way are those of the intermediates: there is generally a monotonic decrease in reactant concentrations and a monotonic, but not necessarily smooth, increase in those of the products. The free energy even of oscillatory systems decreases continuously during the course of the reaction: ΔG does not oscillate. Nor are there specific individual reactions which proceed forwards at some stages and backwards at others: in fact our simplest models will comprise reactions in which the reverse reactions are neglected completely.

1.1. Observed phenomena: oscillations, ignitions, etc.

A classic example of behaviour 'far from equilibrium' in closed systems is that of long-lived oscillations. The Belousov–Zhabotinskii (B–Z) reaction comprises the reduction of bromate ion BrO_3^- by bromide Br^- to bromine Br_2 in the presence of a metal redox catalyst (usually Ce^{3+}/Ce^{4+} or Fe^{2+}/Fe^{3+}). The system also contains an organic substrate, e.g. malonic or citric acid, capable of reacting with the intermediate product Br_2 and then being subsequently oxidized to return Br^-. Under quite a wide range of experimental conditions the reaction shows a short induction period before entering a period of oscillatory behaviour. The oscillations are characterized by a switching in the bromide ion concentration and in the colour of the mixture as shown in Fig. 1.1. The excursions have a period of about 2 min and the train of oscillations may continue for 3 or more hours with more than 100 switches in colour. Oscillatory behaviour must, however, eventually disappear before the reaction is complete and is followed by the required final decay to equilibrium over a much longer timescale. In a strict mathematical sense the oscillations are only a transient phenomenon, but there is clearly a difference between the long-lived excursions which appear undamped for much of their existence and a simple damped oscillatory settling-down, e.g. of a pendulum to its rest state.

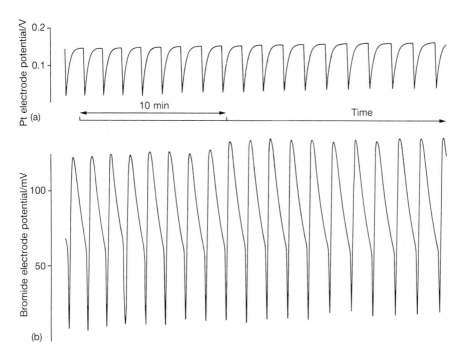

FIG. 1.1. Typical experimental records of oscillatory behaviour in the Belousov–Zhabotinskii reaction: (a) platinum electrode which responds primarily to the Ce^{3+}/Ce^{4+} couple; (b) bromide-sensitive electrode measuring $\ln[Br^-]$.

We are thus, in many instances, more interested in the transient behaviour 'early' in a reaction than we are in the more easily studied final or equilibrium state. With this in mind, we shall be concerned in our early chapters with simple models of chemical reaction that can satisfy all thermodynamic requirements and yet still show oscillatory behaviour of the kind described above in a well-stirred closed system under isothermal or non-isothermal conditions.

In open, or flow, reactors chemical equilibrium need never be approached. The reaction is kept away from that state by the continuous inflow of fresh reactants and a matching outflow of product/reactant mixture. The reaction achieves a 'stationary state', where the rates at which all the participating species are being produced are exactly matched by their net inflow or outflow. This stationary-state composition will depend on the reaction rate constants, the inflow concentrations of all the species, and the average time a molecule spends in the reactor—the mean residence time or its inverse, the flow rate. Any oscillatory behaviour may now, under appropriate operating conditions, be sustained indefinitely, becoming a stable response even in the strictest mathematical sense.

An additional phenomenon is that of multistability. Here, for the same set of rate constants, inflow concentrations, and flow rate, the reaction has more than one stationary-state composition from which to choose. As shown in Fig. 1.2, at the lowest flow rates there will be a single stationary state open to the system. The composition of this will be close to that of the chemical equilibrium mixture. At very high flow rates the molecules do not, on average, spend long in the reactor, so little reaction is possible. There is again only a single stationary state to choose for any of these high flow rates. In Fig. 1.2(a), the system shows a smooth variation in stationary-state composition over the whole range of flow rates. This is known as monostability. In Fig. 1.2(b), however, the stationary-state locus folds back on itself in the middle of the range. Between the two fold (or turning) points, the system has three stationary-state compositions above each flow rate. Actually, only two of these are accessible, as we explain later, and so experimentalists have often termed this type of response 'bistability'.

If we consider an experiment in which the system is set up at low flow rates, the system will settle to the lower branch (sometimes called the thermodynamic branch because it approaches the thermodynamic equilibrium state at the lowest flow rates). If the flow rate is now increased, the system moves along this branch, through the region of multistability, until it reaches the turning point. Beyond the fold, the system must jump suddenly to the

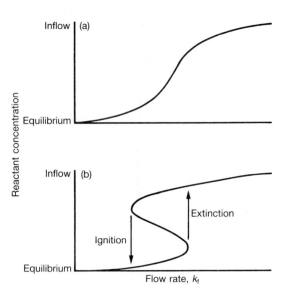

FIG. 1.2. Two of the possible dependences of stationary-state concentration of reactant on flow rate k_f for a well-stirred reactor: (a) monotonic variation (monostability); (b) multistability, with ignition and extinction.

upper branch ('flow branch'). This jump is known variously as extinction or washout and the corresponding downwards jump, at the other fold point, is called ignition.

Conditions at which there are qualitative changes in behaviour, such as the jumping between different branches or the onset of oscillations, as the flow rate is varied for instance, are known as 'bifurcations', with various subclassifications into different types. We will be very much concerned with understanding and learning how to predict these different bifurcation phenomena and whether the different types can be related.

1.2. Feedback: isothermal autocatalysis and non-isothermal self-heating

A vital constituent of any chemical process that is going to show oscillations or other bifurcations is that of 'feedback'. Some intermediate or product of the chemistry must be able to influence the rate of earlier steps. This may be a positive 'catalytic process', where the feedback species enhances the rate, or an 'inhibition' through which the reaction is poisoned. This effect may be chemical, arising from the mechanistic involvement of species such as radicals, or thermal, arising because chemical heat released is not lost perfectly efficiently and the consequent temperature rise influences some reaction rate constants. The latter is relatively familiar; most chemists are aware of the strong temperature dependence of rate constants through, e.g. the Arrhenius law,

$$k(T) = A \exp(-E/RT)$$

where A is a (temperature-independent) pre-exponential factor and E is the activation energy.

Chemical feedback may work, for example, by chain branching or 'autocatalysis'. The mechanism by which hydrogen and oxygen react spontaneously involves a cycle of three elementary steps:

$$H + O_2 \rightarrow OH + O \qquad (1.1)$$

$$OH + H_2 \rightarrow H_2O + H \qquad (1.2)$$

$$O + H_2 \rightarrow OH + H. \qquad (1.3)$$

The first and third of these steps are chain-branching processes as there is a net increase in the number of radicals in each of these. If one H atom enters this cycle, and all OH and O radicals produced react to give back H, then three are produced, i.e. taking $(1.1) + 2 \times (1.2) + (1.3)$ we have stoichiometrically

$$\underline{H} + 3H_2 + O_2 \rightarrow 3\underline{H} + 2H_2O. \qquad (1.4)$$

As reaction (1.1) is the rate-determining step in this cycle, the rate at which the

H atom concentration increases is directly proportional to itself, i.e.

$$d[H]/dt = + 2k_1[H][O_2] \tag{1.5}$$

where the $+$ sign emphasizes that there is positive feedback through this cycle.

A similar 'autocatalysis' occurs in the Belousov–Zhabotinskii reaction, through the species $HBrO_2$. At low concentrations of Br^-, $HBrO_2$ reacts with bromate to produce two molecules of the radical species $BrO_2 \cdot$. Each of these is then reduced by cerium(III) back to the bromine(III) state of $HBrO_2$. The first step is rate determining, so $[HBrO_2]$ increases through the cycle in an autocatalytic fashion:

$$HBrO_2 + BrO_3^- + H^+ \rightarrow 2BrO_2 \cdot + H_2O \tag{1.6}$$

$$BrO_2 \cdot + Ce(III) + H^+ \rightarrow HBrO_2 + Ce(IV) \tag{1.7}$$

so the stoichiometry $(1.6) + 2 \times (1.7)$ gives a net production of $HBrO_2$:

$$\underline{HBrO_2} + BrO_3^- + 3H^+ + 2Ce(III) \rightarrow 2\underline{HBrO_2} + 2Ce(IV) + H_2O$$

with a rate law

$$d[HBrO_2]/dt = + k[HBrO_2][BrO_3^-][H^+]. \tag{1.8}$$

Again the $+$ sign emphasizes that $HBrO_2$ catalyses its own production.

1.3. Skeleton schemes: model representation of isothermal autocatalysis

In this book we do not wish to limit ourselves to any specific reaction, so we will use a more general representation of a prototype autocatalytic sequence. If we have a reaction, or a sequence of elementary steps which provides a net conversion of a species A to B, we can use the representation

uncatalysed $A \rightarrow B$ rate $= k_u a$

for the uncatalysed process. Imagine that the same stoichiometry is now achieved in an autocatalytic manner, so that the rate at which B is produced from A depends on some power of the concentration of B. If empirically this power is found to be n, we can use a shorthand representation for the autocatalytic process of the form

autocatalysis $A + nB \rightarrow (n + 1)B$ rate $= k_1 ab^n$.

In the hydrogen–oxygen and B–Z reactions considered above, the autocatalytic cycles correspond to a value for n of unity. The resulting rate law, rate $= k_1 ab$, involves the product of two concentrations and is known as 'quadratic autocatalysis'. In the reaction between iodate and iodide ions, I^- is produced through an autocatalytic cycle which, at its simplest, corresponds to $n = 2$. The rate law in this case, rate $= k_1 ab^2$, involves the third power of concentration, and so is known as 'cubic autocatalysis'. We shall be particularly concerned with cubic autocatalytic systems, as these often are the easiest and simplest to analyse and understand. The results are also typical of more complex systems which involve combinations of quadratic autocatalysis with more complex effects.

We also should emphasize here that the idea of these prototype representations is not of autocatalysis occurring through a single elementary reaction step, but as the consequence of a series of such steps which may or may not have a single rate-determining reaction.

1.4. Characteristics of autocatalytic growth

Autocatalysis will play a central role in driving the oscillations and other non-linear phenomena of interest in this book. Usually, an autocatalytic process will be combined in a larger mechanism with other steps. Before considering such systems, however, we investigate the properties and behaviour of autocatalysis on its own—in particular how the concentrations and rate vary with time and with respect to each other. We start with quadratic autocatalysis, and then look at the cubic form.

1.4.1. Quadratic autocatalysis in closed vessels

The prototype representation of quadratic autocatalysis is

$$A + B \rightarrow 2B \qquad \text{rate} = k_q ab \qquad (1.9)$$

so the kinetic rate law governing the rate at which the concentration of A changes in a closed vessel is

$$da/dt = - k_q ab. \qquad (1.10)$$

If we know the initial concentrations of the reactants (let us call these a_0 and b_0) then because of the reaction stoichiometry, we can express the concentration of B as

$$b = a_0 + b_0 - a \qquad (1.11)$$

so eqn (1.10) becomes

$$da/dt = - k_q a(a_0 + b_0 - a). \qquad (1.12)$$

Equation (1.12) can be rearranged and integrated using partial fractions to give the explicit dependence of a on time, $a(t)$. A special, but particularly simple, case occurs if there is no autocatalyst B present initially (i.e. $b_0 = 0$).

Then we find

$$a(t) = \frac{a_0 \exp(-k_q a_0 t)}{1 + \exp(-k_q a_0 t)}. \tag{1.13}$$

From this it follows that the reaction rate can be expressed in a similar way as a function of time:

$$\frac{da}{dt} = -\frac{k_q a_0^2 \exp(-k_q a_0 t)}{[1 + \exp(-k_q a_0 t)]^2}. \tag{1.14}$$

Figure 1.3 shows how the concentration and rate vary in time for typical values of a_0 and k_q. Note that the time has to be extrapolated to $-\infty$. This is a particular twist in the behaviour of simple autocatalytic rate laws. If no autocatalyst is present initially, $b_0 = 0$ and hence the rate of production of B is also zero from eqn (1.10). The reaction thus takes an infinite time to get going. This unphysical effect is removed either by including a non-zero initial concentration of B (no matter how small) or by invoking an extra uncatalysed reaction (no matter how slow) converting A directly to B as discussed in the previous section. In the former case, with $b_0 \neq 0$, the resulting integrated forms become:

$$a(t) = \frac{a_0(a_0 + b_0) \exp[-k_q(a_0 + b_0)t]}{b_0 + a_0 \exp[-k_q(a_0 + b_0)t]} \tag{1.15}$$

$$\frac{da}{dt} = -k_q(a_0 + b_0)^2 \frac{a_0 \exp[-k_q(a_0 + b_0)t]}{\{b_0 + a_0 \exp[-k_q(a_0 + b_0)t]\}^2}. \tag{1.16}$$

Typical variations of concentration and rate for this case, with $b_0 = 10^{-6}$ mol dm^{-3}, are shown in Fig. 1.4. There is now a finite *induction*

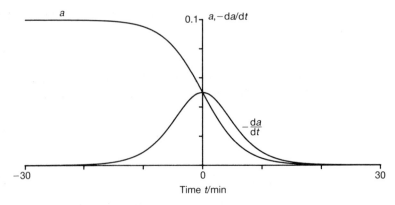

FIG. 1.3.   The variation of reactant concentration, a, and reaction rate, $-da/dt$, with time for a system obeying quadratic autocatalysis: initial reactant concentration $a_0 = 0.1$ mol dm^{-3}; $k_q = 3$ dm^3 mol^{-1} s^{-1}.

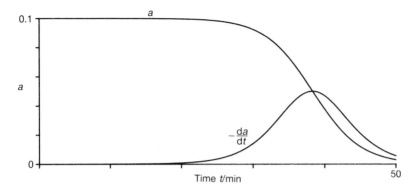

FIG. 1.4. The variation of reactant concentration, a, and reaction rate, $-\,da/dt$, with time for a system obeying quadratic autocatalysis with non-zero initial catalyst concentration: $a_0 = 0.1\ \mathrm{mol\,dm}^{-3}$; $b_0 = 10^{-6}\ \mathrm{mol\,dm}^{-3}$; $k_q = 3\ \mathrm{dm}^3\,\mathrm{mol}^{-1}\,\mathrm{s}^{-1}$.

period before the reaction rate becomes at all noticeable. This is followed by a rapid acceleration of the reaction, with the rate achieving a sharp maximum close to which the majority of the conversion of A to B occurs. The rate then slows as the concentration of A falls to zero and there is a long tail to the reaction.

Figure 1.5 shows how the reaction rate depends on the extent of conversion in the system. The extent of conversion can be represented by the difference between the concentration of A at a given time and its initial value, i.e. by $a_0 - a$. The resulting curves are parabolae—symmetric, with a maximum at 50 per cent conversion if $b_0 = 0$, but slightly lopsided with non-zero b_0.

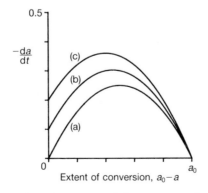

FIG. 1.5. The dependence of reaction rate, $-\,da/dt$, on extent of conversion for quadratic autocatalysis; $a_0 = 0.1\ \mathrm{mol\,dm}^{-3}$: (a) $b_0 = 0$; (b) $b_0 = 0.01\ \mathrm{mol\,dm}^{-3}$; (c) $b_0 = 0.02\ \mathrm{mol\,dm}^{-3}$.

1.4.2. Cubic autocatalysis and clock reactions in closed vessels

The cubic rate law has the overall stoichiometry

$$A + 2B \rightarrow 3B \qquad \text{rate} = k_c ab^2 \qquad (1.17)$$

so the equation for the rate of reaction in a closed vessel has the form

$$da/dt = - k_c ab^2 = - k_c a(a_0 + b_0 - a)^2. \qquad (1.18)$$

This can be integrated in closed form, and the simplest expressions again emerge if the initial concentration of the autocatalyst is zero. The reaction shows a typical autocatalytic induction period, followed by an acceleration through a period of rapid consumption. Figure 1.6 shows the variation of concentration and rate with time for a system with $a_0 = 0.1 \text{ mol dm}^{-3}$ and $b_0 = 0.001 \text{ mol dm}^{-3}$.

The induction period, followed by a sharp increase in rate is, however, the most characteristic feature of autocatalysis in closed vessels. One manifestation of this behaviour is the 'clock reaction'. An experimental system which is a typical chemical clock and which also exhibits cubic autocatalysis is the iodate–arsenite reaction. In the presence of excess iodate, the system which is initially colourless eventually undergoes a sudden colour change to brown (or blue in the presence of starch). The potential of an iodide-sensitive electrode shows a barely perceptible change during most of the induction period, but then rises rapidly, reaching a peak at the point of colour change.

The chemistry underlying the clock can be thought of in the following terms. Iodate ion is reduced by iodide through the overall stoichiometric

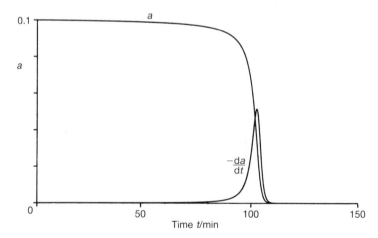

FIG. 1.6. The variation of reactant concentration, a, and reaction rate, $- da/dt$, with time for cubic autocatalysis: $a_0 = 0.1 \text{ mol dm}^{-3}$; $b_0 = 10^{-3} \text{ mol dm}^{-3}$; $k_c = 100 \text{ min}^{-1}$.

process

$$IO_3^- + 5I^- + 6H^+ \rightarrow 3I_2 + 3H_2O.$$

The iodine produced in this reaction is rapidly reduced back to iodide by the arsenite:

$$AsO_3^{3-} + H_2O + I_2 \rightarrow AsO_4^{3-} + 2H^+ + 2I^-.$$

Thus the overall stoichiometry becomes

$$IO_3^- + 5I^- + 3AsO_3^{3-} \rightarrow 3AsO_4^{3-} + 6I^-.$$

Under the simplest circumstances, the net rate of iodate removal in this process is found to follow a cubic rate law of the form

$$-d[IO_3^-]/dt = k_c[I^-]^2[IO_3^-]$$

so iodate can be identified with the reactant A in the general model above and iodide with B.

Clock behaviour can be observed provided $[IO_3^-]_0 > \frac{1}{3}[AsO_3^{3-}]_0$, the brown/blue colour change signalling the build-up of permanent I_2 after the arsenite has been exhausted. The concentration of iodate at the end of the induction period will thus be

$$[IO_3^-]_{cl} = [IO_3^-]_0 - \frac{1}{3}[AsO_3^{3-}]_0.$$

If we assume that the concentration of I_2 is negligible up to t_{c1}, we can use the stoichiometric condition

$$[IO_3^-]_0 + [I^-]_0 = [IO_3^-] + [I^-].$$

Substituting for $[I^-]$ in the rate law and integrating from $[IO_3^-]_0$ to $[IO_3^-]_{cl}$ yields a general expression for the time taken for the iodate concentration to fall to any given value

$$t = \frac{1}{k_c(a_0 + b_0)^2}\left[\ln\left(\frac{a_0 + b_0 - a}{b_0}\right) + \frac{(a_0 + b_0)(a_0 - a)}{b_0(a_0 + b_0 - a)}\right]$$

where $a_0 = [IO_3^-]_0$, $a = [IO_3^-]$, and $b_0 = [I^-]_0$. Clearly this equation holds also for the prototype scheme (1.17). The clock time t_{cl} can now be obtained by substituting for $[IO_3^-]_{cl}$.

Returning to the simple cubic autocatalysis model above, we shall be more interested later in the relationship between the rate and the extent of conversion. This is shown for various values of b_0 in Fig. 1.7. If $b_0 = 0$ (Fig. 1.7(a)), the rate curve is both a minimum and zero at no conversion (i.e. there is a double root at the origin) and has a further zero at complete conversion ($a = 0$, $a_0 - a = a_0$). The rate has a maximum value of $(4/27)k_c a_0^3$ occurring two-thirds of the way across the diagram ($a = \frac{1}{3}a_0$). There is also a point of inflection at 50 per cent conversion, $a_0 - a = \frac{1}{2}a_0$.

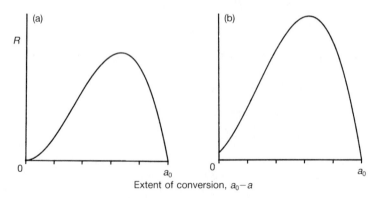

FIG. 1.7. The dependence of reaction rate, $-da/dt$, on the extent of conversion for cubic autocatalysis; $a_0 = 0.1 \ \mathrm{mol \ dm^{-3}}$: (a) $b_0 = 0$; (b) $b_0 = 0.01 \ \mathrm{mol \ dm^{-3}}$.

When the initial concentration of the autocatalyst is not zero, the rate curve becomes significantly different near to the origin. Figure 1.7(b) shows the rate curve for a system with $b_0 = \frac{1}{10}a_0$. Now the reaction rate at zero extent of reaction is also non-zero, being $k_c a_0 b_0^2 = \frac{1}{100}k_c a_0^3$. This lifting-up of the curve near to the beginning of the reaction has important consequences and will be considered at many stages in this book. The maximum rate is enhanced by adding autocatalyst, being now $(4/27)k_c(a_0 + b_0)^3$, and occurs at lower extents of conversion when $a_0 - a = (2a_0 - b_0)/3$.

1.5. Cubic autocatalysis from bimolecular steps

The cubic nature of the 'empirical' rate law discussed in the previous section, and the representation in eqn (1.17), is not at all meant to imply that we are thinking of a single, termolecular, elementary step. There are various ways in which a combination of simple bimolecular steps can combine together to give an overall rate law with this cubic form. For instance, in the two-step mechanism involving an intermediate X

$$\mathrm{B + B \rightleftharpoons X} \quad \left\{ \begin{array}{l} \text{forward rate} = k_3 b^2 \\ \text{reverse rate } = k_{-3}x \end{array} \right. \quad (1.19)$$

$$\mathrm{A + X \rightarrow B + X} \qquad \text{rate} = k_4 ax \qquad (1.20)$$

the two autocatalyst participants form a 'dimer', which then plays the actual catalytic role. (In fact X need not simply be $\mathrm{B_2}$; reaction (1.19) may also involve such species as $\mathrm{H^+}$ or $\mathrm{H_2O}$ whose concentrations are virtually constant in a buffered aqueous environment.) If the reverse rate constant is particularly high, so that $k_{-3} \gg k_3 a_0$ and $k_4 a_0$, then reaction (1.19) becomes a 'pre-equilibrium' process, and the overall rate of production of B (or the rate of removal of A) appears to leading order in the form of (1.18).

Another possible 'decomposition' of the cubic step is

$$A + B \rightleftharpoons X \qquad \begin{cases} \text{forward rate} = k_3 ab \\ \text{reverse rate} \ = k_{-3} x \end{cases} \tag{1.21}$$

$$X + B \rightarrow 3B \qquad \text{rate} = k_4 bx. \tag{1.22}$$

Again if reaction (1.21) is held in a pre-equilibrium state, the overall rate of conversion of A to B can show a cubic form. This realization of cubic autocatalysis seems to be of importance for the iodate–arsenite and iodate–hydrogen sulphite reactions. There the corresponding elementary steps include

$$IO_3^- + I^- + 2H^+ \rightleftharpoons HIO_2 + HOI \tag{1.23}$$

$$HIO_2 + I^- + H^+ \rightarrow 2HOI. \tag{1.24}$$

The species HOI is then rapidly converted to iodide, by reaction first with I^- to produce I_2 which is then reduced by the arsenite of hydrogen sulphite. Thus, identifying A with the reactant iodate and B with iodide, the system shows a cubic autocatalysis with rate proportional to $[IO_3^-][I^-]^2$ at constant pH.

1.6. Consecutive first-order reactions

Kinetic schemes often involve a number of mechanistic steps coupled together. As a simple introduction to a model which we will elaborate subsequently, we can illustrate how the concentrations evolve in a series of consecutive first-order steps:

$$P \xrightarrow{k_0} A \xrightarrow{k_u} B \xrightarrow{k_2} C.$$

The rather peculiar choice of subscripts for the rate constants is made for consistency with what follows in later sections and chapters. The reaction rate equations for this system are

$$dp/dt = -k_0 p \tag{1.25}$$

$$da/dt = k_0 p - k_u a \tag{1.26}$$

$$db/dt = k_u a - k_2 b \tag{1.27}$$

and we will assume that initially the system contains only P, with an initial concentration p_0 (i.e. $p = p_0$, $a = 0$ and $b = 0$ at $t = 0$). Integrating these equations (the details are given in an appendix to this chapter), we find the following expressions for $p(t)$, $a(t)$, and $b(t)$:

$$p(t) = p_0 \exp(-k_0 t) \tag{1.28}$$

$$a(t) = \frac{k_0 p_0}{k_u - k_0} [\exp(-k_0 t) - \exp(-k_u t)] \tag{1.29}$$

$$b(t) = \frac{k_0 k_u p_0}{k_u - k_0}$$
$$\times \left(\frac{\exp(-k_0 t) - \exp(-k_2 t)}{k_2 - k_0} - \frac{\exp(-k_u t) - \exp(-k_2 t)}{k_2 - k_u} \right). \tag{1.30}$$

Taking a typical numerical example, with $k_0 = 1 \times 10^{-4}\,\mathrm{s}^{-1}$, $k_u = 5 \times 10^{-3}\,\mathrm{s}^{-1}$, $k_2 = 0.01\,\mathrm{s}^{-1}$, and $p_0 = 0.1\,\mathrm{mol\,dm}^{-3}$, gives Fig. 1.8. The concentration of the reactant P falls exponentially, whilst the concentrations

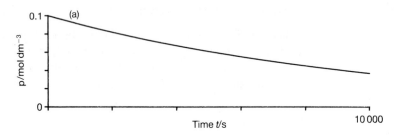

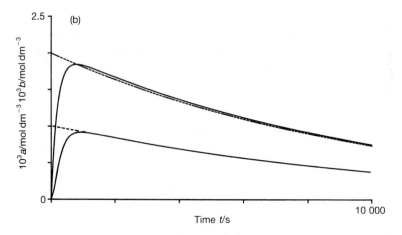

FIG. 1.8. Consecutive first-order reactions with $p_0 = 0.1\,\mathrm{mol\,dm}^{-3}$, $k_0 = 10^{-4}\,\mathrm{s}^{-1}$, $k_u = 5 \times 10^{-3}\,\mathrm{s}^{-1}$, and $k_2 = 10^{-2}\,\mathrm{s}^{-1}$: (a) the exponential decay of precursor reactant concentration, p; (b) growth and decay of intermediate concentrations $a(t)$ and $b(t)$. Also shown in (b), as broken curves, are the pseudo-stationary-state loci, $a_{ss}(t)$ and $b_{ss}(t)$, given by eqns (1.31) and (1.32).

of the intermediates A and B successively show a rise to a maximum followed by a decay to zero. This single maximum is the closest a system of isothermal first-order reactions can get to oscillation, but in the next section we consider the effect of including an additional autocatalytic route between the intermediates A and B.

Also shown in Fig. 1.8 are the pseudo-steady-state solutions for a and b. These are obtained by setting da/dt and db/dt equal to zero, giving

$$a_{ss}(t) = (k_0/k_u)p = (k_0/k_u)p_0 \exp(-k_0 t) \tag{1.31}$$

$$b_{ss}(t) = (k_u/k_2)a = (k_0/k_2)p_0 \exp(-k_0 t). \tag{1.32}$$

For some values of the rate constants, these approximate forms give a good prediction of the intermediate concentrations, at long enough times. Equations (1.31) and (1.32) do not match the initial conditions, i.e. a_{ss} tends to $(k_0/k_u)p_0$ and b_{ss} to $(k_0/k_2)p_0$ rather than to zero as $t \to 0$. Thus there must always be some initial evolution of a and b even if the system eventually settles close to the pseudo-steady states. Equations (1.31) and (1.32) can in fact be seen as a long time limit of the exact solutions (1.29) and (1.30), provided k_u is greater than k_0 and k_2 is greater than k_u: the terms $\exp(-k_u t)$ and $\exp(-k_2 t)$ then tend to zero more quickly than $\exp(-k_0 t)$ which becomes the dominant term.

In the next section we will employ a similar approximate treatment to a slightly elaborated model for which exact analytical expressions of the form (1.29) and (1.30) cannot be obtained.

1.7. Simple models and oscillations in closed systems

Chapter 2 will discuss in some detail a simple isothermal autocatalytic model whereby a reactant P is converted to a product C through two intermediates A and B. The simple kinetic model can be written as

$$P \to A \qquad \text{rate} = k_0 p$$
$$A \to B \qquad \text{rate} = k_u a$$
$$A + 2B \to 3B \qquad \text{rate} = k_1 ab^2$$
$$B \to C \qquad \text{rate} = k_2 b.$$

The rates of the first and last steps (the initial production of A from P and the final conversion of B to C) are taken simply to be first order.

From the above scheme we may easily write down the law of mass action equations for the rates of change of the concentrations p, a, and b:

$$dp/dt = -k_0 p \tag{1.33}$$

$$da/dt = k_0 p - k_1 ab^2 - k_u a \tag{1.34}$$

$$db/dt = k_1 ab^2 + k_u a - k_2 b. \tag{1.35}$$

Applying a steady-state analysis to the intermediates A and B, we set

$$k_0 p - k_1 a_{ss} b_{ss}^2 - k_u a_{ss} = 0 \tag{1.36}$$

$$k_1 a_{ss} b_{ss}^2 + k_u a_{ss} - k_2 b_{ss} = 0. \tag{1.37}$$

These equations can be rearranged to give the stationary-state concentrations

$$b_{ss} = k_0 p / k_2 \tag{1.38}$$

$$a_{ss} = k_2^2 k_0 p / (k_1 k_0^2 p^2 + k_2^2 k_u) \tag{1.39}$$

which depend on the rate constants and the concentration of the reactant P.

We can also find the simple expression for the concentration of the reactant as a function of time by integrating eqn (1.33) to give

$$p(t) = p_0 \exp(-k_0 t). \tag{1.40}$$

We can thus estimate the way in which the concentrations of the intermediates A and B will change during an experiment. Simply by substituting into eqns (1.38) and (1.39) for p, from (1.40) we get

$$a(t) = \frac{k_2^2 k_0 p_0 \exp(-k_0 t)}{k_1 k_0^2 p_0^2 \exp(-2k_0 t) + k_2^2 k_u} \tag{1.41}$$

$$b(t) = \frac{k_0}{k_2} p_0 \exp(-k_0 t). \tag{1.42}$$

These time-dependent or pseudo-steady-state results are shown in Fig. 1.9. The numerical values correspond to the data in Table 1.1, but the curves have the same qualitative form for any values of the rate constants, etc.: $b(t)$ falls exponentially throughout the reaction whilst $a(t)$ builds up to a maximum before decaying.

If, however, we actually integrate the reaction rate equations numerically using the rate constants in Table 1.1 we find that the system does not always stick to, or even stay close to, these pseudo-steady loci. The actual behaviour is shown in Fig. 1.10. There is a short initial period during which a and b grow from zero to their appropriate pseudo-steady values. After this the evolution of the intermediate concentrations is well approximated by (1.41) and (1.42), but only for a while. After a certain time, the system moves spontaneously away from the pseudo-steady curves and oscillatory behaviour develops. We may think of the steady state as being unstable or, in some sense 'repulsive', during this period in contrast to its stability or 'attractiveness' beforehand. Thus we have met a 'bifurcation to oscillatory responses'. The oscillations

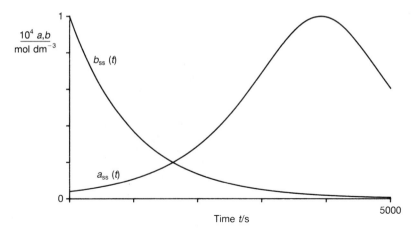

FIG. 1.9. Consecutive first-order reactions and cubic autocatalysis, showing pseudo-steady-state predictions for the intermediate concentrations. Initial concentrations and rate constants are given in Table 1.1.

Table 1.1
*Typical values for rate constants
and initial conditions*

k_0	$1 \times 10^{-3}\,\mathrm{s}^{-1}$
k_u	$1 \times 10^{-2}\,\mathrm{s}^{-1}$
k_1	$2.5 \times 10^9\,\mathrm{dm^6\,mol^{-2}\,s^{-1}}$
k_2	$1\,\mathrm{s}^{-1}$
p_0	$0.1\,\mathrm{mol\,dm}^{-3}$

grow in amplitude and lengthen in period. Eventually, however, reactant consumption must become important and the oscillations eventually die out. The pseudo-steady state becomes attractive again (there is a second bifurcation) and there is a damped oscillatory approach back to the loci (1.41) and (1.42).

There are many questions we wish to answer. Why and when does the steady evolution of the system become unstable? Why do other values of the rate constants or of the initial conditions yield oscillations either immediately (without the induction period) or, in some cases, not at all? How quickly will the oscillations grow? How long will they last? How many will there be? Surprisingly, perhaps, the answers to these and other features emerge in simple analytic results and in many cases require only some simple algebra.

Very similar behaviour is observed for the simplest chemistry in closed vessels if the heat produced during the reaction is not lost with perfect

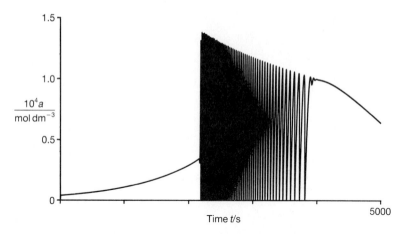

Fig. 1.10. Actual time-dependent concentration of intermediate A for consecutive first-order reactions with cubic autocatalysis showing pseudo-steady-state behaviour, pre-oscillatory evolution, an oscillatory period, and then the return to pseudo-steady-state behaviour.

efficiency. Non-isothermal reactions can also display long trains of oscillations and will be investigated in chapter 3.

1.8. Bistability in flow reactors

The simplest form of flow system is the continuously fed well-stirred tank reactor or CSTR, represented schematically in Fig. 1.11. The behaviour of typical autocatalytic systems in a CSTR will be considered in chapters 4 and 5, but here we may quickly examine how multistability can arise, even with only one overall chemical reaction. We will take a CSTR in which just the

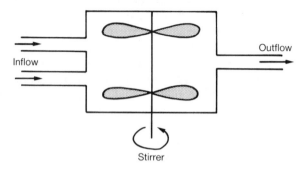

Fig. 1.11. Schematic diagram of a continuous-flow well-stirred tank reactor (CSTR), the simplest open system.

cubic autocatalytic process

$$A + 2B \rightarrow 3B \qquad \text{rate} = k_1 ab^2$$

occurs. The reactor will be fed by a stream, or separate streams, of A and B with concentrations a_0 and b_0 respectively. The concentrations of the two species in the reactor are a and b.

The concentration of A is determined by three rates: the rate of inflow of A, $r_{in} = qa_0$, where q is the volume flow rate of the feed; the rate of outflow of A, $r_{out} = qa$; and the total chemical rate of conversion of A to B, $r_{ch} = - Vk_1 ab^2$. Thus we may write

$$V \, da/dt = qa_0 \quad - \quad qa \quad - \quad Vk_1 ab^2. \qquad (1.43)$$

$$\text{net rate} \qquad \text{inflow} \qquad \text{outflow} \qquad \text{reaction}$$
$$\text{rate} \qquad\quad \text{rate} \qquad\quad \text{rate}$$

The concentrations of A and B are related by the reaction stoichiometry to the inflow concentrations, so

$$a_0 + b_0 = a + b. \qquad (1.44)$$

We can use this to eliminate b from eqn (1.43). If we also divide throughout by the reactor volume V and call q/V the flow rate k_f (it has units of $(\text{time})^{-1}$ and so is like a first-order rate constant)

$$da/dt \quad = \quad k_f(a_0 - a) \quad - \quad k_1 a(a_0 + b_0 - a)^2. \qquad (1.45)$$

$$\text{net} \qquad\qquad \text{net} \qquad\qquad\qquad \text{chemical}$$
$$\text{accumulation} \qquad \text{inflow} \qquad\qquad\qquad \text{reaction}$$

The concentration of A now tends to a value which makes the net inflow rate exactly balance the chemical reaction rate. When this has happened, $da/dt = 0$, so the concentration becomes steady. This 'stationary-state' concentration can be maintained indefinitely.

Stationary states are given by the solutions of the equation

$$k_f(a_0 - a_{ss}) - k_1 a_{ss}(a_0 + b_0 - a_{ss})^2 = 0 \qquad (1.46)$$

so a_{ss} is given by the roots of a cubic involving k_1, a_0, b_0, and the flow rate k_f. Cubic equations have either one or three real roots. The physical quantities we are dealing with—concentrations, flow rates, etc.—mean that we are only interested in positive solutions. If for a given set of k_1, a_0, b_0, and k_f the cubic has only one real positive solution a_{ss}, we have a unique stationary state at the flow rate; if the cubic has three real, positive solutions we are in a region of multistability since all three solutions correspond to $da/dt = 0$ and hence are stationary states.

1.8.1. Flow diagrams

A convenient way of distinguishing between unique and multiple solutions is to use a 'flow diagram', in which the two rates—net inflow and chemical reaction—are plotted as functions of concentration. Figure 1.12 shows a number of flow diagrams, corresponding to different values for k_1, a_0, b_0, and flow rate k_f. The straight line L is the net rate of inflow

$$L = k_f(a_0 - a) \tag{1.47}$$

and so its gradient depends solely on the flow rate. The cubic chemical reaction rate curve R is that seen in §1.4.2, and is given by

$$R = k_1 a(a_0 + b_0 - a)^2. \tag{1.48}$$

This depends on k_1, a_0, and b_0. Points of intersection of these two curves on the flow diagram correspond to conditions where $R = L$, and hence to stationary-state solutions. If R and L have just one intersection, as shown in Fig. 1.12(a) or (e), there is a unique stationary state. If L cuts R three times, as

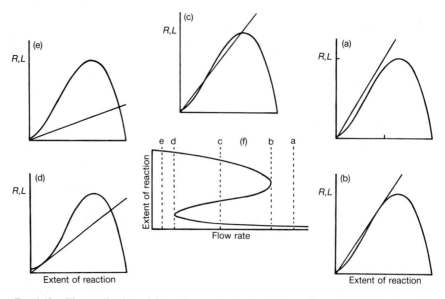

FIG. 1.12. The application of flow diagram to illustrate the origins of bistability for cubic autocatalysis in a CSTR: (a) R and L have just one intersection, close to the origin and hence corresponding to a low stationary-state extent of reaction; (c) R and L have three intersections corresponding to multiple stationary-state solutions; (e) R and L again have only one intersection, now corresponding to a high stationary-state extent of reaction; (d) R and L intersect tangentially at low extents of reaction (ignition tangent); (b) R and L intersect tangentially at high extents of reaction (extinction tangent); (f) the stationary-state locus for extent of reaction as a function of flow rate (the flow rates a–e correspond to the other parts of this figure).

in Fig. 1.12(c), there are three stationary states and we have multistability. An easy way of moving the system between Figs 1.12(a), (c), and (e) is to vary the flow rate, keeping k_1, a_0, and b_0 constant. Increasing the flow rate increases the gradient of L, without effecting R. The way in which the stationary-state solutions vary with k_f is shown in Fig. 1.12(f), revealing the range of multistability at intermediate flow rates.

The jump points in Fig. 1.12(f) correspond to conditions where two stationary states come together and merge. In terms of the roots of the cubic equation, two real roots merge and become a complex pair. In terms of the flow diagram, the curves R and L become tangential. The two tangencies are represented in Figs 1.12(b) and (d).

1.8.2. Tangency conditions for flow diagrams

This correspondence between tangency and jump points is quite universal in chemical models and so it is instructive to work the present simple case in some detail here. The condition for tangency is

$$R = L \qquad \mathrm{d}R/\mathrm{d}a = \mathrm{d}L/\mathrm{d}a. \tag{1.49}$$

In terms of the rate constants etc. for the present model, this becomes

$$k_1 a(a_0 + b_0 - a)^2 = k_f(a_0 - a) \tag{1.50}$$

$$k_1(a_0 + b_0 - a)(a_0 + b_0 - 3a) = -k_f \tag{1.51}$$

to be satisfied simultaneously. Dividing one equation by the other and rearranging gives a quadratic equation in a:

$$2a^2 - 3a_0 a + a_0(a_0 + b_0) = 0. \tag{1.52}$$

At tangency, therefore,

$$a_\pm = \tfrac{1}{4}\{3a_0 \pm [a_0(a_0 - 8b_0)]^{1/2}\}. \tag{1.53}$$

The upper root gives the tangency at low extents of conversion shown in Fig. 1.12(d), the lower root gives Fig. 1.12(b). The corresponding flow rates can then be evaluated by substitution from eqn (1.53) to (1.51).

A response showing multiple stationary states requires that the inflow concentration of B be significantly less than that of A. Multiple intersections and tangencies are only possible if

$$a_0 > 8b_0. \tag{1.54}$$

For larger inflow concentrations of the autocatalyst B, i.e. with $b_0 > a_0/8$, the curves R and L can only intersect once, whatever the flow rate. As shown in Fig. 1.13, we then have a monotonic dependence of a_{ss} on k_f.

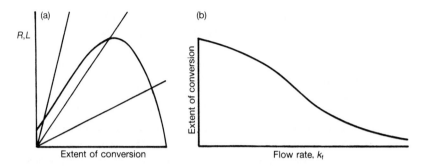

FIG. 1.13. Cubic autocatalysis with relatively high inflow concentration of the catalyst, $b_0 = \frac{1}{6}a_0$: (a) flow diagram; (b) corresponding stationary-state locus showing monostability.

The stationary-state response curves, or 'bifurcation diagrams' shown in Figs 1.13(b) and 1.12(f), represent two of the simplest possible patterns: monotonic variation and a single hysteresis loop respectively. These are the only qualitatively different responses possible for the cubic autocatalytic step on its own. They are also found for a first-order exothermic reaction in an adiabatic flow reactor (see chapter 6). With only slightly more complex chemical mechanisms a whole array of extra exotic patterns can be found, such as those displayed in Fig. 1.14. The origins of these shapes will be determined in chapter 4.

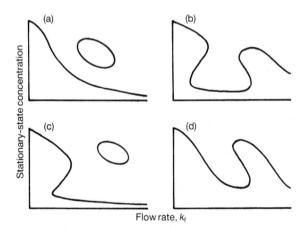

FIG. 1.14. Four more of the possible stationary-state bifurcation diagrams for chemical systems (see also Fig. 1.2) in flow reactors: (a) isola; (b) mushroom; (c) isola + hysteresis loop; (d) breaking wave.

1.8.3. Stability of stationary states

Let us return to Fig. 1.12(c), where there are multiple intersections of the reaction rate and flow curves R and L. The details are shown on a larger scale in Fig. 1.15. Can we make any comments about the stability of each of the stationary states corresponding to the different intersections? What, indeed, do we mean by stability in this case? We have already seen one sort of instability in §1.6, where the pseudo-steady-state evolution gave way to oscillatory behaviour. Here we ask a slightly different question (although the possibility of transition to oscillatory states will also arise as we elaborate on the model). If the system is sitting at a particular stationary state, what will be the effect of a very small perturbation? Will the perturbation die away, so the system returns to the same stationary state, or will it grow, so the system moves to a different stationary state? If the former situation holds, the stationary state is stable; in the latter case it would be unstable.

The term R is the chemical rate of removal of A from the reactor: this process tends to increase the extent of conversion. The flow term L is the net rate of inflow of fresh reactant A, and this tends to decrease the extent of conversion. If $L > R$, the extent of conversion decreases and we move to the left along the abscissa in Fig. 1.15; conversely, if $R > L$, the system moves to the right (and, of course, if $L = R$ we have a stationary-state solution and so we do not move at all). Let us concentrate first on the intersection furthest to the right, corresponding to the highest extent of conversion. If the system is somehow disturbed from this state, so that the extent of conversion increases further, then L will exceed R so the system will automatically adjust itself back towards lower extents of conversion, back to the original intersection. Similarly, if the extent of conversion momentarily drops from this stationary-state value, then R will exceed L and the system will move back to the right. This is a stable situation. Exactly the same arguments apply at the lowest of the three intersections, which is therefore also a stable state.

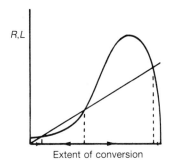

Extent of conversion

FIG. 1.15.   Flow diagram for cubic autocatalysis showing different stabilities of multiple stationary-state intersections.

If we ask the same questions of the middle intersection, however, we find a very different situation. Consider first a momentary increase in the extent of conversion from this state. To the right of this point, the curve R lies above L, so the system will tend to move further to the right. The perturbation will grow and, in fact, the extent of conversion will increase until it approaches the uppermost stationary state. Similarly, a momentary decrease in the extent of conversion from the middle intersection also tends to diverge, and the system moves to the lowest stationary state. The middle solution is thus unstable to perturbations. It is like a pendulum balanced perfectly upright: in theory it can maintain its state (upright), but the slightest nudge will send it to a different state (in this case falling to either of two different states, one to the right or one to the left, depending on which way it is pushed). The middle solution can be said to play a role separating the two stable stationary states—it is sometimes known as a saddle point.

1.9. Reaction and diffusion

Chemical reactions with autocatalytic or thermal feedback can combine with the diffusive transport of molecules to create a striking set of spatial or temporal patterns. A reactor with permeable wall across which fresh reactants can diffuse in and products diffuse out is an open system and so can support multiple stationary states and sustained oscillations. The diffusion processes mean that the stationary-state concentrations will vary with position in the reactor, giving a 'profile', which may show distinct banding (Fig. 1.16). Similar patterns are also predicted in some circumstances in closed vessels if stirring ceases. Then the spatial dependence can develop spontaneously from an initially uniform state, but uniformity must always return eventually as the system approaches equilibrium.

Travelling wavefronts are familiar in non-isothermal gas-phase systems, as flames. Once established, these generally propagate along a tube into a stationary reactant mixture at a steady velocity or can be stabilized on a burner,

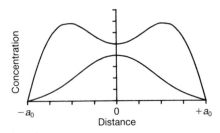

FIG. 1.16. Stationary-state concentration profiles for reaction coupled with diffusion.

with the reactant flowing at a steady velocity towards the flame zone. Similar fronts also occur widely in autocatalytic solution-phase reactions. Again there is a relatively sharp band of reaction passing through the solution into fresh reactants and leaving products behind. Figure 1.6 could describe the variation in concentration through such a front, with the time axis replaced by a distance coordinate: the wave is propagating from right to left in the diagram, towards the region of high reactant concentration. Typically the front velocity is much slower in solution reactions, perhaps of the order of a few millimetres per minute.

Other travelling waveforms include target patterns, fronts, or pulses as expanding and concentric circles and which may be broken to form spirals (which may also arise apparently spontaneously under some conditions) as shown in Fig. 1.17.

Many of these various spatio-temporal features emerge in the simple models to be treated in this book and are discussed in chapters 9–11.

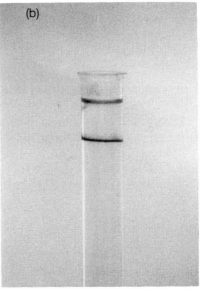

Fig. 1.17(a, b)

FIG. 1.17.   Different forms of travelling wavefronts: (a) travelling wavefront (propagating down tube) in iodate–arsenous acid reaction with excess iodate; (b) travelling wavefront or pulse in iodate–arsenous acid reaction with excess arsenite; (c) target patterns in Belousov–Zhabotinskii reaction.

1.10. Complex oscillations and chaos

The simplest types of oscillations are those for which each peak is the same as the previous one. These simple 'period-1' responses are the most commonly observed form for chemical reactions. However, particularly when working with flow reactors, more complex periodicities can emerge.

The simplest extension comes when a period-1 oscillation changes so that it then repeats every other maximum. The time interval between each peak need not change much, but as the repeating unit involves two peaks, the oscillatory period will double. Such a period doubling may occur as, say, the residence time is varied, just as the transition from steady reaction to period-1 oscillations occurs. Further period doublings may follow if the experimental conditions are changed again, giving rise to a sequence of period-4, period-8, etc., as shown in Fig. 1.18. These higher period doublings come closer and closer together, and eventually we may pass a point at which all periodicity vanishes. The concentrations now vary continuously in time, but the traces never repeat themselves. Such behaviour is aperiodic, or 'chaotic'.

Other complex oscillatory waveforms include 'bursting', in which large-amplitude oscillations are interspersed with periods of non-oscillatory evolution, or by small oscillations (Fig. 1.19).

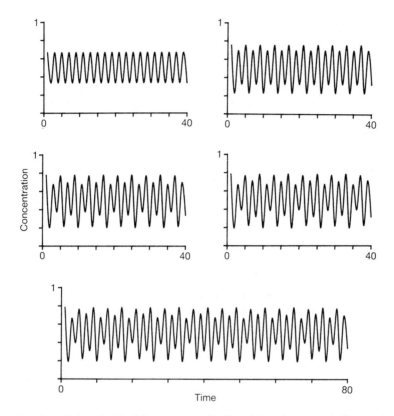

FIG. 1.18. A typical period-doubling sequence leading ultimately to aperiodic waveforms (chemical chaos).

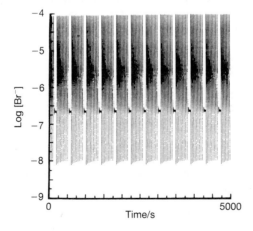

FIG. 1.19. Complex, but strictly periodic, oscillations in a chemical reaction showing 'bursting' in a model of the Belousov–Zhabotinskii reaction. (Reprinted with permission from Bar-Eli, K. and Noyes, R. M. (1988). *J. Chem. Phys.*, **88**, 3636–54. © American Institute of Physics.)

Appendix. Evaluation of concentrations for consecutive first-order reactions

The purpose of this appendix, in giving detail of the derivation of eqns (1.22)–(1.24), is to demonstrate a method of analysis which will be of particular use in later chapters when we discuss the local stability of a stationary-state solution. We will see here concentrations of different species evolving as the sum of a series of exponential terms which involve first-order rate constants. Later we will see similar sums of exponential terms, where the exponents, although more complicated can also be interpreted as pseudo-first-order rate coefficients.

Let us now turn to the specific example provided by the rate equations (1.19)–(1.21):

$$dp/dt = -k_0 p \tag{1.19}$$

$$da/dt = k_0 p - k_u a \tag{1.20}$$

$$db/dt = k_u a - k_2 b. \tag{1.21}$$

This set of equations involves only terms which depend on a single concentration. Mathematically they are known as a set of first-order differential equations. It is this simple structure which tells us that the solutions (i.e. how the concentrations vary in time) will have the appearance of various exponential terms, $\exp(-k_0 t)$, $\exp(-k_u t)$, and $\exp(-k_2 t)$, added together.

Equation (1.19) for p is the easiest: it only involves a rate constant, so the solution will have the form

$$p = \Pi_0 \exp(-k_0 t) \tag{A1.1}$$

where Π_0 is some constant to be determined. In this trivial case we can integrate (1.19) directly anyway and use the initial condition $p = p_0$ at $t = 0$ to determine the pre-exponential factor. So $\Pi_0 = p_0$ and

$$p = p_0 \exp(-k_0 t). \tag{A1.2}$$

For the concentration of A, eqn (1.20) involves k_0 and k_1, but not k_2. The latter can have no effect on a, so the solution must have the form

$$a = \alpha_0 \exp(-k_0 t) + \alpha_1 \exp(-k_1 t) \tag{A1.3}$$

where α_0 and α_1 are the coefficients to be determined. We can eliminate one, α_1 say, using the initial condition $a = 0$ at $t = 0$. Substituting these into eqn (A1.3), we require

$$0 = \alpha_0 + \alpha_1 \quad \text{so} \quad \alpha_1 = -\alpha_0 \tag{A1.4}$$

and then

$$a = \alpha_0 [\exp(-k_0 t) - \exp(-k_1 t)]. \tag{A1.5}$$

To find α_0 we can differentiate eqn (A1.5) and compare the resulting expression for da/dt with (1.20). Differentiation gives

$$da/dt = -\alpha_0[k_0 \exp(-k_0 t) - k_1 \exp(-k_1 t)] \qquad (A1.6)$$

whilst substitution for p and a from (A1.2) and (A1.5) into (1.20) gives

$$da/dt = k_0 p_0 \exp(-k_0 t) - k_1 \alpha_0[\exp(-k_0 t) - \exp(-k_1 t)]. \qquad (A1.7)$$

As the right-hand sides of (A1.6) and (A1.7) represent the same quantity, da/dt, they must be equal. In the resulting equation all the exponential terms cancel and the remaining terms can be rearranged to yield α_0:

$$\alpha_0 = k_0 p_0/(k_1 - k_0) \qquad (A1.8)$$

so that

$$a(t) = \frac{k_0 p_0}{k_1 - k_0} [\exp(-k_0 t) - \exp(-k_u t)] \qquad (A1.9)$$

To find $b(t)$, we follow a similar procedure. Equation (1.21) involves k_u and k_2 directly, and k_0 indirectly through the concentration of a. We thus look for a solution involving all three exponential terms

$$b(t) = \beta_0 \exp(-k_0 t) + \beta_1 \exp(-k_u t) + \beta_2 \exp(-k_2 t) \qquad (A1.10)$$

with three coefficients to be determined. Again we can use the initial conditions to find one, β_2 say. We require $b = 0$ at $t = 0$, so

$$0 = \beta_0 + \beta_1 + \beta_2 \qquad \text{giving} \quad \beta_2 = -(\beta_0 + \beta_1). \qquad (A1.11)$$

We now have for $b(t)$:

$$b(t) = \beta_0[\exp(-k_0 t) - \exp(-k_2 t)] + \beta_1[\exp(-k_u t) - \exp(-k_2 t)] \qquad (A1.12)$$

and hence by differentiation

$$db/dt = -\beta_0[k_0 \exp(-k_0 t) - k_2 \exp(-k_2 t)] \\ - \beta_1[k_1 \exp(-k_u t) - k_2 \exp(-k_2 t)]. \qquad (A1.13)$$

The right-hand side of (A1.13) must be equal to the right-hand side of (1.21), and substituting for $a(t)$ and $b(t)$ into the latter from (A1.9) and (A1.12) we find, after some rearrangement,

$$\beta_0(k_2 - k_0)\exp(-k_0 t) + \beta_1(k_2 - k_u)\exp(-k_u t) \\ = \frac{k_0 k_u p_0}{(k_u - k_0)} [\exp(-k_0 t) - \exp(-k_u t)]. \qquad (A1.14)$$

This time the exponential terms have not cancelled and we require condition (A1.14) to give us both β_0 and β_1. In fact we can split equation (A1.14) into two, one equation involving terms containing $\exp(-k_0 t)$ and the other with those terms containing $\exp(-k_u t)$:

terms in $\exp(-k_0 t)$: $\quad \beta_0(k_2 - k_0) = k_0 k_u p_0 / (k_u - k_0)$ $\hspace{2cm}$ (A1.15)

terms in $\exp(-k_u t)$: $\quad \beta_1(k_2 - k_u) = -k_0 k_u p_0 / (k_u - k_0)$. $\hspace{1.5cm}$ (A1.16)

These rearrange simply to give β_0 and β_1 and the final expression for $b(t)$ then becomes

$$b(t) = \frac{k_0 k_u p_0}{k_u - k_0} \left(\frac{\exp(-k_0 t) - \exp(-k_2 t)}{k_2 - k_0} - \frac{\exp(-k_u t) - \exp(-k_2 t)}{k_2 - k_u} \right)$$

References

Specific references will be given at the end of each chapter. Here we list a series of general or introductory references.

The thermodynamics of systems 'far from equilibrium' are discussed in a number of texts, such as

Prigogine, I. (1962). *Introduction to nonequilibrium thermodynamics.* Wiley-Interscience, New York.

Glansdorff, P. and Prigogine, I. (1971). *Thermodynamics of structure, stability and fluctuations.* Wiley-Interscience, New York.

Nicolis, G. and Prigogine, I. (1977). *Self-organization in nonequilibrium systems.* Wiley-Interscience, New York.

Prigogine, I. (1980). *From being to becoming.* Freeman, San Francisco.

Haken, H. (1983). *Advanced synergetics.* Springer, Berlin.

Prigogine, I. and Stengers, I. (1983). *Order out of chaos.* Heinemann, London.

Babloyantz, A. (1986). *Molecules, dynamics and life.* Wiley-Interscience, New York.

A selection of monographs covering chemical and mathematical aspects of oscillations etc. include

Faraday Symposia of the Chemical Society (1975). *Physical chemistry of oscillatory phenomena.* The Royal Society of Chemistry, London.

Gray, B. F. (1974). Kinetics of oscillatory reactions. In *Reaction kinetics: specialist periodical reports,* (ed. P. G. Ashmore), pp. 309–86. The Chemical Society, London.

Murray, J. D. (1979). *Lecture notes on nonlinear differential equation models in biology.* Clarendon Press, Oxford.

Cooke, D. O. (1979). *Inorganic reaction mechanisms*. The Chemical Society, London.

Tyson, J. J. (1976). *The Belousov–Zhabotinskii reaction*. Springer, Berlin.

Field, R. J. and Burger, M. (eds) (1985). *Oscillations and travelling waves in chemical systems*. Wiley-Interscience, New York.

Various journals publish material from this area, including: *Chemical Engineering Science, The Journal of Chemical Physics, The Journal of Physical Chemistry, Journal of the Chemical Society Faraday Transactions, Physica D, Proceedings of the Royal Society, Journal of The American Chemical Society.*

Part 1

The Techniques

OSCILLATIONS IN A CLOSED
ISOTHERMAL SYSTEM

This chapter introduces the simplest chemical kinetic scheme for an iso-thermal oscillatory reaction in a closed system. This model scheme is used to illustrate concepts of very general importance and applicability. A mathematically deeper analysis is given in chapter 3.

After a study of this chapter it should be possible to:

(1) appreciate the fundamental importance of equations of conservation and of the reaction kinetic rate laws;

(2) recognize the usefulness of the stationary-state approximation when the precursor decays slowly whilst the autocatalytic step proceeds rapidly;

(3) recognize the limitation of the stationary-state approximation and the need to establish the stability of stationary states;

(4) understand how oscillatory behaviour is intimately connected with un-stable stationary states;

(5) realize how a very simple kinetic model can generate quite richly varied transient oscillations on its way to the final state of complete conversion or chemical equilibrium;

(6) note the connections between a particular set of reaction rate coefficients and initial concentrations and the size, shape, duration, and number of oscillations.

In this chapter we concentrate on the simplest chemical model of oscillations introduced briefly in the previous chapter. Our example involves the irreversible conversion of a precursor reactant P to a final product C through two intermediate species A and B. The intermediates are supposed to be much more reactive than the relatively stable reactant P, so that their concentrations will always be relatively low compared with the initial concentration of P.

We allow *two* chemical pathways by which intermediate A is converted to B. First, there is a direct or uncatalysed step

uncatalysed $\quad$ A $\rightarrow$ B $\quad$ rate $= k_u a.$

Secondly, there is an autocatalytic contribution for which we use the

Table 2.1

*Typical values for rate constants
and initial conditions*

k_0	$1 \times 10^{-3} \, \text{s}^{-1}$
k_u	$1 \times 10^{-2} \, \text{s}^{-1}$
k_1	$2.5 \times 10^9 \, \text{dm}^6 \, \text{mol}^{-2} \, \text{s}^{-1}$
k_2	$1 \, \text{s}^{-1}$
p_0	$0.1 \, \text{mol} \, \text{dm}^{-3}$

short-hand notation

cubic autocatalysis $A + 2B \rightarrow 3B$ rate $= k_1 ab^2$.

This 'step' provides a mechanism for feedback within the model, since a product of the reaction has an accelerating influence on the reaction rate. The feedback is also non-linear with a b^2 factor leading to an overall dependence of the reaction rate on concentration that is cubic in form.

The full kinetic scheme can now be written as

$$P \xrightarrow{k_0 p} A \underset{k_u a}{\overset{k_1 ab^2}{\rightleftarrows}} B \xrightarrow{k_2 b} C$$

or

$$P \rightarrow A \qquad \text{rate} = k_0 p$$

$$A \rightarrow B \qquad \text{rate} = k_u a$$

$$A + 2B \rightarrow 3B \qquad \text{rate} = k_1 ab^2$$

$$B \rightarrow C \qquad \text{rate} = k_2 b.$$

The rates of the first and last steps (the initial production of A from P and the final conversion of B to C) are taken simply to be first order. As a typical example we can consider a set of reaction rate constants such as that given in Table 2.1.

2.1. Kinetic rate equations

From the above scheme we may easily write the kinetic equations for the rates of change of the concentrations p, a, and b:

$$dp/dt = -k_0 p \tag{2.1}$$

$$da/dt = k_0 p - k_1 ab^2 - k_u a \tag{2.2}$$

$$db/dt = k_1 ab^2 + k_u a - k_2 b. \tag{2.3}$$

The concentration of the product C can be evaluated at any time from the initial concentrations p_0 etc. and the instantaneous concentrations p, a and b, from the stoichiometry, i.e.

$$p_0 + a_0 + b_0 + c_0 = p + a + b + c. \qquad (2.4)$$

Thus the value of c cannot vary independently and we do not need to consider a reaction rate equation for this species as well.

2.2. Chemical equilibrium

We first examine the 'final state' of the reaction, i.e. the chemical equilibrium composition. This is not of great relevance to oscillatory behaviour but is an important first check that the model is 'chemically reasonable'. Equilibrium arises when all three rates of change become zero simultaneously. Equations (2.1)–(2.3) have a unique point satisfying this condition, as required chemically, given by

$$p_e = a_e = b_e = 0 \qquad c_e = p_0 + a_0 + b_0 + c_0. \qquad (2.5)$$

For the present model the equilibrium state corresponds to complete conversion. It is also relatively easy to show that the final approach of each of the concentrations to this state is given by a sum of three exponentially decaying terms $\exp(-k_0 t)$, $\exp(-k_u t)$, and $\exp(-k_2 t)$ (this forms the basis of the relaxation technique in chemical kinetics for measuring the rate constants for fast reactions).

2.3. Behaviour 'far from equilibrium'

Consider a beaker, or some other reaction vessel, filled with a solution of pure P at some initial concentrations $p(t = 0) = p_0$. The initial concentrations of all other species are zero, so $a_0 = b_0 = c_0 = 0$ in the equations above. We know that after an infinite time we achieve the equilibrium state, so p, a, and b will approach zero, and that c will tend to the value p_0. However, we know little else of the course of the reaction and what happens to these concentrations on the way from their initial to their final states. We turn to this question now.

Equation (2.1) is of a particularly simple form and can be integrated directly to give the concentration of the reactant P at any time:

$$p(t) = p_0 \exp(-k_0 t). \qquad (2.6)$$

With $k_0 = 10^{-3} \, \text{s}^{-1}$, the reaction half-life is approximately 2 h; after 6 h, only 1 per cent of P remains. Thus we also know that the concentration of P shows a monotonic exponential decay from its initial value to zero. If k_0 is small, this is a slow decay.

We are left with the behaviour of the intermediates A and B. A common approach to kinetic models involving relatively reactive species is to apply the pseudo-stationary-state (PSS) hypothesis.

2.4. Pseudo-stationary states

The stationary-state concentrations of a and b are given by setting their rates of change simultaneously to zero. Thus

$$k_0 p - k_1 a_{ss} b_{ss}^2 - k_u a_{ss} = 0 \tag{2.7}$$

$$k_1 a_{ss} b_{ss}^2 + k_u a_{ss} - k_2 b_{ss} = 0. \tag{2.8}$$

Adding these two equations eliminates all terms containing a_{ss} and leads to the simple result

$$b_{ss} = k_0 p / k_2. \tag{2.9}$$

Substituting this into (2.8) we then get for a_{ss}

$$a_{ss} = k_2^2 k_0 p / (k_1 k_0^2 p^2 + k_2^2 k_u). \tag{2.10}$$

Thus the intermediate B has a concentration that is directly proportional to that of the reactant P. The dependence of a_{ss} on p is slightly more complicated but can still be evaluated explicitly. As an example, for the values of the reaction rate constants and the initial value of p from Table 2.1, we find $b_{ss} = 10^{-4} \text{ mol dm}^{-3}$ and $a_{ss} = 4 \times 10^{-6} \text{ mol dm}^{-3}$.

2.4.1. Dependence of a_{ss} and b_{ss} on p

In a real experiment the concentration of the reactant falls in time. We should, therefore, establish the way in which b_{ss} and a_{ss} vary with p. This, in fact, will turn out to be the basis of a particularly convenient approach in which we regard the pseudo-stationary-state concentrations as relatively simple functions of p, rather than the more complex functions of time which we derive later. The stationary-state loci are shown in Fig. 2.1.

At high concentrations of the reactant, the stationary-state concentration of B is higher than that of A: at low p, a_{ss} is greater than b_{ss}. The two loci must, therefore, cross. This occurs when

$$a_{ss} = b_{ss} = [(k_2 - k_u)/k_1]^{1/2}. \tag{2.11}$$

and when this happens

$$p = (k_2/k_0)[k_2 - k_u)/k_1]^{1/2}. \tag{2.12}$$

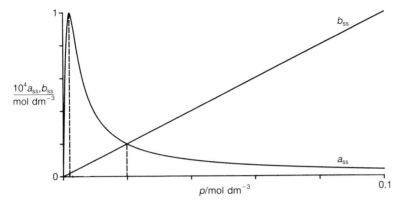

FIG. 2.1.   Pseudo-stationary-state dependences of the intermediate concentrations a_{ss} and b_{ss} on the instantaneous concentration of the precursor, p. For the rate data in Table 2.1, the locus for intermediate A shows a maximum of $a_{ss} = 2 \times 10^{-4}$ mol dm^{-3} at $p = 2 \times 10^{-3}$ mol dm^{-3} and the two loci cross with $a_{ss} = b_{ss} = 2 \times 10^{-5}$ mol dm^{-3} with $p = 0.02$ mol dm^{-3}.

For the data in Table 2.1, this corresponds to $a_{ss} = b_{ss} \approx 2 \times 10^{-5}$ mol dm^{-3} and would occur when $p = 0.02$ mol dm^{-3}.

The stationary-state concentration of A also shows a maximum. Differentiating eqn (2.10) with respect to p, we find the condition for this as

$$(a_{ss})_{max} = \tfrac{1}{2} k_2/(k_1 k_u)^{1/2} \tag{2.13}$$

and is attained when

$$p = (k_2^2 k_u/k_1 k_0^2)^{1/2}. \tag{2.14}$$

Again, using the data in Table 2.1 as an example, $(a_{ss})_{max} = 2 \times 10^{-4}$ mol dm^{-3} when $p = 2 \times 10^{-3}$ mol dm^{-3}.

The locations of the crossing and maximum points in the figure are of more than just mathematical interest. These give an upper and lower bound, respectively, on the range of conditions over which we may expect to see oscillations. We can also see from the equations above that the different rate constants in this model have a habit of combining together to produce significant quantities with the units of concentration. We will make use of this later on.

2.4.2. Time dependence of pseudo-stationary states

Because we also know exactly how the concentration of the reactant varies in time, we can estimate the way in which the concentrations of the intermediates A and B will change during an experiment. Simply by substituting into eqns (2.9) and (2.10) for p, from (2.6) we get

$$a(t) = \frac{k_2^2 k_0 p_0 \exp(-k_0 t)}{k_1 k_0^2 p_0^2 \exp(-2k_0 t) + k_2^2 k_u}$$ (2.15)

$$b(t) = \frac{k_0}{k_2} p_0 \exp(-k_0 t).$$ (2.16)

These time-dependent, or pseudo-steady-state, results are shown in Fig. 2.2. The numerical values correspond to the data in Table 2.1, but the curves are qualitatively correct for any values of the rate constants etc.: $b(t)$ falls exponentially throughout the reaction whilst $a(t)$ builds up to a maximum before decaying.

The time at which the curves for a and b cross will be that at which the concentration p has fallen from its initial value to that given by eqn (2.12), i.e. t_{cross} satisfies

$$p_0 \exp(-k_0 t_{cross}) = \frac{k_2 (k_2 - k_u)^{1/2}}{k_0 k_1^{1/2}}$$ (2.17)

so that

$$t_{cross} = k_0^{-1} \ln \left(\frac{k_0 k_1^{1/2} p_0}{k_2 (k_2 - k_u)^{1/2}} \right).$$ (2.18)

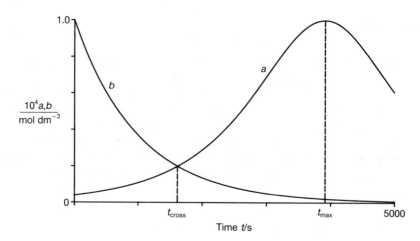

FIG. 2.2. Predicted pseudo-stationary-state evolution of the intermediate species concentrations $a(t)$ and $b(t)$, as given by eqns (2.15) and (2.16). Specific numerical values correspond to the rate data in Table 2.1. The time at which the two concentrations become equal and that at which $a(t)$ attains its maximum are indicated.

In a similar way, the maximum in $a(t)$ will occur at time t_{max}, given by

$$t_{max} = (\tfrac{1}{2}) k_0^{-1} \ln \left(\frac{k_1 k_0^2 p_0^2}{k_2^2 k_u} \right). \tag{2.19}$$

Using our example data (Table 2.1), we find $t_{cross} = 1614\,s$ and $t_{max} = 3912\,s$.

2.4.3. Predicted behaviour

We may now make a prediction that if we integrate the full reaction rate equations (2.1)–(2.3) then while the assumptions involved in the PSS hypothesis hold, the actual concentrations of A and B will be well approximated

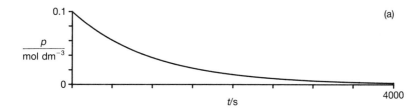

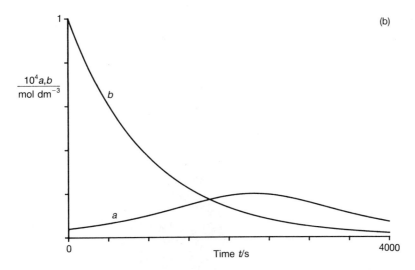

FIG. 2.3. Computed concentration histories for the cubic autocatalytic model with rate data from Table 2.1 *except* for the uncatalysed reaction rate constant $k_u = \tfrac{1}{4}$: (a) the exponential decay of the precursor reactant P; (b) intermediate species concentrations $a(t)$ and $b(t)$ showing typical pseudo-stationary evolution over the whole course of the reaction.

by the above results. There will need to be a short initial period during which a and b grow from zero to their appropriate pseudo-steady values, but once the concentrations are close to eqns (2.15) and (2.16) we may expect steady evolution. A typical computer run showing exactly this agreement is shown in Fig. 2.3: the values used for the rate constants are given in the legend and correspond to those in Table 2.1 except that we take $k_u = \frac{1}{4}$.

If the computations are made with all the data taken from Table 2.1, however, a remarkable difference appears (Fig. 2.4). There is an initial transient behaviour during which the concentrations of the intermediates move quickly to the appropriate pseudo-stationary values. The PSS curves

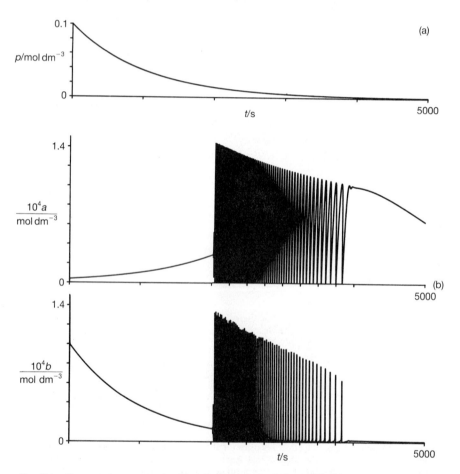

FIG. 2.4.   Computed concentration histories for autocatalytic model with rate constants given exactly as in Table 2.1: (a) exponential decay of precursor; (b) intermediate concentrations $a(t)$ and $b(t)$, showing initial pseudo-stationary-state behaviour but subsequent development of an oscillatory period of finite duration, $1752\ \text{s} < t < 3940\ \text{s}$.

Table 2.2

Analytical expressions relating to oscillatory behaviour

Birth of oscillations:

     oscillations begin at time $t_1^* = (1/k_0) \ln(p_0/p_1^*)$

     where $p_1^* = (k_2^2/2k_1 k_0^2)^{1/2} \{k_2 - 2k_u + [k_2(k_2 - 8k_u)]^{1/2}\}$

     initial oscillatory period $t_p = 2\pi/(k_2 - k_u)$

Death of oscillations:

     oscillations cease at time $t_2^* = (1/k_0)\ln(p_0/p_2^*)$

     where $p_2^* = (k_2^2/2k_1 k_0^2)^{1/2} \{k_2 - 2k_u - [k_2(k_2 - 8k_u)]^{1/2}\}$

     final oscillatory period $t_p = 2\pi/(k_2 k_u)^{1/2}$

Number of oscillations:

$$N_{osc} = \frac{1}{4\pi} \frac{k_2}{k_0} \left(\frac{2k_u}{k_2}\right)^{1/4} \ln\left(\frac{k_2}{k_u}\right)$$

are then followed for a while and the curves cross close to the conditions specified in eqns (2.11), (2.12), (2,17), and (2.18). After this, however, the system begins to move away from the predicted curves. The concentrations of A and B begin to oscillate at $t \approx 1752$ s. The amplitude of the oscillations grows at first and the period lengthens. In all, 81 excursions can be seen for this particular example. Oscillatory behaviour dies away at $t \approx 3940$ s, i.e. approximately $36\frac{1}{2}$ minutes after it began. The concentrations return to the pseudo-stationary curves and decay to zero.

There is a period of time or, if we prefer, a range of reactant concentration over which the system *spontaneously* moves away from the pseudo-stationary state. The idea that stationary states may be *unstable* is not widely appreciated in chemical kinetics but it is fundamental to the analysis of oscillatory systems.

Can we understand this instability and, if so, can we predict when it will occur? We would like to map out the experimental conditions under which oscillatory behaviour can be expected in terms of the rate constants k_0 etc. and the concentration of the reactant. In fact we can do even better than this. We will see that much can be said about the details of the oscillations: when they start, how they grow in period and amplitude, how long they will last, how and when they die out, and how many we can expect to see. Some typical results are given in Table 2.2.

2.5. Stability of stationary states

We introduced in the previous section the idea that stationary states need not always be stable. Here we wish to consider further what we mean by stability in this context. Stationary states can only correspond to resting states that can be achieved in practice if they are (locally) stable.

Stability relates to the behaviour of a system when it is subjected to a small perturbation away from a given stationary state (or if fluctuations occur naturally). If the perturbation decays to zero, the system has some in-built tendency to return back to the same state. In this case it is described as *locally stable* . (The qualification 'local' means that very large perturbations may have different consequences.) We will introduce the relatively simple mathematical techniques required to determine this local stability of a given state in Chapter 3. It will also be useful before then to reduce the reaction rate equations (2.1)–(2.3) to their simplest possible form by introducing dimensionless variables and quantities.

At this stage, however, we may proceed to the important features of the model's behaviour qualitatively from Fig. 2.4. For some range of time or of reactant concentration the pseudo-stationary state described by eqns (2.15) and (2.16) become locally unstable. Let this range be denoted

$$p_2^* \leqslant p_0 \leqslant p_1^* . \tag{2.20}$$

If we recall that the concentration of P is decreasing in time, we can note that this instability starts when p reaches p_1^* and stability will be regained when $p = p_2^*$.

We will soon show that the upper and lower bounds of the oscillatory region are given approximately by

$$p_{1,2}^* = (k_2^2/2k_1 k_0^2)^{1/2} \{k_2 - 2k_u \pm [k_2(k_2 - 8k_u)]^{1/2}\} . \tag{2.21}$$

For the data in Table 2.1, eqn (2.21) gives $p_1^* = 0.0197 \text{ mol dm}^{-3}$ and $p_2^* = 2.04 \times 10^{-3} \text{ mol dm}^{-3}$.

The time taken for the reactant concentration to fall from its initial value to p_1^* gives the induction period for oscillations, t_1^*. The decay of the reactant concentration is exponential, as given by eqn (2.6), so

$$p_1^* = p_0 \exp(-k_0 t_1^*) \Rightarrow t_1^* = k_0^{-1} \ln(p_0/p_1^*). \tag{2.22}$$

Using the value for p_1^* just derived and the data from Table 2.1, we predict an induction time of 1625 s (≈ 27 min).

In a similar way, the system will reach the end of the oscillatory period when the reactant concentration falls to p_2^* at time t_2^* given by

$$t_2^* = k_0^{-1} \ln(p_0/p_2^*). \tag{2.23}$$

Again for our example data we have 3892 s, so the oscillations should cease 1 h and 5 min from the start of the reaction. The difference between these two times $t_2^* - t_1^*$ gives the duration of the oscillatory epoch:

$$t_2^* - t_1^* = k_0^{-1} \ln(p_1^*/p_2^*) \tag{2.24}$$

so for our example we expect the oscillations to cease approximately 38 min after they begin.

These formulae actually tend to underestimate t_1^* and t_2^* (they are in fact strict lower bounds). Comparing these figures with the corresponding numerical computation in Fig. 2.4, we see that the onset of oscillatory behaviour is delayed slightly until $t \approx 1752$ s (i.e. by a further 2 min). The oscillations also survive slightly longer than predicted, the final vestiges dying away at $t \approx 4000$ s, but the difference is again less than 2 min. The reasons for these discrepancies can be understood, and even quantified, but that will be done in the next chapter.

During the period of instability, the system will move spontaneously away from the stationary state. For the present model there is only ever one stationary state, so there is no other resting state for the system to move to. The concentrations of A and B must vary continuously in time. They eventually tend to a periodic oscillatory motion around the unstable state. We thus see oscillations over the range of conditions described by (2.20).

We can characterize the oscillations in terms of their size (amplitude) and the period between successive peaks. It is particularly useful to establish how the amplitude and period vary with the reactant concentration. One way of doing this is artificially to hold p constant and then integrate the rate equations until $a(t)$ and $b(t)$ settle down to a steady oscillation. Figure 2.5 shows the stable oscillatory response obtained from eqns (2.2) and (2.3) with the reactant concentration held constant at the value $p = 0.01 \text{ mol dm}^{-3}$, inside the range of instability. The concentration of species A varies between a maximum value of $1.36 \times 10^{-4} \text{ mol dm}^{-3}$ and a minimum of $2.77 \times 10^{-7} \text{ mol dm}^{-3}$. The difference between the maximum and minimum gives the amplitude of the oscillation appropriate to this value of p (and to the particular values of the rate constants used from Table 2.1), $1.36 \times 10^{-4} \text{ mol dm}^{-3}$. The period can easily be read off from the figure as the time between successive maxima: $t_p = 19.0$ s. Similarly, $b(t)$ has a maximum of $1.235 \times 10^{-4} \text{ mol dm}^{-3}$ and a minimum of $6.48 \times 10^{-7} \text{ mol dm}^{-3}$, so the oscillatory amplitude is $1.229 \times 10^{-4} \text{ mol dm}^{-3}$.

As well as showing how a and b vary in time, the numerical traces can also be plotted in a different way. If we plot the two concentrations against each other (Fig. 2.5(c)), we find that they draw out a closed curve or 'limit cycle' around which the system circulates. This limit cycle surrounds the unstable pseudo-stationary state appropriate to p. The amplitude of the oscillations is a measure of the size of this limit cycle.

Repeating such computations for other values of p within the range of instability, we can see how the waveform, amplitude, and period vary with the reactant concentration. Figure 2.6(a) plots the maximum and minimum values of the concentration of A as a function of p. This representation emphasizes how the system oscillates about the unstable stationary state

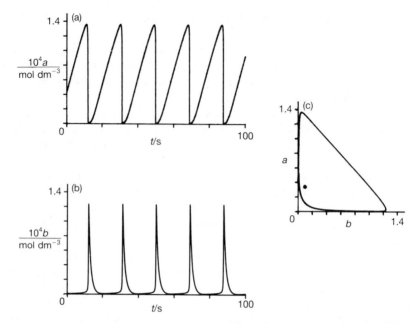

FIG. 2.5. Sustained oscillations in intermediate concentrations $a(t)$ and $b(t)$ for pool chemical model with precursor reactant concentration $p = 0.01 \, \text{mol dm}^{-3}$. Also shown is the corresponding limit cycle in the a–b phase plane.

a_{ss} between p_2^* and p_1^* (where the stationary-state locus is shown as a broken curve). Figures 2.6(b)–(d) then show the actual oscillations in time for selected reactant concentrations $p = 0.0195$, 0.01, and 0.005. The actual amplitude and period are shown as functions of p in Fig. 2.7.

Close to the upper end of the range of instability, the oscillations have small amplitude and a short period: near p_1^*, the waveform is close to sinusoidal. As p is decreased the excursions increase in amplitude, quite quickly, attaining a maximum at $p \approx 0.015 \, \text{mol dm}^{-3}$ with the particular values of the rate constants used here. The period is now longer and the waveform less symmetric. At yet lower reactant concentrations, the amplitude decreases slightly: the period continues to increase smoothly as p decreases over most of the oscillatory range, and the oscillations become more and more 'sawtooth' in form. Finally, extremely close to p_2^*, the oscillatory amplitude and the period decrease rapidly again. The amplitude tends to zero, although the period remains finite.

For a full understanding of the changes in the oscillations and the associated limit cycle as p varies, we need a three-dimensional representation, such as Fig. 2.8. This shows the stationary-state locus as a line passing through space in p–a–b coordinates (the concentration 'phase space'). The

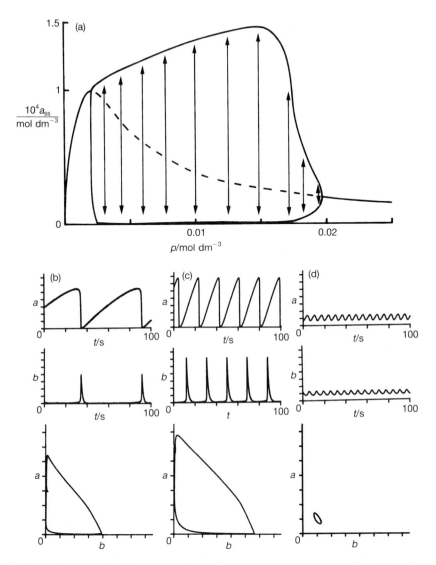

FIG. 2.6. Variation of oscillatory waveform with precursor reactant concentration: (a) stationary-state locus for intermediate A, $a(p)$, showing region of instability for $p_2^* \leqslant p \leqslant p_1^*$ indicating the magnitude of the oscillations; (b) oscillations in a and b and the corresponding limit cycle for $p = 0.005$; (c) oscillations in a and b and the corresponding limit cycle for $p = 0.010$; (d) oscillations in a and b and the corresponding limit cycle for $p = 0.0195$.

locus is a curved path: where the line is solid the stationary state is stable; where it is broken it is unstable. The 'cocoon' around the unstable portion represents the varying size of the limit cycle corresponding to the oscillatory behaviour. As p decreases in time we can imagine the concentrations follow-

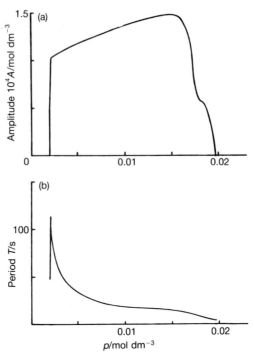

FIG. 2.7.   Variation in (a) absolute amplitude and (b) oscillatory period across region of instability for the pool chemical model with rate data from Table 2.1. (The qualitative form is appropriate for all combinations of rate constants giving oscillatory behaviour with this model.)

ing the locus, moving from high p to lower values. As it reaches the region of instability it moves from this path and winds on to the limit cycle envelope, returning to the stable path again at the other end.

As k_u increases, the range of reactant concentrations over which the system is unstable decreases. From eqn (2.21) we can see that oscillatory behaviour will only be possible provided $k_2 > 8k_u$, otherwise there are no real solutions for p_1^* and p_2^*. This means that the uncatalysed step converting A to B must not proceed too quickly compared with the rate at which B can decay to the final product C.

If we consider a system in which k_u is very small compared with k_2, so $k_u/k_2 \ll 1$, we can expand (2.21) to obtain for the points of onset and death of oscillatory behaviour:

upper root $$p_1^* \approx (k_2/k_1 k_0^2)^{1/2}(k_2 - \tfrac{3}{2}k_u) \qquad (2.25)$$

lower root $$p_2^* \approx (k_u/k_1 k_0^2)^{1/2}(k_2 + 2k_u). \qquad (2.26)$$

To compare these results with the point where the two stationary-state loci

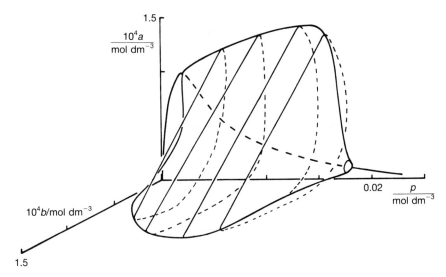

FIG. 2.8. Three-dimensional representation of stationary-state locus, with the growth and development of the surrounding limit cycle, in a–b–p space for the pool chemical model.

cross and with the maximum in a_{ss} we can expand (2.12) and (2.14) in the same way. This gives

for crossing: $p \approx (k_2/k_1 k_0^2)^{1/2} (k_2 - \tfrac{1}{2}k_u)$ (2.27)

for maximum: $p = (k_u/k_1 k_0^2)^{1/2}k_2$.   (2.28)

Thus the upper end of the oscillatory region p_1^* lies close to the conditions at which the two stationary-state loci cross, but at slightly lower reactant concentration. The lower end of the unstable region p_2^* lies at slightly higher reactant concentrations than that at which the locus $a_{ss}(p)$ has its maximum.

For larger values of the uncatalysed reaction rate constant, i.e. as k_u approaches $\tfrac{1}{8}k_2$, the two solutions of (2.21) move closer together and so the region of instability and oscillations shrinks. The two points merge when $k_u = \tfrac{1}{8}k_2$, when

$$p_1^* = p_2^* = (\tfrac{3}{8}k_2^3/k_1 k_0^2)^{1/2}.$$ (2.29)

For the data in Table 2.1, this gives $p_1^* = p_2^* = 0.0122 \text{ mol dm}^{-3}$.

2.6. Simple tests for oscillatory instability

The general philosophy of our test for oscillatory instability can be summarized as follows. At the pseudo-steady state, the net rates of change of the intermediate concentrations are zero: $da/dt = db/dt = 0$. If we make a small addition of A, increasing a, how does da/dt respond? If the net rate of production of A decreases from zero because of this increase in a, then da/dt becomes negative and the concentration of A will decay back towards its unperturbed pseudo-steady value. This has the hallmarks of stability. If, instead, da/dt were to be increased to a positive value by the addition of A, the concentration would then tend to grow further, moving away from the pseudo-steady state. This has implications of instability. Similar considerations apply to how db/dt changes with a small addition of B.

To quantify this idea in mathematical terms, we can recognize that we are really talking about the partial derivative quantities $\partial(da/dt)/\partial a$ and $\partial(db/dt)/\partial b$. Stability has been associated in some sense with these two quantities being negative (i.e. da/dt decreases as a increases, so $\partial(da/dt)/\partial a < 0$), instability with these being positive. In most 'normal' chemical systems, e.g. those with deceleratory kinetics, the two partial derivatives will be negative. It is a characteristic of autocatalysis, however, that at least one of these may become positive — at least over some ranges of composition and experimental conditions.

In the present case, the partial derivatives can be obtained from eqns (2.2) and (2.3) as

$$\partial(da/dt)/\partial a = -k_1 b^2 - k_u$$

$$\partial(db/dt)/\partial b = +2k_1 ab - k_2.$$

Clearly $\partial(da/dt)/\partial a$ will always be negative. For the intermediate B (the autocatalyst) the situation is less clear: $\partial(db/dt)/\partial b$ contains both positive and negative contributions. In order to determine the sign of this partial derivative we need to specify the values of a and b. As we are testing the stability of the pseudo-steady state, we should use a_{ss} and b_{ss}. It is easiest to work in terms of the reactant concentration p rather than explicitly in time t, so we take (2.9) and (2.10). Then

$$\frac{\partial}{\partial a}\left(\frac{da}{dt}\right) = -[k_1(k_0 p/k_2)^2 + k_u] \tag{2.30}$$

$$\frac{\partial}{\partial b}\left(\frac{db}{dt}\right) = k_2 \frac{k_0^2 k_1 p^2 - k_u k_2^2}{k_0^2 k_1 p^2 + k_u k_2^2}. \tag{2.31}$$

There may thus be a range of reactant concentrations for which the numerator of eqn (2.31) is positive. The system is stable to perturbations of A but unstable to perturbations in B.

When one partial derivative is positive and the other negative, stability of the system as a whole is determined by which of the two has the larger magnitude or, equivalently, the sign of the sum $\partial(da/dt)/\partial a + \partial(db/dt)\partial b$. If both partial derivatives are negative (we know then that the system will be stable), their sum will be negative. If both were positive (and the system is totally unstable), then the sum would be positive. If the partial derivatives have opposite sign, but their sum is negative, then the system will still be stable; if the sum is positive, the system will be unstable. For the above equations,

$$\frac{\partial}{\partial a}\left(\frac{da}{dt}\right) + \frac{\partial}{\partial b}\left(\frac{db}{dt}\right) = k_2 \frac{k_0^2 k_1 p^2 - k_u k_2^2}{k_0^2 k_1 p^2 + k_u k_2^2} - [k_1 (k_0 p/k_2)^2 + k_u]. \quad (2.32)$$

The pseudo-steady state changes stability whenever the sum of the two partial derivatives becomes equal to zero. The right-hand side of eqn (2.32) can be viewed as a quadratic in p^2 whose roots we have given earlier as eqn (2.21).

2.7. Behaviour of model in absence of uncatalysed reaction

In the previous section we saw that oscillatory behaviour will be favoured if the uncatalysed reaction converting A to B is slow compared with the decay of B to C. It is interesting, therefore, to consider briefly the system in which the uncatalysed reaction is omitted, i.e. where k_u is set equal to zero. This special case actually has a few problems associated with it, but also has some important features. (The problems are related to the anomalous induction periods of autocatalytic reactions discussed in §1.4 and will be discussed later in this chapter.)

The crossing point in the stationary-state loci then occurs at

$$a_{ss} = b_{ss} = (k_2/k_1)^{1/2} \qquad \text{at} \quad p = k_2^{3/2}/k_0 k_1^{1/2}.$$

Instability, and hence potential oscillatory behaviour, also sets in exactly at this point. Taking the limit of eqn (2.25) also gives

$$p_1^* = k_2^{3/2}/k_0 k_1^{1/2}.$$

The oscillatory period at this point is simply related to the value of the rate constant for the step B → C, with $t_p(p_1^*) = 2\pi/k_2$. If $k_2 = 1 \text{ s}^{-1}$, the period is approximately 6 s.

With $k_u = 0$, p_2^* is also zero, so we expect oscillations to survive until all the reactant has been consumed. In fact this behaviour is not found numerically and surprisingly the model ceases to represent 'reasonable chemistry'.

We can recognize that because k_1 and k_2 have respectively the units of first- and third-order rate constants, the ratio $(k_2/k_1)^{1/2}$ has the units of concentration. (For the data in Table 2.1 it has the value $2 \times 10^{-5} \text{ mol dm}^{-3}$.) This quotient also turns out to be a sensible reference concentration, offering a natural scale on which to measure concentrations in this model. Some concentrations may be judged as being small compared with $(k_2/k_1)^{1/2}$, others as being large. For example, the concentration of reactant P in eqn (2.14) exceeds $(k_2/k_1)^{1/2}$ by a factor of k_2/k_0, or two orders of magnitude with the data from Table 2.1. Thus the reactant has a 'high concentration'.

2.8. Influence of reversibility

So far all the reaction steps have been considered as being totally irreversible. This choice has been made in the interests of keeping the model at its simplest possible level. The fact that the model shows oscillations under such conditions is revealing, as it clearly demonstrates that oscillatory behaviour does not correspond to particular elementary steps sometimes proceeding forwards, at other times running backwards. We should also show, on the other hand, that oscillations are not a consequence of our simplification. All the qualitative results derived above should be seen in the model with reverse reactions included, and the quantitative relationships for these more general forms should clearly reduce to those already obtained in the limit of high values for the equilibrium constants for the various steps.

We will be particularly interested in the influence of reversibility on the two 'propagation' steps involving the intermediates A and B:

$$A \rightleftharpoons B \qquad k_u, k_{-u}$$

$$A + 2B \rightleftharpoons 3B \qquad k_1, k_{-1}.$$

Both of these steps have the net stoichiometry $A \rightarrow B$ and therefore have the same free energy change $\Delta G^0 = \Delta G_f^0(B) - \Delta G_f^0(A)$. This, in turn, means that both reactions must have the *same* value for their equilibrium constant K_e. This latter point can also be seen from the equilibrium requirement that the forward and reverse reaction rate should be equal:

$$k_u a_{eq} = k_{-u} b_{eq} \Rightarrow b_{eq}/a_{eq} = k_u/k_{-u} = K_e$$

$$k_1 a_{eq} b_{eq}^2 = k_{-1} b_{eq}^3 \Rightarrow b_{eq}/a_{eq} = k_1/k_{-1} = K_e.$$

In this latter case, although we have made use of the equality between forward and reverse rate, which only holds at equilibrium, the resulting relationship $k_u/k_{-u} = k_1/k_{-1} = K_e$ only involves (rate and equilibrium) constants and therefore holds in general.

The reaction rate equation for the intermediate A can now be written as

$$da/dt = k_0 p - k_1 ab^2 + k_{-1} b^3 - k_u a + k_{-u} b$$

$$= k_0 p - (k_1 b^2 + k_u)(a - K_e^{-1} b). \tag{2.33}$$

Similarly, for the autocatalytic species B, we find

$$db/dt = (k_1 b^2 + k_u)(a - K_e^{-1} b) - k_2 b. \tag{2.34}$$

These equations differ from (2.2) and (2.3) by extra terms involving $K_e^{-1}b$. Clearly, for reactions which lie well to the right at equilibrium, so that K_e is large, this can remain only a small correction. In the limit of irreversible steps $K_e \to \infty$ and so the inverse K_e^{-1} tends to zero and we regain our previous rate equations.

The stationary states of the reversible system are given by

$$a_{ss} = \frac{k_0 k_2^2 p}{k_1 k_0^2 p^2 + k_u k_2^2} + \frac{k_0 p}{k_2 K_e} \tag{2.35}$$

$$b_{ss} = \frac{k_0 p}{k_2} . \tag{2.36}$$

The second expression, for b_{ss}, is independent of the equilibrium constant and has exactly the same form as that derived for the irreversible system (eqn (2.9)). For the intermediate A, the stationary-state concentration is increased by the reversibility of the steps: the first term in eqn (2.30) is that corresponding to the irreversible solution (eqn (2.10)), the second is proportional to the inverse of the equilibrium constant. Thus, as $K_e \to \infty$, a_{ss} tends smoothly to our previous result.

The new stationary-state locus showing the dependence of a_{ss} on the reactant concentration p is shown in Fig. 2.9 and now has both a maximum and, at high concentrations, a minimum. The locations of these turning points can be obtained by differentiating eqn (2.35) with respect to p. For large values of K_e, the leading order expressions are as follows.

For the maximum

$$p = \frac{k_u^{1/2} k_2}{k_1^{1/2} k_0} \left(1 + \frac{2k_u}{k_2 K_e} \cdots \right) \tag{2.37}$$

$$a_{ss} = \frac{1}{2} \frac{k_2}{(k_1 k_u)^{1/2}} \left(1 + \frac{2k_u}{k_2 K_e} \cdots \right) . \tag{2.38}$$

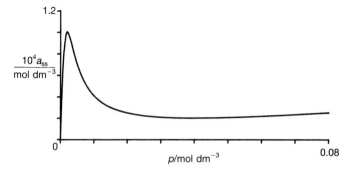

FIG. 2.9. The influence of reversibility on the form of the stationary-state locus for intermediate concentration $a(p)$, showing the appearance of a minimum at high reactant concentrations.

For the minimum

$$p = \frac{k_2 K_e^{1/2}}{k_1^{1/2} k_0} + \cdots \tag{2.39}$$

$$a_{ss} = 2 \left(\frac{k_2}{K_e k_1} \right)^{1/2} \left(1 - \frac{1}{2} \frac{k_u}{k_2 K_e} \cdots \right). \tag{2.40}$$

Equations (2.37) and (2.38) tend smoothly to the previous results (eqns (2.13) and (2.14)) as the system approaches irreversibility ($K_e \to \infty$): for finite K_e, the maximum stationary-state concentration of A increases as K_e decreases and occurs at a higher reactant concentration. The minimum in the locus lies at high reactant concentrations, $p \propto K_e^{1/2}$, and corresponds to a low concentration of the intermediate $a_{ss} \propto K_e^{-1/2}$.

The loci for a_{ss} and b_{ss} still have a crossing point. Again for large K_e, the leading-order expression for the location of this point is

$$p = \frac{k_1^{3/2}}{k_1^{1/2} k_0} (1 + \tfrac{1}{2} K_e^{-1} \cdots). \tag{2.41}$$

The conditions under which the above stationary-state solution loses its stability can be determined following the recipe of §2.6. Again we find that instability may arise, and hence oscillatory behaviour is possible, in this reversible case. The condition for the onset of instability can be expressed in terms of the reactant concentration $p_2^* \leqslant p \leqslant p_1^*$, where

$$(p_{1,2}^*)^2 = \frac{k_2^3}{2k_1 k_0^2 (1 + K_e^{-1})} \left[1 - \frac{2k_u}{k_2} (1 + K_e^{-1}) \right.$$

$$\left. \pm \left(1 - \frac{8k_u}{k_2} (1 + K_e^{-1}) \right)^{1/2} \right]. \tag{2.42}$$

This is very similar to eqn (2.21), to which it reduces in the limit $K_e \to \infty$ when the terms $(1 + K_e^{-1})$ tend to unity.

Again we require real roots to this expression if oscillations are to be observable, and this, in turn, puts certain limits on the relative sizes of k_2 and k_u and now also on K_e. If the uncatalysed rate does not occur, so $k_u = 0$, reversibility has no effect on the possibility of oscillations. Otherwise, the region of instability increases as the reactions become more irreversible. The general condition for instability over some range of reactant concentrations is now

$$k_u < \tfrac{1}{8} k_2 / (1 + K_e^{-1}) \qquad \text{i.e.} \quad K_e > 8 k_u / (k_2 - 8 k_u). \tag{2.43}$$

Whenever this condition is satisfied, the resulting region of unstable solutions lies strictly between the crossing point and the maximum in the stationary-state loci.

We can see from the above discussion that oscillatory behaviour can be found if the reactions are made irreversible and also, with only small and smooth changes, if the reversibility is recognized. Oscillations are neither the consequence of reversibility nor of its neglect.

2.9. Summary

In addition to the general aims set out at the beginning of this chapter we have discovered a wealth of specific detail about the behaviour of the simple kinetic model introduced here. Most results have been obtained analytically, despite the non-linear equations involved, with numerical computation reserved for confirmation, rather than extension, of our predictions. Much of this information has been obtained using the idea of a pseudo-stationary state, and regarding this as not just a function of time but also as a function of the reactant concentration. Stationary states can be stable or unstable.

Though reduced to the barest of essentials, the scheme shows many features observed in real examples of oscillatory reactions: a pre-oscillatory period, a period of oscillatory behaviour, and then a final monotonic decay of reactant and intermediate concentrations to their equilibrium values. We can identify from the model such features as the dependence of the length of the pre-oscillatory period on the initial reactant concentration and the rate constants, an estimate for the number of oscillations, and the length of the oscillatory phase. By tuning the parameters we can obtain as many oscillations as we wish.

References

The application of cubic autocatalysis to model chemical and biochemical schemes in thermodynamically closed systems is dealt with in the following references.

Goldbeter, A. and Nicolis, G. (1976). An allosteric enzyme model with positive feedback applied to glycolytic oscillations. *J. Theor. Biol.*, **4**, 65–160.

Gray, P. and Scott, S. K. (1986). A new model for oscillatory behaviour in closed systems: the autocatalator. *Ber. Bunsenges. Phys. Chem.*, **90**, 985–96.

Hanusse, P. (1973). Etude des systemes dissipatifs chimiques à deux et trois especes intermediare. *C. R. Acad. Sci., Paris*, **C 277**, 263.

Merkin, J. H., Needham, D. J., and Scott, S. K. (1986). Oscillatory chemical reactions in closed vessels. *Proc. R. Soc.*, **A 406**, 299–323.

Prigogine, I. and Lefever, R. (1968). Symmetry breaking instabilities in dissipative systems. *J. Chem. Phys.*, **48**, 1695–700.

Selkov, E. E. (1968). Self-oscillations in glycolysis. *Eur. J. Biochem.*, **4**, 79–86.

Tracqui, P., Perault-Staub, A. M., Milhaud, G., and Staub, J. F. (1987). Theoretical study of a two-dimensional autocatalytic model for calcium dynamics at the extracellular fluid-bone interface. *Bull. Math. Biol.*, **49**, 597–613.

Tyson, J. J. (1973). Some further studies of non-linear oscillations in chemical systems. *J. Chem. Phys.*, **58**, 3919–30.

Tyson, J. J. and Kauffman, S. (1975). Control of mitosis by a continuous biochemical oscillation. *J. Math. Biol.*, **1**, 289–310.

Tyson, J. J. and Light, J. C. (1973). Properties of two-component bimolecular and trimolecular chemical reaction schemes. *J. Chem. Phys.*, **59**, 4164–73.

OSCILLATIONS IN A CLOSED ISOTHERMAL
SYSTEM: MATHEMATICAL ANALYSIS

This chapter presents the systematic analysis of the scheme $P \rightarrow A \rightarrow B \rightarrow C$ where the middle reaction has both direct and self-catalysed components. The aims are to enable the reader:

(1) to exploit the benefits of compact dimensionless equations and to recognize the physical significance of terms appearing in them;

(2) to construct a stationary-state scaffolding (with $[P]$ = constant) for the non-stationary 'real' case and to determine the locus of stationary states;

(3) to establish from first principles the local stability of stationary states and the conditions for changes in the character of stationary states;

(4) to deduce, for the particular model scheme, conditions for Hopf bifurcation;

(5) to relate all the foregoing to the prediction and interpretation of behaviour in a closed system.

The previous chapter has provided some indication of the behaviour which can be exhibited by the simple cubic autocatalysis model. In order to make a full analysis, it is convenient both for algebraic manipulation and as an aid to clarity to recast the rate equations in dimensionless terms. This is meant to be a painless procedure (and beloved of chemical engineers even though traditionally mistrusted by chemists). We aim wherever possible to make use of symbols which can be quickly identified with their most important constituents: thus for the dimensionless concentration of A we have α, with β for the dimensionless concentration of B. Once this transformation has been achieved, we can embark on a quite detailed and comprehensive analysis of the behaviour of this prototype chemical oscillator.

3.1. Dimensionless equations

The reaction rate equations for the cubic autocatalysis model of the previous chapter are

$$dp/dt = -k_0 p \tag{3.1}$$

$$da/dt = k_0 p - k_1 ab^2 - k_u a \tag{3.2}$$

$$db/dt = k_1 ab^2 + k_u a - k_2 b. \tag{3.3}$$

These involve three concentration variables (a, b, p) and four parameters given by the different rate constants $(k_0, k_1, k_u, \text{and } k_2)$. In fact there are other parameters involved implicitly as we must also specify the initial concentrations of P, A, and B before we can do any computation or analysis.

The main purpose in transforming the above set of dimensional equations to an equivalent dimensionless form is to enable us to identify any terms which may be particularly important in determining the behaviour of the system. Generally, such terms will be revealed by some dimensionless parameters being much larger than others. An additional bonus of the process, however, is often that we can also reduce the number of parameters in the equations. Thus it may turn out that the behaviour is not determined primarily by the absolute values of all the rate constants, but only by one or two ratios of them.

(a) Dimensionless concentrations

How do we make concentrations etc. dimensionless? We do so by dividing the physical concentration by some characteristic measure or concentration scale. In the previous chapter we found that the stationary-state loci cross at a particular point where, in the limit $k_u = 0$, $a_{ss} = b_{ss} = (k_2/k_1)^{1/2}$. With this quotient, the model is actually suggesting a concentration scale to us, so we now divide the concentration of A by this quantity to obtain the dimensionless form α. We must use the same reference scale for all concentrations, so our dimensionless forms will be

for reactant P: $\pi = (k_1/k_2)^{1/2} p$ (3.4)

for intermediate A: $\alpha = (k_1/k_2)^{1/2} a$ (3.5)

for intermediate B: $\beta = (k_1/k_2)^{1/2} b.$ (3.6)

For example, with the data in Table 2.1 the dimensionless initial concentration of the reactant $\pi_0 = 5000$. (Recall that this concentration is meant to be 'large'.) Also, the maximum concentration of a_{ss} is given by eqn (2.13), so

$$(\alpha_{ss})_{max} = (k_1/k_2)^{1/2} a_{max} = \tfrac{1}{2}(k_2/k_u)^{1/2}. \tag{3.7}$$

For our example data, with $k_2 = 1\,\text{s}^{-1}$ and $k_u = 10^{-2}\,\text{s}^{-1}$, we have $(\alpha_{ss})_{max} = 5$.

(b) Dimensionless time

In order to make time dimensionless we again make use of the results at the simple crossing point. When $k_u = 0$ we found that the natural frequency of the oscillations at the crossing point (where they begin) is k_2. This is a first-order rate constant and has units of (time)$^{-1}$, so we can use its inverse k_2^{-1} as a natural timescale for this model. Instead of timing events in minutes or seconds the system times them in multiples of $1/k_2$. We may now define our dimensionless time τ as

$$\tau = k_2 t. \tag{3.8}$$

For the example data we have been considering $k_2 = 1 \text{ s}^{-1}$, so for this case τ just becomes the time in seconds: $\tau \equiv t/\text{s}$. If we had taken $k_2 = 10^3 \text{ s}^{-1}$, the dimensionless system would be working on a millisecond timescale.

(c) Dimensionless rate constant ratios

Two ratios of rate constants will emerge in our equations: k_0/k_2 and k_u/k_2. As k_0 and k_u are also pseudo-first-order rate constants these two groups are dimensionless and we will denote their values as follows:

$$\varepsilon = k_0/k_2 \quad \text{and} \quad \kappa_u = k_u/k_2. \tag{3.9}$$

The use of the symbol κ_u for the second of these quantities allows its meaning to remain clear. Rather than taking the equivalent symbol κ_0 for the first group we have chosen to use ε, which will declare in advance to mathematicians our prejudice that this will in general be a small quantity. For our example data, $\varepsilon = 1 \times 10^{-3}$ and $\kappa_u = 1 \times 10^{-2}$. We may also note that there will be no need for a dimensionless measure of the rate constant k_1. This is because we have already involved it in the concentration scale.

The values of the above dimensionless groups for the kinetic data from Table 2.1 are listed in Table 3.1, together with some of the specific results we will derive in the following sections.

3.1.1. Dimensionless forms of rate equations

The reaction rate equations for the present model, eqns (3.1)–(3.3), can be cast in terms of the above dimensionless groups as follows:

for P: $\quad d\pi/d\tau = -\varepsilon\pi$ \hfill (3.10)

for A: $\quad d\alpha/d\tau = \varepsilon\pi - \alpha\beta^2 - \kappa_u\alpha$ \hfill (3.11)

for B: $\quad d\beta/d\tau = \alpha\beta^2 + \kappa_u\alpha - \beta$ \hfill (3.12)

and the initial conditions become

$$\pi = \pi_0 \quad \alpha = \beta = 0 \quad \text{at} \quad \tau = 0. \tag{3.13}$$

Equation (3.10) can be integrated to give

$$\pi = \pi_0 e^{-\varepsilon\tau}. \tag{3.14}$$

Table 3.1

Typical values for dimensionless groups

$\pi_0 = 5000$	$\varepsilon = 1 \times 10^{-3}$	$\kappa_u = 1 \times 10^{-2}$	$\mu_0 = \varepsilon\pi_0 = 5$
$(\alpha_{ss})_{max} = 5$	when $\mu = 0.1$		
$\alpha_{ss} = \beta_{ss} = 0.995$	when $\mu = 0.995$		
$\mu_1^* = 0.9847$	$\mu_2^* = 0.1021$		

We have said that the reactant P will decay slowly. This is equivalent to requiring ε to be a small number, very much less than unity. Thus, by requiring a slow decay, we mean slow compared with the timescale of the reaction B → C: k_0 must be very small compared with k_2.

Substituting eqn (3.14) into the rate equation for A:

$$d\alpha/d\tau = \varepsilon\pi_0 e^{-\varepsilon\tau} - \alpha\beta^2 - \kappa_u\alpha. \tag{3.15}$$

We can see that this equation involves ε twice. On its first occurrence it is multiplied by the dimensionless initial concentration of the reactant π_0. It has already been mentioned that π_0 will generally be a very large number, $\pi_0 \gg 1$. The first term in eqn (3.15) therefore involves the multiple of a small number ε and a large number π_0 as well as the exponential term involving ε. If π_0 is of the same order of magnitude as the *inverse* of ε (we say if π_0 is of the order of ε^{-1}, or write $\pi_0 \approx O(\varepsilon^{-1})$) then their product in (3.15) will be neither large nor small. We can thus express their product as another dimensionless group μ_0 which will then be defined by

$$\mu_0 = \varepsilon\pi_0 = k_0 k_1^{1/2} p_0/k_2^{3/2}. \tag{3.16}$$

We can think of μ_0 as another dimensionless measure of the initial reactant concentration.

Introducing μ_0 into eqn (3.15) for $d\alpha/d\tau$, we have

$$d\alpha/d\tau = \mu_0 e^{-\varepsilon\tau} - \alpha\beta^2 - \kappa_u\alpha \tag{3.17}$$

$$= \mu - \alpha\beta^2 - \kappa_u\alpha \tag{3.18}$$

where we have used a shorthand notation

$$\mu = \mu_0 e^{-\varepsilon\tau}. \tag{3.19}$$

This term represents a measure of the reactant concentration at any time τ or, more correctly, of the dimensionless rate at which P is being converted to A. Using the data in Table 2.1 we have $\mu_0 = 5$. Because the exponential term involves ε, the decay of μ is slow: for $\tau = 0.693\,\varepsilon^{-1}$ ($= 693$ for our example data) $\mu = \frac{1}{2}\mu_0$ and so the reactant concentration and the rate have fallen to half their initial values. We will make use of this slow variation in μ when we assess the pseudo-stationary-state behaviour and local stability of the model. Just as we investigated a_{ss} and b_{ss} in terms of p rather than of time, we will often prefer to look at the dependences of α_{ss} and β_{ss} on μ rather than on τ.

The pair of dimensionless rate equations for the two intermediate concentrations are now

$$d\alpha/d\tau = \mu - \alpha\beta^2 - \kappa_u\alpha \qquad (3.20)$$

$$d\beta/d\tau = \alpha\beta^2 + \kappa_u\alpha - \beta. \qquad (3.21)$$

These equations involve two variables, α and β, and two parameters, μ and κ_u. This is at least a more economical representation than the full dimensional forms where we had five parameters—the rate constants and p_0.

We now turn again to the evaluation of the pseudo-stationary-state responses.

3.1.2. Stationary-state solutions

The stationary states are found by setting $d\alpha/d\tau = d\beta/d\tau = 0$, so that

$$\mu - \alpha_{ss}\beta_{ss}^2 - \kappa_u\alpha_{ss} = 0 \qquad (3.22)$$

$$\alpha_{ss}\beta_{ss}^2 + \kappa_u\alpha_{ss} - \beta_{ss} = 0. \qquad (3.23)$$

These give

$$\alpha_{ss} = \mu/(\mu^2 + \kappa_u) \qquad (3.24)$$

$$\beta_{ss} = \mu. \qquad (3.25)$$

Substituting for μ and κ_u from (3.16) and (3.09) we recover the dimensional forms presented earlier in (2.9) and (2.10).

In a given experiment μ is, in effect, being varied for us (by the exponential decay in (3.19)). The dimensionless uncatalysed reaction rate constant κ_u would remain constant during a given experiment, but might be varied from one run to another. In our analysis we seek first to determine how α_{ss} and β_{ss} depend on μ for a fixed value of κ_u. We then investigate how this dependence changes as κ_u varies.

The stationary-state concentration β_{ss} of the autocatalyst shows a linear dependence on the reactant concentration μ, as shown in Fig. 3.1. The locus for $\alpha_{ss}(\mu)$ shows a maximum:

$$(\alpha_{ss})_{max} = \tfrac{1}{2}\kappa_u^{-1/2} \qquad \text{at} \quad \mu = \kappa_u^{1/2}. \qquad (3.26)$$

Recalling that κ_u will usually be small, this can be a relatively large value, occurring at low values of the reactant concentration. For the example data in Table 2.1, $\kappa_u = 10^{-2}$ and hence $(\alpha_{ss})_{max} = 5$ and is achieved when $\mu = 0.1$. Somewhat surprisingly, the maximum stationary-state concentration of the

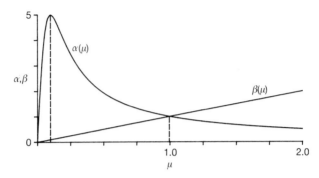

FIG. 3.1.   Stationary-state loci $\alpha(\mu)$ and $\beta(\mu)$ for the dimensionless concentrations of the inter-mediate species A and B: numerical values correspond to the parameter values in Table 3.1.

intermediate A does not depend on the initial concentration of the reactant P from which it is produced. This independence can also be seen from eqn (2.13).

The loci $\alpha_{ss}(\mu)$ and $\beta_{ss}(\mu)$ cross when

$$\alpha_{ss} = \beta_{ss} = (1 - \kappa_u)^{1/2} \qquad \text{when} \quad \mu = (1 - \kappa_u)^{1/2} \qquad (3.27)$$

i.e. close to unity. With the rate constants in Table 2.1, the crossing occurs at $\alpha_{ss} = \beta_{ss} = \mu = 0.995$.

Turning briefly to the special case of no uncatalysed reaction $\kappa_u = 0$, the various results above take even simpler forms. For the stationary states we then have

$$\alpha_{ss} = 1/\mu \qquad \beta_{ss} = \mu \qquad (3.28)$$

and these loci cross (Fig. 3.2) when

$$\alpha_{ss} = \beta_{ss} = \mu = 1. \qquad (3.29)$$

However, we may also see an immediate problem: as the reactant concentration μ approaches zero, the value of the stationary-state concentration of A becomes infinite, which is physically unrealistic. This problem does not occur if κ_u is given any non-zero value, no matter how small.

Equations (3.20) and (3.21) with their stationary-state solutions (3.24) and (3.25) are simple enough to provide a good introduction to some of the mathematical techniques which can serve us so well in analysing these sorts of chemical models. In the next sections we will explain the ideas of local stability analysis (§3.2) and then apply them to our specific model (§3.3). After that we introduce the basic aspects of a technique known as the Hopf bifurcation analysis (§3.4) which enables us to locate the conditions under which oscillatory states are likely to appear. We set out only those aspects that are required within this book, without any pretence at a complete

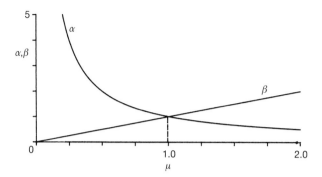

FIG. 3.2.   Stationary-state loci $\alpha(\mu)$ and $\beta(\mu)$ for the dimensionless concentrations of the intermediate species A and B for the scheme without the uncatalysed reaction $A \rightarrow B$, $\kappa_u = 0$.

coverage. Various excellent mathematical monographs can be found which cover these techniques in detail and with rigour; some which we have found particularly helpful are listed at the end of this chapter.

3.2. Local stability analysis: mathematics and recipe

To determine the local stability of a given stationary state we need to know whether infinitesimally small perturbations of the system about that state will grow (in which case it is unstable) or decay (stability). This idea is analogous to the principles of relaxation experiments for measuring fast reaction rate constants when a system is disturbed from its stationary or equilibrium state and the return is followed. Rather than present the theory in general terms, we will concentrate here on two-variable systems. These show sufficient complexity to permit a wide range of behaviour, including sustained oscillations, whilst retaining almost complete algebraic tractability.

An additional approximation to be made concerns the concentration of the reactant, i.e. the value of μ. We have seen that in many cases of interest this will only be varying very slowly. If we can regard this as a very slow process, then the concentration of P will appear constant on the much faster timescale with which the concentrations of the intermediates change. In other words, we are hoping that while we are watching the decay or growth of a small perturbation we can regard μ as having a constant value.

The general philosophy has already been rehearsed in §2.6 of the previous chapter. There are, however, some additional intricacies which must be considered in a full analysis, and other ways in which a stationary state can become unstable. These aspects are considered now. Those readers less interested in this mathematical background can move straight to §3.3 where the method is applied—consulting Table 3.2 and Fig. 3.3 as appropriate.

3.2.1. Time dependence of small perturbations

Equations (3.20) and (3.21) are typical examples of a general class of equations which can be written in the shorthand form

$$d\alpha/d\tau = f(\alpha, \beta) \tag{3.30}$$

$$d\beta/d\tau = g(\alpha, \beta). \tag{3.31}$$

Here the notation $f(\alpha, \beta)$ and $g(\alpha, \beta)$ is taken to represent that the derivatives are given by functions of the variables α and β. For the present model we have

$$f(\alpha, \beta) = \mu - \alpha\beta^2 - \kappa_u \alpha \tag{3.32}$$

$$g(\alpha, \beta) = \alpha\beta^2 + \kappa_u \alpha - \beta. \tag{3.33}$$

The stationary states are defined generally as the solutions of

$$f(\alpha_{ss}, \beta_{ss}) = 0 \qquad g(\alpha_{ss}, \beta_{ss}) = 0. \tag{3.34}$$

If we now imagine that the system is given a sudden small perturbation, so that the new concentrations of A and B become

$$\alpha = \alpha_{ss} + \Delta\alpha \qquad \beta = \beta_{ss} + \Delta\beta \tag{3.35}$$

with $\Delta\alpha \ll 1$ and $\Delta\beta \ll 1$, then the rate of change of the perturbations can be obtained by substituting these forms into eqns (3.30) and (3.31) and expanding as a Taylor series. This gives, quite simply,

$$d\,\Delta\alpha/d\tau = d\alpha/d\tau = f(\alpha_{ss}, \beta_{ss}) + (\partial f/\partial\alpha)_{ss}\Delta\alpha$$
$$+ (\partial f/\partial\beta)_{ss}\Delta\beta + \cdots \tag{3.36}$$

$$d\,\Delta\beta/d\tau = d\beta/d\tau = g(\alpha_{ss}, \beta_{ss}) + (\partial g/\partial\alpha)_{ss}\Delta\alpha$$
$$+ (\partial g/\partial\beta)_{ss}\Delta\beta + \cdots \tag{3.37}$$

where the higher-order terms involve $(\Delta\alpha)^2$, $(\Delta\beta)^2$, or $(\Delta\alpha\Delta\beta)$ and higher-order partial derivatives. The subscripts 'ss' on $\partial f/\partial\alpha$ etc. indicate that these partial derivatives are evaluated at the stationary state under consideration. Because the perturbations have been assumed to be infinitesimally small, the squares and higher-order powers can usually be neglected. We may also use eqn (3.34) to eliminate the first terms on the right-hand side. Thus, to leading order, the time dependence of the perturbations is governed by a pair of linear equations

$$d\,\Delta\alpha/d\tau = (\partial f/\partial\alpha)_{ss}\Delta\alpha + (\partial f/\partial\beta)_{ss}\Delta\beta \tag{3.38}$$

$$d\,\Delta\beta/d\tau = (\partial g/\partial\alpha)_{ss}\Delta\alpha + (\partial g/\partial\beta)_{ss}\Delta\beta. \tag{3.39}$$

Linear ordinary differential equations are relatively easily manipulated and yield solutions which are simply the sum of exponential terms:

$$\Delta\alpha(\tau) = c_1 e^{\lambda_1 \tau} + c_2 e^{\lambda_2 \tau} \tag{3.40}$$

$$\Delta\beta(\tau) = c_3 e^{\lambda_1 \tau} + c_4 e^{\lambda_2 \tau}. \tag{3.41}$$

The coefficients c_1–c_4 are merely constants which depend on the size and sign of the initial perturbation. More important here, however, are the exponents λ_1 and λ_2. It is the sign (positive or negative) and character (real or complex) of these two quantities which determine whether and how the perturbations decay or grow. These exponents can be determined as the eigenvalues of the Jacobian matrix of eqns (3.38) and (3.39):

$$\mathbf{J} = \begin{pmatrix} \partial f/\partial\alpha & \partial f/\partial\beta \\ \partial g/\partial\alpha & \partial g/\partial\beta \end{pmatrix}_{ss} \tag{3.42}$$

where the subscript 'ss' again signifies that the elements are evaluated at the stationary state. The eigenvalues λ_1 and λ_2 are given by the roots of the quadratic equation

$$\lambda^2 - \mathrm{tr}(\mathbf{J})\lambda + \det(\mathbf{J}) = 0 \tag{3.43}$$

where $\mathrm{tr}(\mathbf{J})$ is the trace of the Jacobian matrix (the sum of the terms on the leading diagonal)

$$\mathrm{tr}(\mathbf{J}) = (\partial f/\mathrm{d}\alpha)_{ss} + (\partial g/\partial\beta)_{ss} \tag{3.44}$$

and $\det(\mathbf{J})$ is the determinant

$$\det(\mathbf{J}) = (\partial f/\mathrm{d}\alpha)_{ss}(\partial g/\partial\beta)_{ss} - (\partial f/\mathrm{d}\beta)_{ss}(\partial g/\partial\alpha)_{ss}. \tag{3.45}$$

Note that the first constituent of the trace is the partial derivative of $\mathrm{d}\alpha/\mathrm{d}\tau$ with respect to α and the second is that of $\mathrm{d}\beta/\mathrm{d}\tau$ with respect to β.

We may thus write

$$\lambda_{1,2} = \tfrac{1}{2}\{\mathrm{tr}(\mathbf{J}) \pm [\mathrm{tr}(\mathbf{J})^2 - 4\det(\mathbf{J})]^{1/2}\}. \tag{3.46}$$

There are a number of possibilities for the sign and character of λ_1 and λ_2 depending on the signs and relative magnitudes of $\mathrm{tr}(\mathbf{J})$ and $\det(\mathbf{J})$. These are summarized in Table 3.2 and Fig. 3.3, and discussed in turn below.

(a) $tr(J) < 0$, $det(J) > 0$, $tr(J)^2 - 4det(J) > 0$

With these inequalities, both λ_1 and λ_2 are real and have negative values. All of the exponential terms in eqns (3.40) and (3.41) will then decrease monotonically to zero and the perturbed system will return to the same stationary state. The stationary state is stable, because the perturbations decay, and is called a *node* because of the final monotonic approach.

Table 3.2

Classification of local stability and character in terms of Jacobian matrix and its eigenvalues

tr($\mathbf{J}$)	det($\mathbf{J}$)	tr($\mathbf{J}$)^2 − 4det($\mathbf{J}$)	$\lambda_{1,2}$	Character and stability
−	+	+	Real, both − ve	Stable node (monotonic approach)
−	+	−	Complex, real parts − ve	Stable focus (damped oscillatory approach)
0	+	−	Imaginary, (real parts = 0)	Hopf bifurcation point or centre
+	+	−	Complex, real parts + ve	Unstable focus (oscillatory divergence)
+	+	+	Real, both + ve	Unstable node (monotonic divergence)
±	0	+	One zero, one − ve or one + ve	Saddle–node bifurcation point
±	−	+	Real, one − ve, one + ve	Saddle point (unstable)
0	0	0	both zero	Double-zero eigenvalue bifurcation point

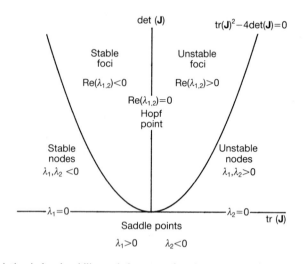

FIG. 3.3. Variation in local stability and character of stationary-state solutions with the values of the trace and determinant of the corresponding Jacobian matrix (see also Table 3.2).

(b) $tr(\boldsymbol{J}) < 0$, $det(\boldsymbol{J}) > 0$, $tr(\boldsymbol{J})^2 - 4det(\boldsymbol{J}) < 0$

The last inequality here means that the eigenvalues λ_1 and λ_2 will be complex numbers

$$\lambda_{1,2} = \text{Re}(\lambda) \pm i\text{Im}(\lambda) \tag{3.47}$$

with

$$\text{Re}(\lambda) = \tfrac{1}{2}\text{tr}(\boldsymbol{J}) \qquad \text{Im}(\lambda) = \tfrac{1}{2}[4det(\boldsymbol{J}) - \text{tr}(\boldsymbol{J})^2]^{1/2}. \tag{3.48}$$

The real parts will have the same sign as each other and as the trace of the Jacobian, so in this case they will be negative. The sum of the exponentials of such a complex conjugate pair is equivalent to the product of an exponential decay and a cosine function, i.e.

$$\Delta\alpha = c_1 e^{\text{Re}(\lambda t)} \cos\left[\text{Im}(\lambda t) + \phi_1\right] \tag{3.49}$$

$$\Delta\beta = c_2 e^{\text{Re}(\lambda t)} \cos\left[\text{Im}(\lambda t) + \phi_2\right] \tag{3.50}$$

where ϕ_1 and ϕ_2 are arbitrary phase angles. The decaying terms ensure a return to the original stationary state (i.e. stability), but because of the cosine functions this is a damped oscillatory approach, known as *focal* behaviour.

(c) $tr(\boldsymbol{J}) > 0$, $det(\boldsymbol{J}) > 0$, $tr(\boldsymbol{J})^2 - 4det(\boldsymbol{J}) < 0$

Again λ_1 and λ_2 are complex eigenvalues of the form $\text{Re}(\lambda) \pm i\text{Im}(\lambda)$, but now they have positive real parts. Thus the exponential terms in (3.47) and (3.48) grow in time. The perturbation grows away from the unstable stationary state in a divergent oscillatory or unstable focal manner.

(d) $tr(\boldsymbol{J}) > 0$, $det(\boldsymbol{J}) > 0$, $tr(\boldsymbol{J})^2 - 4det(\boldsymbol{J}) > 0$

Now λ_1 and λ_2 are real and both positive. The exponential terms in (3.40) and (3.41) all increase monotonically in time. The perturbations grow exponentially and the system moves directly away from the unstable nodal state.

(e) $det(\boldsymbol{J}) < 0$

When the determinant of the Jacobian is negative, the roots of eqn (3.43) will be real and have opposite sign irrespective of $\text{tr}(\boldsymbol{J})$. One of the exponential terms in each of (3.41) and (3.42) (that corresponding to the negative eigenvalue) decreases exponentially; the other, with the positive root, will increase in time. Only in the very special case that the c_i multiplying the growing exponential in (3.40) and (3.41) are exactly zero will the perturbation decay back to the stationary state. This requires a very special perturbation. For general perturbations, such that c_1 to c_4 are all non-zero, the growing terms will eventually dominate and the system will move away from the stationary state. This is known as *saddle point* behaviour.

As well as the five possibilities for λ_1 and λ_2 discussed above, there are two special cases which will be particularly important in locating and classifying the conditions under which the behaviour exhibited by our models changes ('bifurcates').

(f) tr(J) = 0, det(J) > 0 (Hopf bifurcation)

Under these conditions, when the trace of the Jacobian is changing sign, the eigenvalues become imaginary numbers:

$$\lambda_{1,2} = \pm i\det(\mathbf{J})^{1/2} = \pm i\omega_0. \tag{3.51}$$

The time dependence of the perturbations in (3.40) and (3.41) would then be equivalent to an undamped cosine function of frequency ω_0, leading to indefinitely sustained oscillations.

In fact we must work a little harder under these conditions as higher-order terms in the full expressions for $d\,\Delta\alpha/d\tau$ and $d\,\Delta\beta/d\tau$ have now to be studied, and this simple oscillatory response may be modified or not realized at all in practice. Nevertheless, the change in local stability associated with the condition

$$\text{tr}(\mathbf{J}) = 0 \qquad \det(\mathbf{J}) > 0 \tag{3.52}$$

is characteristic of the conditions under which oscillatory behaviour first appears. These Hopf bifurcations are of sufficient importance to deserve a separate section to themselves and will be a recurrent theme throughout this book.

(g) det(J) = 0

When the determinant of the Jacobian matrix becomes zero, one of the roots of (3.43) also becomes zero. This represents the point at which a node (stable or unstable depending on the sign of tr(**J**)) is just changing to a saddle point or vice versa. Such saddle–node bifurcations, characterized by

$$\det(\mathbf{J}) = 0 \tag{3.53}$$

frequently (but not always) involve the coming together and mutual destruction of two different stationary states, as we shall see in later chapters when we deal with systems which can have multiple stationary solutions.

3.2.2. Phase plane representation of local stabilities

The concentrations α and β vary in time as they approach or move away from any particular stationary state. Often it is convenient to visualize the time-dependent behaviour another way, by plotting the variation of one concentration against that of the other, in what is known as the α–β phase plane.

In the phase plane we are really considering the solutions of the differential equation

$$d\alpha/d\beta = f(\alpha, \beta)/g(\alpha, \beta). \qquad (3.54)$$

As α and β vary in time, or equivalently as eqn (3.54) is solved from some given initial set of conditions, the α–β locus marks out a trajectory on the phase plane.

Stationary-state solutions correspond to conditions for which both numerator and denominator of (3.54) vanish, giving $d\alpha/d\beta = 0/0$, and so are singular points in the phase plane. There will be one singular point for each stationary state: each of the different local stabilities and characters found in the previous section corresponds to a different type of singularity. In fact the terms node, focus, and saddle point, as well as limit cycle, come from the patterns on the phase plane made by the trajectories as they approach or diverge. Stable stationary states or limit cycles are often refered to as 'attractors', unstable ones as 'repellors' or 'sources'. The different phase plane patterns are shown in Fig. 3.4.

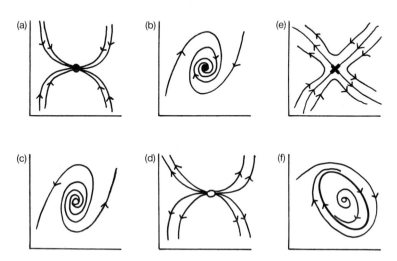

Fig. 3.4. Representations of the different singular points in the concentration phase plane: (a) stable node, sn; (b) stable focus, sf; (c) unstable focus uf; (d) unstable node, un; (e) saddle point, sp.

3.3. Local stability analysis: application

We can now return to our autocatalytic reaction scheme which is waiting to have the local stability of its stationary state analysed. Although we only have one solution for any given μ and κ_u, we can expect that just as the value of this state varies with the parameters, so might its local stability.

In terms of the previous section, the functions f and g are given by eqns (3.32) and (3.33), with stationary states given by eqns (3.24) and (3.25):

$$f(\alpha, \beta) = \mu - \alpha\beta^2 - \kappa_u \alpha \tag{3.32}$$

$$g(\alpha, \beta) = \alpha\beta^2 + \kappa_u \alpha - \beta \tag{3.33}$$

$$\alpha_{ss} = \mu/(\mu^2 + \kappa_u) \tag{3.24}$$

$$\beta_{ss} = \mu. \tag{3.25}$$

Evaluating the four partial derivatives we obtain:

$$(\partial f/\partial \alpha)_{ss} = -(\beta_{ss}^2 + \kappa_u) \quad = -(\mu^2 + \kappa_u) \tag{3.55}$$

$$(\partial f/\partial \beta)_{ss} = -2\alpha\beta_{ss} \quad = -2\mu^2/(\mu^2 + \kappa_u) \tag{3.56}$$

$$(\partial g/\partial \alpha)_{ss} = (\beta_{ss}^2 + \kappa_u) \quad = (\mu^2 + \kappa_u) \tag{3.57}$$

$$(\partial g/\partial \beta)_{ss} = 2\alpha\beta_{ss} - 1 \quad = (\mu^2 - \kappa_u)/(\mu^2 + \kappa_u). \tag{3.58}$$

The determinant of $\mathbf{J}$ is given in this example simply by

$$\det(\mathbf{J}) = \mu^2 + \kappa_u \tag{3.59}$$

and hence is always positive. In this model, therefore, we can expect neither saddle points (case (e)) nor saddle–node bifurcations (case (g)).

The trace of $\mathbf{J}$, defined by eqn (3.44), becomes

$$\text{tr}(\mathbf{J}) = -(\mu^2 + \kappa_u) + (\mu^2 - \kappa_u)/(\mu^2 + \kappa_u)$$

$$= -[\mu^4 - (1 - 2\kappa_u)\mu^2 + \kappa_u(1 + \kappa_u)]/(\mu^2 + \kappa_u). \tag{3.60}$$

For $\kappa_u = 10^{-2}$ and $\mu = 5$ the trace is negative, $\text{tr}(\mathbf{J}) = -24.01$, indicating stability. Later, in an experiment, when the precursor concentration has fallen so that $\mu = 0.5$, we find that $\text{tr}(\mathbf{J}) = +0.66$: the trace has changed sign and become positive, so the stationary state will have become locally unstable.

Before considering the consequences of these expressions, we will take the simple case in which the uncatalysed reaction makes no contribution, with $\kappa_u = 0$, as an example.

3.3.1. Model without uncatalysed reaction

With $\kappa_u = 0$, the trace and determinant have the simpler forms

$$\text{tr}(\mathbf{J}) = (1 - \mu^2) \tag{3.61}$$

$$\det(\mathbf{J}) = \mu^2. \tag{3.62}$$

The discriminant $\Delta = \text{tr}(\mathbf{J})^2 - 4\det(\mathbf{J})$ is given by

$$\Delta = \mu^4 - 6\mu^2 + 1. \tag{3.63}$$

The character and local stability of the stationary-state solution, eqns (3.24) and (3.25), now depend only on μ, the dimensionless initial concentration of the reactant. There are four different patterns in this system:

1. $\mu > 1 + \sqrt{2}$. Here $\text{tr}(\mathbf{J})$ is negative, $\det(\mathbf{J})$ is positive, and the discriminant Δ is also positive. We thus have case (a) of §3.2.1. The stationary state is a stable node for these values of μ, and small perturbations decay monotonically.

2. $1 < \mu < 1 + \sqrt{2}$. In this range $\text{tr}(\mathbf{J})$ is negative and $\det(\mathbf{J})$ is positive but Δ is now negative. The eigenvalues λ_1 and λ_2 are complex conjugates with negative real parts (case (b) of §3.2.1). The stationary state is a stable focus and perturbations decay in a damped oscillatory way.

3. $\sqrt{2} - 1 < \mu < 1$. Over this range the trace of the Jacobian is positive and the discriminant is negative. The stationary state is now an unstable focus (case (c) of §3.2.1). Small perturbations grow in a divergent oscillatory way and the system can no longer sit at the stationary state.

4. $0 < \mu < \sqrt{2} - 1$. Here the trace is still positive and the discriminant has changed sign and become positive too. The stationary state is an unstable node (case (d) of §3.2.1). Perturbations now show exponential growth.

These changes in stability and character are marked on the stationary-state loci in Fig. 3.5. This also shows clearly that the loss of stability occurs as the two loci cross, when $\mu = 1$. At this point eqn (2.84) is also satisfied with $\text{tr}(\mathbf{J}) = 0$, corresponding to the special case (f) of §3.2.1 and holding the promise of the onset of oscillatory behaviour. We return to this point later.

3.3.2. Model with uncatalysed reaction

The behaviour of the system with κ_u non-zero shows many similarities, but also some important differences, to the simpler case just examined. The more complex expression for the trace, eqn (3.60), involves both parameters, is

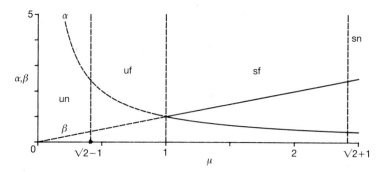

FIG. 3.5. Stationary-state loci $\alpha(\mu)$ and $\beta(\mu)$ for the dimensionless concentrations of the inter-mediate species A and B and with no uncatalysed step ($\kappa_u = 0$), showing changes in local stability and character at $\mu = \sqrt{2} + 1$ (sn to sf), $\mu = 1$ (sf to uf, i.e. Hopf bifurcation), $\mu = \sqrt{2} - 1$ (uf to un).

a quartic (although in a particularly helpful biquadratic form), and leads to an equation for Δ which has eighth powers and which is not algebraically tractable in any way. Thus we can obtain no explicit form for the change from node to focus (real to complex eigenvalues). Nevertheless, the condition

$$\Delta = 0 \qquad (3.64)$$

is easily computed numerically and we know, from the discussion above, that the roots of this equation must tend to $\sqrt{2} \pm 1$ as κ_u tends to zero. Figure 3.6 shows the boundary obtained from eqn (3.64) separating nodal from focal states in the μ–κ_u parameter plane.

The quartic expression for $\text{tr}(\mathbf{J})$ conveniently only involves even powers of μ. The condition for the change of local stability, $\text{tr}(\mathbf{J}) = 0$, therefore is a quadratic equation in μ^2 with roots given by

$$(\mu_{1,2}^*)^2 = \tfrac{1}{2}[(1 - 2\kappa_u) \pm (1 - 8\kappa_u)^{1/2}]. \qquad (3.65)$$

Because of the importance of points where the trace of the Jacobian matrix vanishes, we will denote such values of the parameters by a superscript asterisk. Equation (3.65) has two real roots, provided the dimensionless rate constant for the uncatalysed step has a value less than $\tfrac{1}{8}$. In terms of the original rate constants this requires $k_2 > 8k_u$ as presented previously (§2.5). For the data in Table 2.3, $\mu_1^* = 0.9847$ and $\mu_2^* = 0.1021$. As κ_u increases, these two bifurcation points move closer together: for $\kappa_u = 0.1$, $\mu_1^* = 0.790$ and $\mu_2^* = 0.420$, so the oscillatory range is smaller.

For very small values of κ_u, we may expand (3.65) to give

$$\mu_1^* \approx 1 - \tfrac{3}{2}\kappa_u - \cdots \qquad (3.66)$$

$$\mu_2^* \approx \kappa_u^{1/2}(1 + 2\kappa_u) + \cdots \qquad (3.67)$$

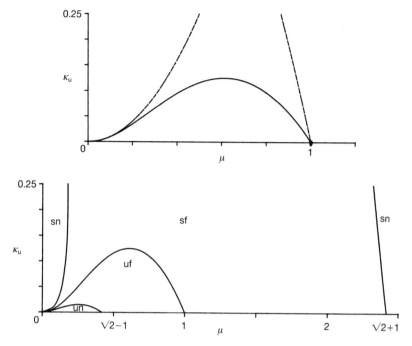

FIG. 3.6.   The dependence of local stability and character of the stationary-state solution on the parameters μ amd κ_u. (a) The locus of Hopf bifurcation points $\mu^*(\kappa_u)$ with $\text{tr}(\mathbf{J}) = 0$; beneath this locus the stationary state is unstable, above it is stable. Also shown (as broken curves) are the loci corresponding to the maximum in the $\alpha_{ss}(\mu)$ locus and the crossing point where $\alpha_{ss}(\mu) = \beta_{ss}(\mu)$. The region of instability lies completely between these points. (b) Division of the parameter plane into regions of stable node (sn), stable focus (sf), unstable focus (uf), and unstable node (un) the Hopf curve is the same as that shown in (a).

With our example data, these give 0.9850 and 0.1020 respectively. These locations may be compared with two other values of μ: (i) the crossing point for the stationary-state loci, which for small κ_u is given by $\mu \sim 1 - \frac{1}{2}\kappa_u$ and which therefore lies slightly above μ_1^*; and (ii) the maximum in the α_{ss} locus which occurs at $\mu = \kappa_u^{1/2}$ and therefore lies just below μ_2^*. These conclusions about the relative positions of the various features just described in fact hold for all values of κ_u for which unstable stationary states can occur, not just for the limit $\kappa_u \rightarrow 0$. The curves in the μ–κ_u parameter plane corresponding to each of these conditions are also shown in Fig. 3.6.

Figure 3.7 shows the changes of stability and character occurring along the stationary-state loci for a typical value of κ_u. For μ slightly greater than μ_2^*, the solution is a stable focus; for μ slightly less than μ_1^* it is an unstable focus. The stationary state is unstable over the whole range of μ between μ_1^* and μ_2^*, but may change its character to that of an unstable node within this region if the discriminant can change sign. This change has no immediate effect for us,

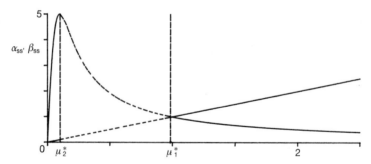

FIG. 3.7.   Stationary-state loci $\alpha(\mu)$ and $\beta(\mu)$ showing changes in local stability when the uncatalysed step is included in the model and $\kappa_u < \frac{1}{8}$, showing two Hopf bifurcation points μ_1^* and μ_2^*. Particular numerical values correspond to parameter values in Table 3.1.

but in a later chapter we will see that it is important for the possibility of spatial pattern formation.

When the dimensionless reactant concentration is slightly greater than the lower root of eqn (3.65), i.e. $\mu > \mu_2^*$, the stationary state must be an unstable focus, which becomes stable as μ passes through μ_2^*. Below this the state is first a stable focus then, as μ approaches zero, a stable node.

If the uncatalysed reaction rate increases with respect to the rate of catalyst decay, so that k_u becomes larger than $\frac{1}{8}k_2$, there are no real solutions to eqn (3.60). The stationary state can no longer become unstable as μ is varied. Damped oscillatory responses can still be observed when we have a stable focus, but undamped oscillations will not be found.

The question of what happens to the system in the range of instability, and how the concentrations of A and B vary as they move away from the unstable stationary state, leads us to the study of sustained oscillatory behaviour. Before a full appreciation of the latter can be obtained, however, we must rehearse the relevant theoretical background. Fortunately the autocatalytic model is again an exemplary system with which to introduce at least the basic aspects of the Hopf bifurcation, and we will do this in the next section.

3.4. Oscillatory behaviour: Hopf bifurcation analysis

In the previous sections we have implied that the loss of local stability which occurs for a stationary-state solution as the real part of the eigenvalues changes from negative to positive is closely linked to the conditions under which sustained oscillatory responses are born.

For a two-variable system, the condition for the real part to vanish is simply that the trace of the Jacobian matrix should become zero, with the determinant being positive, as discussed in case (f) of §3.2.1. These requirements are met for the present model at the points μ_1^* and μ_2^* and we have seen that $\det(\mathbf{J})$ is always positive. We can thus make use of a mathematical result, known as the Hopf bifurcation theorem, which guarantees that as the parameter μ is varied through μ_1^* or μ_2^* a non-stationary or periodic state emerges (or disappears if we go in the opposite direction). This periodic solution will exist over some non-zero range of experimental conditions. These periodic solutions appear as sustained oscillations in the concentrations $\alpha(\tau)$ and $\beta(\tau)$ and as closed loops or limit cycles in the α–β phase plane.

Before we can conclude, in general, that a given system will begin to show oscillatory behaviour between two Hopf bifurcation points we must attend to a few additional requirements of the theorem.

3.4.1. Degeneracy

First we must check that the real part of the eigenvalues, $\mathrm{Re}(\lambda) = \frac{1}{2}\mathrm{tr}(\mathbf{J})$, actually passes through zero and becomes positive for some range. It will not do that if $\mathrm{Re}(\lambda)$ has a maximum at zero. To avoid a maximum, we require that at the points μ_1^* and μ_2^* the derivative of $\mathrm{tr}(\mathbf{J})$ with respect to μ should be non-zero, i.e.

$$\mathrm{d}\,\mathrm{tr}(\mathbf{J})/\mathrm{d}\mu = -2\mu[\mu^4 + 2\kappa_u\mu^2 - \kappa_u(2 - \kappa_u)]/(\mu^2 + \kappa_u)^2 \neq 0 \quad (3.68)$$

when $\mathrm{tr}(\mathbf{J}) = 0$. In fact this inequality is satisfied for all the Hopf bifurcation points μ_1^* and μ_2^* provided the uncatalysed reaction rate constant $\kappa_u < \frac{1}{8}$. For $\kappa_u = \frac{1}{8}$, the Hopf points merge and there is then no range of μ over which $\mathrm{tr}(\mathbf{J})$ can be positive.

If we have conditions for which

$$\mathrm{tr}(\mathbf{J}) = 0 \quad \text{and} \quad \mathrm{d}\,\mathrm{tr}(\mathbf{J})/\mathrm{d}\mu = 0 \quad\quad (3.69)$$

simultaneously, the Hopf bifurcation is said to be 'degenerate'. At this sort of degeneracy, typically, two Hopf bifurcation points merge and cancel each other out. These are often the conditions for which oscillatory responses cease to be possible in a given system. Other types of degeneracy can occur and will be introduced in the next chapter. They are of no concern with the present model.

3.4.2. Stability of oscillations

We must also examine the stability of the periodic solution and its limit cycle as it emerges from the bifurcation point. Just as stationary states may be stable or unstable, so may oscillatory solutions. If they are stable they may be observable in practice; if they are unstable they will not be directly observable although their existence still has some physical relevance. We will give the recipe for evaluating the stability and character of a Hopf bifurcation in the next chapter when we discuss a different model for which the algebra is slightly more straightforward. As a reward for the extra effort required for this analysis we gain a great deal of additional information. Not only can we tell whether the limit cycle is born stable or unstable, we can find out whether it grows in size as μ is reduced or as μ is increased; we can also find how quickly it grows and how the oscillatory period varies.

3.4.3. Growth of oscillations

The size and period of the oscillations, or of the corresponding limit cycle, varies with the dimensionless reactant concentration μ. We may determine this dependence in a similar way to that used in §2.5. Close to the Hopf bifurcation points we can in fact determine the growth analytically, but in general we must employ numerical computation. For now we will merely present the basic result for the present model. The qualitative pattern of response is the same for all values of $\kappa_u < \frac{1}{8}$.

As the dimensionless concentration of the reactant decreases so that μ just passes through the upper Hopf bifurcation point μ_1^* in Fig. 3.8, so a stable limit cycle appears in the phase plane to surround what is now an unstable stationary state. Exactly at the bifurcation point, the limit cycle has zero size. The corresponding oscillations have zero amplitude but are born with a finite period. The limit cycle and the amplitude grow smoothly as μ is decreased. Just below the bifurcation, the oscillations are essentially sinusoidal. The amplitude continues to increase, as does the period, as μ decreases further, but eventually attains a maximum somewhere within the range $\mu_2^* < \mu < \mu_1^*$. As μ approaches the lower bifurcation point μ_2^* from above, the oscillations decrease in size and period. The amplitude falls to zero at this lower bifurcation point, but the period remains non-zero.

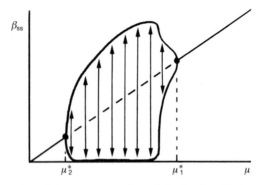

FIG. 3.8.   Representation of the onset, growth, and death of oscillations in the isothermal autocatalytic model as μ varies for reaction with the uncatalysed step included, showing emergence of the stable limit cycle at μ_1^* and its disappearance at μ_2^*.

There are no unstable limit cycles in this model, and the oscillatory solution born at one bifurcation point exists over the whole range of stationary-state instability, disappearing again at the other Hopf bifurcation. Both bifurcations have the same 'character' (stable limit cycle emerging from zero amplitude), although they are mirror images, and are called 'supercritical' Hopf bifurcations.

3.4.4. Poincaré–Bendixson theorem

Another useful rule which can frequently guide us to situations where oscillatory solutions will be found is the Poincaré–Bendixson theorem. This states that if we have a unique stationary state which is unstable, or multiple stationary states all of which are unstable, but we also know that the concentrations etc. cannot run away to infinity or become negative, then there must be some other non-stationary atractor to which the solutions will tend. Basically this theorem says that the concentrations cannot just wander around for an infinite time in the finite region to which they are restricted; they must end up somewhere. For two-variable systems, the only other type of attractor is a stable limit cycle. In the present case, therefore, we can say that the system must approach a stable limit cycle and its corresponding stable oscillatory solution for any value of μ for which the stationary state is unstable.

This quick test does not, however, tell us that there will be only one stable limit cycle, or give any information about how the oscillatory solutions are born and grow, nor whether there can be oscillations under conditions where the stationary state is stable. We must also be careful in applying this theorem. If we consider the simplified version of our model, with no uncatalysed step, then we know that there is a unique unstable stationary state for all reactant concentrations such that $\mu < 1$. However, if we integrate the mass-balance equations with $\mu = 0.9$, say, we do not find limit cycle behaviour. Instead the concentration of B tends to zero and that for A become infinitely large (growing linearly with time). In fact for all values of μ less than 0.900 32, the concentration of A becomes unbounded and so the Poincaré–Bendixson theorem does not apply.

3.5. Conserving mass: explicit incorporation of reactant consumption

Our approach of regarding α and β as functions of μ has served us well. It has allowed us to identify stationary-state solutions and how they vary with the reactant concentration and the uncatalysed reaction rate. By examining the local stability of these solutions we have also been able to obtain simple formulae which tell us the realms of experimental conditions over which we should expect to see oscillations in the concentrations of the intermediates.

We may now allow for reactant consumption explicitly by restoring the exponential decay, $p = p_0 e^{-k_0 t}$. In dimensionless terms this means recognizing that μ is a time-dependent parameter, $\mu = \mu_0 e^{-\varepsilon \tau}$. The governing rate equations are

$$d\alpha/d\tau = \mu_0 e^{-\varepsilon\tau} - \alpha\beta^2 - \kappa_u \alpha \qquad (3.70)$$

$$d\beta/d\tau = \alpha\beta^2 + \kappa_u \alpha - \beta. \qquad (3.71)$$

These have only one true stationary-state solution, $\alpha_{ss} = \beta_{ss} = 0$ (as $\tau \to \infty$), corresponding to complete conversion of the initial reactant to the final product C. This stationary or chemical equilibrium state is a stable node as required by thermodynamics, but of course that tells us nothing about how the system evolves in time. If ε is sufficiently small, we may hope that the concentrations of the intermediates will follow pseudo-stationary-state histories which we can identify with the results of the previous sections. In particular we may obtain a guide to the kinetics simply by replacing μ by $\mu_0 e^{-\varepsilon\tau}$ wherever it occurs in the stationary-state and Hopf formulae. Thus at any time τ the dimensionless concentrations α and β would be related to the initial precursor concentration by

$$\alpha(\tau) = \mu_0 e^{-\varepsilon\tau}/(\mu_0^2 e^{-2\varepsilon\tau} + \kappa_u) \qquad (3.72)$$

$$\beta(\tau) = \mu_0 e^{-\varepsilon\tau}. \qquad (3.73)$$

These equations will cease to give good approximations if ε becomes large or at intermediate times when τ is about the same size as ε^{-1} (so $t \approx k_0^{-1}$).

Figure 3.9 shows a comparison of the exact numerical solutions of the full equations (3.70) and (3.72) with these leading-order forms for a number of values of ε. In each case there is an initial transient as the concentrations move from their initial values ($\alpha = \beta = 0$ at $\tau = 0$) to the pseudo-stationary curve. For the smallest ε, the exact and approximate histories are virtually indistinguishable over almost all of their course. For larger values of the dimensionless rate constant for the decay of P, a significant departure can develop during the reaction, but the two forms approach each other again at long times.

The traces in Fig. 3.9 were computed for a system with an uncatalysed reaction rate constant such that $\kappa_u > \frac{1}{8}$, and hence there are no oscillatory responses in the corresponding pool chemical equations. For $\kappa_u < \frac{1}{8}$, we may also ask about the (time-dependent) local stability of the pseudo-stationary state. The concentration histories may become unstable to small perturbations for a limited time period. For sufficiently small ε this should occur whilst the group $\mu_0 e^{-\varepsilon\tau}$ lies within the range

$$\mu_2^* < \mu_0 e^{-\varepsilon\tau} < \mu_1^* \qquad (3.74)$$

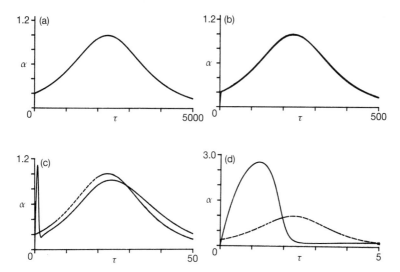

FIG. 3.9. Comparison of exact concentration histories for species A, $\alpha(\tau)$, with pseudo-stationary-state form (broken curves) showing the influence of the precursor decay rate constant ε when the uncatalysed reaction rate is relatively large, $\kappa_u = \frac{1}{4}$: (a) $\varepsilon = 10^{-3}$; (b) $\varepsilon = 10^{-2}$; (c) $\varepsilon = 10^{-1}$; (d) $\varepsilon = 1$.

where the $\mu^*_{1,2}$ are given in terms of κ_u by eqn (3.65). Recasting this in terms of the time τ, we have to leading order that oscillations should be observable over the period

$$\varepsilon^{-1} \ln (\mu_0/\mu^*_2) = \tau^*_2 < \tau < \tau^*_1 = \varepsilon^{-1} \ln (\mu_0/\mu^*_1). \tag{3.75}$$

Note that the dimensionless time τ^*_1, which gives the length of the pre-oscillatory period, will only be positive if the initial concentration μ_0 exceeds the upper Hopf bifurcation value μ^*_1. If we start with a lower initial reactant concentration, so that $\mu_0 < \mu^*_1$ (but still with $\mu_0 > \mu^*_2$), there will be no pre-oscillatory period: the system will jump straight into oscillations which will persist until time τ^*_2.

At the extreme of very low initial reactant concentrations, μ_0 would be less than μ^*_2; in this case τ^*_2 would be negative and hence no oscillations could be observed.

Equation (3.75) shows that the times τ^*_1 and τ^*_2 depend most sensitively on ε, the ratio of the rate constants k_0/k_2: in terms of real time, the corresponding quantities t^*_1 and t^*_2 lengthen inversely with k_0. The dependences on the initial reactant concentration and the uncatalysed reaction rate (through $\mu^*_{1,2}$) are logarithmic and hence less sensitive.

These general quantitative predictions concerning the existence and length of pre-oscillatory and subsequent oscillatory periods depending on μ_0, κ_u, and ε are borne out by numerical computation of the exact equations with

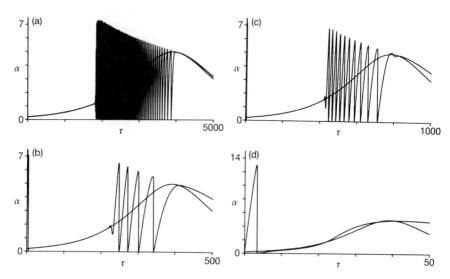

FIG. 3.10.   Comparison of exact concentration histories for species A, $\alpha(\tau)$, with pseudo-station-ary-state form (broken curves) showing the influence of the precursor decay rate constant ε when the uncatalysed reaction rate is small, $\kappa_u = 0.01$: (a) $\varepsilon = 10^{-3}$; (b) $\varepsilon = 5 \times 10^{-2}$; (c) $\varepsilon = 10^{-2}$; (d) $\varepsilon = 0.1$.

small values of ε as shown in Fig. 3.10. There are, however, some significant differences particularly relating to the exact times for the onset and death of the excursions. First, the onset is delayed beyond the value τ_1^*. This arises because the concentrations take a long time to move away from the pseudo-stationary-state locus: the solutions given by (3.72) and (3.73) may correspond to unstable states but nevertheless they do satisfy $d\alpha/d\tau = d\beta/d\tau = 0$ to leading order, so in the absence of perturbations there is no large spontaneous driving force for departure. The growth of small perturbations is governed by the usual exponential forms such as eqns (3.40) and (3.41) or, with complex eigenvalues, (3.49) and (3.50)—although now λ_1 and λ_2 can be thought of as slowly varying functions of time. The rate of growth depends on the magnitude of the real parts of λ and near to the Hopf bifurcation points $\text{Re}(\lambda) \approx 0$, so again this stimulated growth is very slow. If the perturbations have a typical size δ, where δ is small compared even with the decay rate ε, then the increase in the pre-oscillatory period will be of the order $[\varepsilon^{-1} \ln(\delta^{-1})]^{1/2}$. The term δ may represent the round-off error of numerical calculations and for example may be approximately 10^{-9}: if $\varepsilon \approx 10^{-3}$ then the delay between τ_1^* and the first appearance of oscillations will be about 143. This last prediction compares well with the observed result shown in Fig. 3.10(a), where $\tau_1^* = 1625$ but the onset of oscillations is delayed until $\tau = 1782$ in the computed solutions.

The second significant difference between the predictions and the actual results is that oscillations survive beyond the time τ_2^*. This arises because the pseudo-stationary state has focal character just after the second Hopf bifurcation (i.e. the slowly varying eigenvalues $\lambda_{1,2}$ are complex conjugates with now negative real parts) so there is a damped oscillatory return to the locus. In Fig. 3.10(a) this can be seen after $\tau \approx 3966$, whilst $\tau_2^* = 3891$.

Because oscillatory behaviour persists only for a finite length of time, only a finite number of excursions can occur. We can estimate this number by obtaining an approximate value for the mean oscillatory period, τ_m. For this we take a geometric mean of the periods at the two Hopf bifurcation points. These latter quantities can be evaluated from the frequency ω_0 defined by

$$\omega_0 = (\det \mathbf{J})^{1/2} \tag{3.76}$$

with the determinant evaluated at the Hopf bifurcation point, i.e. from eqn (3.59) with μ^2 given by (3.65). The oscillatory period at these points is then given by $\tau_p = 2\pi/\omega_0$. Thus

$$\tau_p(\mu_1^*) = 2\pi/(\mu_1^{*2} + \kappa_u)^{1/2} \qquad \tau_p(\mu_2^*) = 2\pi/(\mu_2^{*2} + \kappa_u)^{1/2}. \tag{3.77}$$

For the data in Table 2.1, we have $\tau_p(\mu_1^*) = 6.35$ and $\tau_p(\mu_1^*) = 43.96$ respectively ($\kappa_u = 10^{-2}$, $\mu_1^* = 0.9847$, $\mu_2^* = 0.1021$).

Now let the mean oscillatory period be $\tau_m = [\tau_p(\mu_1^*)\tau_p(\mu_2^*)]^{1/2}$. For small κ_u we may use eqns (3.66) and (3.67) to obtain

$$\tau_m \approx 2\pi/(2\kappa_u)^{1/4}. \tag{3.78}$$

For our example data, the mean period would thus be 16.7.

The length of time $\Delta\tau$ over which oscillations occur is given roughly by $\Delta\tau \approx (\tau_2^* - \tau_1^*) = \varepsilon^{-1} \ln(\mu_2^*/\mu_1^*) \approx \frac{1}{2}\varepsilon^{-1} \ln(\kappa_u^{-1})$, so then the number of excursions would be

$$N_{\text{osc}} \approx \Delta\tau/\tau_m \approx \varepsilon^{-1}(2\kappa_u)^{1/4} \ln(\kappa_u^{-1})/4\pi. \tag{3.79}$$

The number of oscillations will be independent of the initial reactant concentration and is determined most strongly by the rate constant ratio k_0/k_2: the faster the decay of the reactant the fewer oscillatory excursions one would expect to see. For $\kappa_u = 0.01$ and $\varepsilon \approx 10^{-3}$ we predict aproximately 140 oscillations which is of the correct order of magnitude compared with the results in Fig. 3.10(a) (82 excursions).

If the value of the reactant decay rate ε is not very small, higher-order correction terms will become significant more quickly. Exact (i.e. precisely computed) concentration histories will not be well appproximated by the pseudo-stationary forms (3.72) and (3.73) even when the state is locally stable. During any possible period of oscillatory behaviour, the number of oscillations will naturally decrease as ε increases, as expressed by eqn (3.79). In addition to this, however, the time for the first excursion to develop, which depends on $\varepsilon^{-1/2}$, may become comparable with $(\tau_2^* - \tau_1^*)$, which depends on ε^{-1}. In the extreme case, the system may move through the region of instability so quickly that oscillations cannot develop and hence are not observed. A typical sequence for varying ε is shown in Figs 3.10(a–e).

3.6. Summary

We have now seen how local stability analysis can give us useful information about any given state in terms of the experimental conditions (i.e. in terms of the parameters μ and κ_u for the present isothermal autocatalytic model). The methods are powerful and for low-dimensional systems their application is not difficult. In particular we can recognize the range of conditions over which damped oscillatory behaviour or even sustained oscillations might be observed. The Hopf bifurcation condition, in terms of the eigenvalues λ_1 and λ_2, enabled us to locate the onset or death of oscillatory behaviour. Some comments have been made about the stability and growth of the oscillations, but the details of this part of the analysis will have to wait until the next chapter.

References

Many of the chemistry references appropriate to this chapter have been given in chapter 2. Local stability analysis is covered in most advanced mathematical texts on non-linear ordinary differential equations, for example:

Andronov, A. A., Vitt, E. A., and Khaiken, S. E. (1966). *Theory of oscillators.* Pergamon, Oxford.

Andronov, A. A., Leontovich, E. A., Gordon, I. I., and Maier, A. G. (1973). *Theory of bifurcations of dynamic systems on a plane.* Wiley, New York.

Jordan, D. W. and Smith, P. (1977). *Nonlinear ordinary differential equations.* Clarendon Press, Oxford.

Pippard, A. B. (1985). *Response and stability.* Cambridge University Press.

Thompson, J. M. T. (1982). *Instabilities and catastrophes in science and engineering.* Wiley, Chichester.

THERMOKINETIC OSCILLATIONS IN A CLOSED SYSTEM

This chapter and chapter 5 study the prototypical thermokinetic oscillator. Thermal feedback replaces autocatalysis, and the Arrhenius temperature dependence of rate coefficients supplies non-linearity in the scheme $P \rightarrow A \rightarrow B$ + heat. After careful study of this chapter the reader should be able to:

(1) see clearly the kinetic and thermal foundations of oscillatory behaviour;

(2) see clearly the virtues of dimensionless equations in their most compact form;

(3) recognize and benefit from the links with thermal explosion theory and exploit the exponential approximation to the Arrhenius law;

(4) exploit the stationary state approximation (Salnikov's assumption of constant P) to establish the locus of stationary states and their local stability;

(5) map in parameter space boundaries between types of behaviour, especially the onset of oscillations;

(6) relate all the foregoing to the interpretation and prediction of behaviour in a closed system.

In the preceding chapters we investigated the basic patterns of behaviour which might be exhibited by a reaction scheme which involved a certain form of chemical feedback under isothermal conditions. Here we make a similar analysis for systems with purely first-order chemical reactions but under conditions in which the heat produced by the natural exothermicity can lead to departures from isothermal operation. Feedback is then provided from thermal coupling as the increase in temperature of the reacting mixture leads to an increase in the local value of the reaction rate constant.

Again we are concerned with thermodynamically closed systems, so we may have exchange of heat but not of matter with the surroundings. Our governing equations are those of mass and energy conservation. Again we shall find a pseudo-stationary-state analysis invaluable in providing a guide to the evolution of such systems.

In the next sections we introduce the scheme and show how the conservation equations can be reduced to the simple dimensionless forms

$$d\alpha/d\tau = \mu - \kappa\alpha e^{\theta}$$

$$d\theta/d\tau = \alpha e^{\theta} - \theta$$

where α is the intermediate chemical concentration and θ is the dimensionless temperature rise due to self-heating. Comparing these forms with eqns (3.20) and (3.21) we can see some distinct similarities. In particular, the term e^{θ} which arises above from the Arrhenius temperature dependence of the reaction rate constant appears as an 'exponential autocatalysis' where the term β^2 appears in the isothermal autocatalysis. Again, the group of terms in μ represents the (slowly decaying) concentration of the precursor reactant, and we have a second parameter κ to vary between successive experiments.

4.1. The model

We take the simplest kinetic scheme for our requirements: two consecutive first-order chemical reactions converting an initial reactant P into a final product B through a single intermediate species A

$$P \xrightarrow{k_0 p} A \xrightarrow{k_1 a} B.$$

We will assume that the second of these two steps will be exothermic ($\Delta H_1 < 0$) and the rate constant k_1 will obey an Arrhenius temperature dependence

$$k_1 = A \exp(-E/RT) \tag{4.1}$$

where E is the activation energy and T is the local temperature. For simplicity, however, we can assume that the first step is virtually thermoneutral (i.e. $\Delta H_0 \approx 0$) and has zero activation energy, so k_0 does not vary with temperature.

4.2. Governing equations of mass and energy balance

The model can be written in terms of its constituent steps:

$$P \to A \qquad \text{rate} = k_0 p \quad E_0 = Q_0 = 0$$

$$A \to B \qquad \text{rate} = k_1(T)a \quad E > 0, Q > 0$$

where Q is the reaction exothermicity ($Q = -\Delta H_1$) and the form $k_1(T)$ reminds us that the rate constant for this step varies with the temperature. We imagine a well-stirred closed vessel of volume V sitting in a heat bath (e.g.

a laboratory or perhaps an oven) which has a temperature T_a. The temperature of the reacting mixture within the vessel is denoted by T and this may rise above T_a as the exothermic reaction progresses. The equations for mass balance and energy conservation for this system are

$$dp/dt = -k_0 p \tag{4.2}$$

$$da/dt = k_0 p - k_1(T)a \tag{4.3}$$

$$(V\sigma c)\, dT/dt = QVk_1(T)a - S\chi(T - T_a). \tag{4.4}$$

Here σ is the density of the reactant mixture (perhaps measured in units of kg m^{-3}), c the specific heat capacity ($\text{J K}^{-1}\text{kg}^{-1}$), S the surface area, and χ the surface heat transfer coefficient ($\text{W m}^{-2}\text{K}^{-1}$). We have assumed a Newtonian cooling term for the transfer of heat to the surroundings.

The feedback in this model arises from the thermal coupling represented in the form $k_1(T)$. An increase in the temperature T increases the value of k_1 and hence of the rate of production of heat. The feedback is highly non-linear, following the Arrhenius temperature dependence given by eqn (4.1).

Typical values for the various physico-chemical quantities, as might be appropriate to a gaseous system, are given in Table 4.1. The initial conditions might be

$$p(t = 0) = p_0 \qquad a(t = 0) = 0 \qquad b(t = 0) = 0 \qquad T(t = 0) = T_a. \quad (4.5)$$

Table 4.1

Representative values for dimensional and dimensionless parameters. The quantities E, c, σ, χ, and Q match those for the decomposition of di-tertiarybutyl peroxide in a glass vessel at low pressures

---

Volume, V, $1.0\,\text{dm}^3$
Surface area, S, $5\,\text{dm}^2$
Ambient temperature, T_a, $400\,\text{K}$
Activation energy, E, $166\,\text{kJ mol}^{-1}$
Heat capacity per unit volume, $c\sigma$, $150\,\text{J K}^{-1}\text{m}^{-3}$
Surface heat transfer coefficient, χ, $30\,\text{W m}^{-2}\text{K}^{-1}$
Exothermicity, Q, $400\,\text{kJ mol}^{-1}$
Initial concentration of reactant, p_0, $3 \times 10^{-3}\,\text{mol dm}^{-3}$
Rate constants: k_0, $0.1\,\text{s}^{-1}$; $k_1(T_a)$, $0.5\,\text{s}^{-1}$
Newtonian cooling time, t_N, $0.1\,\text{s}$
Reference concentration, c_{ref}, $6 \times 10^{-5}\,\text{mol dm}^{-3}$
Dimensionless quantities: $\varepsilon = 10^{-2}$; $\kappa = 0.05$; $\mu_0 = 0.5$; $\gamma = 1/50$

4.2.1. Reactant concentration

Equation (4.2) involves only the variable p and so can be integrated to give

$$p(t) = p_0 e^{-k_0 t}. \tag{4.6}$$

We again have a monotonic exponential decrease in the reactant concentration during the course of the reaction. We are left to determine the time dependence of the concentration of intermediate A and the temperature of the reacting mixture T.

4.2.2. Mass balance for intermediate

We can regard the concentration of the intermediate A as a function of P by means of eqn (4.3), or explicitly in terms of the initial reactant concentration and the time. Substituting for p from eqn (4.6), we have

mass balance $da/dt = k_0 p_0 e^{-k_0 t} - k_1(T)a. \tag{4.7}$

4.2.3. Energy balance

When we come to model this system it will be particularly useful to work not in terms of the actual temperature T but in terms of the ambient temperature T_a (which is known) and the temperature rise or degree of self-heating, $\Delta T = T - T_a$. The latter, ΔT, then becomes our second variable, i.e. we look for $d\Delta T/dt$. Equation (4.4) can be written as

energy balance $(V\sigma c)d\Delta T/dt \quad = \quad QVk_1(\Delta T)a - S\chi\Delta T. \tag{4.8}$

                          rate of change   generation   loss

Here the form $k_1(\Delta T)$ expresses the dependence of the reaction rate constant on the temperature rise.

4.3. Example computation of concentration and temperature rise

The evolution in time of the concentration of the species A and of the temperature rise ΔT, for the example data in Table 4.1, is shown in Fig. 4.1. The behaviour is in many ways similar to that of the isothermal cubic autocatalysis model of the previous chapters. The concentration of the precursor P decreases exponentially throughout the reaction. The temperature excess jumps rapidly to approximately 80 K, from which value it begins to decay approximately exponentially. At the same time, the concentration of the intermediate A rises relatively slowly to values of the order of 10^{-5} mol dm^{-3}. After approximately 15 s, the concentration of A and the

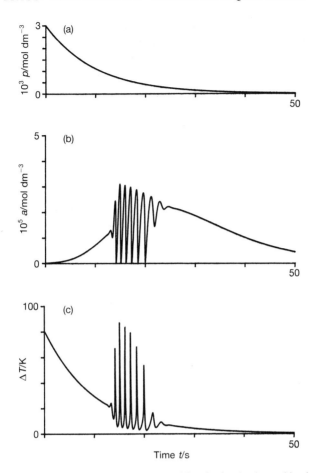

FIG. 4.1.   Computed concentration and temperature histories for the thermokinetic model with parameters given in Table 4.1 showing monotonic decay of precursor reactant p but oscillations in the concentration of intermediate A and the temperature excess ΔT: (a) $p(t)$, (b) $a(t)$, and (c) $\Delta T(t)$.

temperature rise begin to oscillate. For the above data, there are eight oscillatory excursions. Oscillatory behaviour lasts approximately 10 s. The post-oscillatory period sees the temperature rise and the concentration of A fall back to zero.

4.4. Dimensionless form of mass- and energy-balance equations

In the previous chapter we found that the equations could be written in particularly economical forms and in such a way that large or small quantities could be easily recognized, by adopting a number of dimensionless groups. This approach is based on the principle that it is often not necessarily the absolute values of the rate constants and other physico-chemical quantities that are important so much as their relative magnitudes. With the present non-isothermal model we are again well advised to go to dimensionless terms at this early stage. We will need to establish three scales: one for concentration, one for temperature, and one for time.

(a) Dimensionless concentrations

For the moment let us assume that we shall be able to find a natural measure of concentration c_{ref} that will play a similar role to the quotient $(k_2/k_1)^{1/2}$ in the previous chapter. We will define c_{ref} later, and we will then write

$$\alpha = a/c_{ref} \qquad \text{and} \qquad \pi = p/c_{ref}. \tag{4.9}$$

(b) Dimensionless temperature rise

The temperature rise, ΔT, may be made dimensionless by borrowing from thermal explosion theory. There the natural dimensionless temperature rise θ is defined by

$$\theta = E(T - T_a)/RT_a^2. \tag{4.10}$$

From the data in Table 4.1, the group $RT_a^2/E = 8$ K, so a temperature rise of 16 K would give a value $\theta = 2$.

(c) Dimensionless form of Arrhenius rate law

The temperature dependence of the reaction rate constant $k_1(\Delta T)$ can be expressed as the product of two terms: (i) the value of k_1 at T_a and (ii) a factor related to the dimensionless temperature rise:

$$k_1(T) = k_1(T_a)f(\theta) \tag{4.11}$$

where

$$k_1(T_a) = A \exp(-E/RT_a). \tag{4.12}$$

The function $f(\theta)$ giving an exact representation of the Arrhenius form is

$$f(\theta) = \exp[\theta/(1 + \gamma\theta)] \tag{4.13}$$

where the extra dimensionless parameter γ is defined by

$$\gamma = RT_a/E. \tag{4.14}$$

For the activation energy and ambient temperature in Table 4.1, $\gamma = 0.02$, a typically small value.

(d) Dimensionless time

For the natural timescale we take the Newtonian cooling time t_N

$$t_N = c\sigma V/\chi S. \tag{4.15}$$

This is the time taken for a temperature perturbation in a chemically inert mixture with the same physical properties to decay to $1/e$ of its original value. For gases near to atmospheric pressure, t_N will be of the order of 0.1 to 0.5 s. The cooling time is proportional to the density or the pressure of the reacting mixture.

By noting that the temperature rise in this system takes the feedback role that was played by the autocatalytic species B in the previous chapter, we may recognize the resemblance between the rate constant k_2 of the previous chapter (rate constant for decay of B) with $1/t_N$ here (a pseudo-rate constant for decay of the temperature rise). For a system in which heat transfer is slow, t_N will be large; if heat transfer across the surface is rapid, t_N will be small.

The dimensionless time, therefore, becomes

$$\tau = t/t_N. \tag{4.16}$$

(e) Dimensional rate constants

There are two dimensionless rate constants which characterize this system. One is related to the reactant decay step and we use the symbol ε as in the previous chapter:

$$\varepsilon = k_0 t_N. \tag{4.17}$$

The data in Table 4.1 give $\varepsilon = 10^{-2}$. The other rate constant is the dimensionless value for k_1 evaluated at the known ambient temperature:

$$\kappa = k_1(T_a)t_N. \tag{4.18}$$

This quantity in effect compares the timescales for chemical heat release with Newtonian heat transfer. If the chemistry proceeds quickly compared with heat loss, κ will be large; if the chemistry is slow or heat transfer comparatively fast, κ will be small. For the system of Table 4.1, $\kappa = 0.05$.

(f) Reference concentration

Now we must be specific about our choice of the concentration scale c_{ref}. For reasons partly based in the experiences of thermal explosion theory and partly by the desire to reduce the dimensionless equations to their simplest possible forms, we take

$$c_{ref} = \chi S \, RT_a^2/EQV \, k_1(T_a). \tag{4.19}$$

Taking data from Table 4.1, a typical value for c_{ref} might be 6×10^{-5} mol dm^{-3}. This is two orders of magnitude lower than the initial reactant concentration (so the latter is 'large' in the context of this scheme).

(g) Dimensionless rate equations

The mass- and energy-balance equations become

$$d\alpha/d\tau = \varepsilon \pi_0 e^{-\varepsilon\tau} - \kappa \alpha f(\theta) \tag{4.20a}$$

$$d\theta/d\tau = \alpha f(\theta) - \theta. \tag{4.20b}$$

These equations involve the two variables α and θ, the dimensionless time τ, and the four parameters ε, π_0, κ, and γ (which appears in $f(\theta)$). If the rate of decay of the reactant is slow on the Newtonian timescale then ε will be small compared with unity. Conversely, if the initial reactant concentration is large compared with c_{ref}, then the term π_0 will be much larger than unity. We again have a term in eqn (4.20a) which contains the product of these small and large quantities and so we introduce the new parameter μ

$$\mu_0 = \varepsilon \pi_0 = k_0 t_N p_0 / c_{ref} \tag{4.21}$$

which may be of order unity, e.g. $\mu_0 = 0.5$ for the data in Table 4.1.

The full governing equations are now

$$d\alpha/d\tau = \mu_0 e^{-\varepsilon\tau} - \kappa \alpha f(\theta) \tag{4.22}$$

$$d\theta/d\tau = \alpha f(\theta) - \theta. \tag{4.23}$$

If we prefer, we can suppress the explicit time dependence and write the equations in terms of μ:

$$d\alpha/d\tau = \mu - \kappa \alpha f(\theta) \tag{4.24}$$

$$d\theta/d\tau \doteq \alpha f(\theta) - \theta. \tag{4.25}$$

where

$$\mu = \mu_0 e^{-\varepsilon\tau}. \tag{4.26}$$

The parameters μ and κ are dimensionless measures of the reactant concentration and of the size of the reaction rate constant for the step A $\rightarrow$ B evaluated at ambient temperature.

The parameter γ is still involved in these equations, although hidden in the function $f(\theta)$. Before looking at the stationary-state and local stability properties of eqns (4.24) and (4.25) we will introduce a second approximation, concerning $f(\theta)$, which will simplify the model to one with only two parameters.

4.5. The exponential approximation to the Arrhenius rate law

The highly non-linear dependence of a reaction rate constant on temperature, embodied in the Arrhenius rate law (4.1), or its equivalent dimensionless representation (4.13), can lead to problems in both the algebraic tractability and the ease of accurate computations for this model. Again we can learn from thermal explosion theory and make use of a simplification which preserves the strong non-linearity yet will reduce the equations to a more manageable form. The resulting analysis will provide a clear and useful background on which to build our investigation of the full system.

We may note that for most chemical reactions, activation energies are generally large compared with the thermal energy RT_a. This means that the number γ defined by eqn (4.14) will typically be small compared with unity, $\gamma \ll 1$. If the dimensionless temperature rises encountered in the model do not become particularly large, i.e. $\theta \approx O(1)$, then the product $\gamma\theta$ which occurs in the denominator of $f(\theta)$ will also generally be small. Provided these conditions hold, therefore, we can simplify the temperature dependence by neglecting $\gamma\theta$ and using the 'exponential approximation'

$$f(\theta) = e^{\theta}. \tag{4.27}$$

In this form, therefore, the reaction rate constant increases simply by a factor of 2.718 for every unit increase in the dimensionless temperature excess. If we take the data from Table 4.1, $RT_a^2/E = 8$ K, so an 8 K temperature rise causes an increase in the rate by e.

We may represent clearly our required conditions for the exponential approximation in terms of the original quantities, in particular the temperature of the mixture and the ambient temperature. From the definitions for γ and θ, the inequality $\gamma\theta \ll 1$ becomes $T - T_a \ll T_a$, i.e. the temperature rise must be small compared with the absolute ambient temperature. For a system with $T_a = 400$ K in which there is 12 K self-heating, $\gamma\theta = 0.03$, so we would expect the exponential approximation to hold quite well.

In the next few sections we will concentrate on the form of the governing equations (4.24) and (4.25) with the exponential approximation to $f(\theta)$ as given by (4.27). We will determine the stationary-state solution and its dependence on the parameters μ and κ, the changes which occur in the local stability, and the conditions for Hopf bifurcation. Then we shall go on and use the full power of the Hopf analysis, to which we alluded in the previous chapter, to obtain expressions for the growth in amplitude and period of the emerging oscillatory solutions.

Eventually we shall return to the exact Arrhenius form for $f(\theta)$, eqn (4.13), and find that the behaviour of this system can show some additional complexity depending on the parameter γ.

4.6. Exponential approximation: stationary states and local stability

Before proceeding, let us quickly summarize the operations carried out so far in this chapter. We have taken a chemical model consisting of two first-order reactions. The first step, which produces the intermediate A, is essentially a slow process and so we have made the pool chemical approximation for the concentration of the original reactant P. The second step converts A to a final product, releasing heat. The non-linear temperature dependence of the rate of this second step has been approximated by a simple exponential function of the consequent dimensionless temperature rise.

The mass- and energy-balance equations can now be written in the explicit form

mass $d\alpha/d\tau = \mu - \kappa\alpha e^{\theta}$ (4.28)

energy $d\theta/d\tau = \alpha e^{\theta} - \theta.$ (4.29)

Again we may consider that the dimensionless concentration of the reactant μ will automatically vary during the course of a given experiment with the consumption of P, whilst κ might vary between successive experiments.

4.6.1. Pseudo-stationary states

We will regard α and θ as functions of the reactant concentration as expressed in μ and assume that they change on a fast timescale compared with reactant consumption (small ε), i.e. we apply the pseudo-stationary-state hypothesis. The pseudo-stationary-state condition $d\alpha/d\tau = d\theta/d\tau = 0$ yields the following simultaneous equations:

$$\mu - \kappa\alpha_{ss}e^{\theta_{ss}} = 0$$ (4.30)

$$\alpha_{ss}e^{\theta_{ss}} - \theta_{ss} = 0.$$ (4.31)

If we divide the first equation by κ and then add it to the second, we obtain the simple result for the stationary-state temperature rise:

$$\theta_{ss} = \mu/\kappa.$$ (4.32)

Substitution of this result back into eqn (4.31) then gives, for the dimensionless intermediate concentration,

$$\alpha_{ss} = (\mu/\kappa)e^{-(\mu/\kappa)}.$$ (4.33)

Based on the initial reactant concentration μ_0, the data in Table 4.1 give $\theta_{ss} = 10$, $\alpha_{ss} = 10e^{-10} = 4.54 \times 10^{-4}$. In dimensional terms, $\Delta T_{ss} = 10RT_a^2/E = 80$ K and $a_{ss} = 4.54 \times 10^{-4} c_{ref} = 2.72 \times 10^{-8}$ mol dm^{-3}.

Note that the stationary states depend only on the ratio of μ and κ, rather than on their individual values. We can thus draw a 'universal' locus in

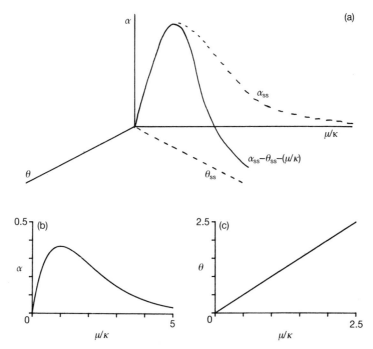

FIG. 4.2. Dependence of dimensionless pseudo-steady-state intermediate concentration and temperature excess on reactant concentration for the model with the exponential approximation: (a) three-dimensional representation of 'universal locus'; (b) projection showing dependence of α_{ss} on μ; (c) projection showing $\theta_{ss}(\mu)$.

$\theta-\alpha-(\mu/\kappa)$ space, as represented in Fig. 4.2(a). Any given system with particular values of μ and κ will start at some point along this locus and then evolve in time in the direction of decreasing μ. Equations (4.32) and (4.33) then give the projections of this curve on to the appropriate planes in Figs 4.2(b) and (c).

(a) Significance of θ_{ss}

The pseudo-stationary temperature excess is clearly linearly proportional to the reactant concentration μ and hence would show an exponential decrease in time:

$$\theta_{ss} = \mu_0 e^{-\varepsilon t/\kappa}. \tag{4.34}$$

For a system with fixed μ, θ_{ss} is inversely proportional to the dimensionless rate constant κ. However, if we look at this result in terms of the original dimensional quantities, the temperature rise does not depend on the value of the rate constant $k_1(T_a)$. Rather we have

$$\theta_{ss} = k_0 p(EQV)/(\chi SRT_a^2) \qquad (4.35a)$$

or, if we substitute in the explicit dependence of p on time,

$$\theta_{ss} = k_0 p_0 e^{-k_0 t}(EQV)/(\chi SRT_a^2) \qquad (4.35b)$$

showing that the dimensionless temperature rise increases, not surprisingly, as the rate of heat transfer decreases and as the exothermicity of the reaction increases.

Equation (4.34) can also be recast in various ways to highlight its physical significance. For instance we may write

$$\theta_{ss} = \varepsilon\theta_{ad}e^{-k_0 t} \qquad (4.36)$$

where θ_{ad} is given by

$$\theta_{ad} = p_0 QE/c\sigma RT_a^2 \qquad (4.37)$$

and is the dimensionless temperature rise which would be achieved if all the reactants were converted to product under adiabatic conditions (no heat transfer). In terms of actual physical quantities, the dimensional stationary-state temperature rise $\Delta T_{ss} = T_{ss} - T_a$ is given by

$$\Delta T_{ss} = k_0 pQV/\chi S = k_0 t_N \Delta T_{ad}e^{-k_0 t} \qquad (4.38)$$

where $\Delta T_{ad} = p_0 Q/c\sigma$ is the adiabatic temperature rise in kelvin. For the data in Table 4.1, $\Delta T_{ad} = 8000$ K and $\theta_{ad} = 1000$.

(b) Significance of α_{ss}

The dependence of the pseudo-stationary-state concentration of the intermediate A on μ/κ shows a maximum value of e^{-1} when $\mu/\kappa = 1$. The time dependence of α_{ss} is given by

$$\alpha_{ss}(\tau) = (\mu_0/\kappa)e^{-\varepsilon\tau} \exp[-(\mu_0/\kappa)e^{-\varepsilon\tau}]. \qquad (4.39)$$

In dimensional terms the maximum value of the intermediate concentration is

$$(a_{ss})_{max} = e^{-1}c\sigma RT_a^2/EQk_1(T_a)t_N$$

and will be attained after a time

$$t_{max} = k_0^{-1} \ln(\mu_0/\kappa).$$

The maximum concentration is enhanced by slow chemistry, efficient heat transfer, and reductions in the reaction exothermicity.

4.6.2. Local stability

In the previous section we found that only the relative values of the parameters μ and κ, through their quotient, are important in determining the stationary state. We now consider the local stability of this solution and will find here that μ and κ play separate roles. We apply the same recipe as that introduced in § 3.2 of the previous chapter beginning from the Jacobian matrix

$$\mathbf{J} = \begin{pmatrix} \partial(d\alpha/d\tau)/\partial\alpha & \partial(d\alpha/d\tau)/\partial\theta \\ \partial(d\theta/d\tau)/\partial\alpha & \partial(d\theta/d\tau)/\partial\theta \end{pmatrix}_{ss}. \tag{4.40}$$

The important features for us are the trace and determinant of $\mathbf{J}$ which determine the eigenvalues $\lambda_{1,2}$ as the solutions of

$$\lambda^2 - \mathrm{tr}(\mathbf{J})\lambda + \det(\mathbf{J}) = 0 \tag{4.41}$$

and the discriminant $\Delta = (\mathrm{tr}(\mathbf{J}))^2 - 4\det(\mathbf{J})$.

For the governing eqns (4.28) and (4.29), evaluated using the stationary-state results (4.32) and (4.33), $\mathrm{tr}(\mathbf{J})$ and $\det(\mathbf{J})$ become

$$\mathrm{tr}(\mathbf{J}) = (\mu/\kappa) - 1 - \kappa e^{\mu/\kappa} \tag{4.42}$$

$$\det(\mathbf{J}) = \kappa e^{\mu/\kappa}. \tag{4.43}$$

For all physically acceptable conditions, the determinant of $\mathbf{J}$ is positive, so we will not find saddle points or saddle–node bifurcations. We can, however, expect to find conditions under which nodal states become focal (damped oscillatory responses), i.e. where $\Delta = 0$, and where focal states lose stability at Hopf bifurcations, i.e. where $\mathrm{tr}(\mathbf{J}) = 0$ and where we shall look for the onset of sustained oscillations.

(a) Transition from nodal to focal character

In order to find the parameter values at which the eigenvalues λ_1 and λ_2 become a complex conjugate pair we need to solve the equation

$$\Delta = 0. \tag{4.44}$$

The direct method of locating these conditions would be to substitute the explicit forms for $\mathrm{tr}(\mathbf{J})$ and $\det(\mathbf{J})$, in terms of μ and κ, from eqns (4.42) and (4.43) into Δ and then try to solve (4.44). However, the particular form of the stationary-state solutions is such that we can most easily proceed by specifying a value for θ_{ss} and then solving for μ and κ. We may replace μ in (4.42) and (4.43) by $\kappa\theta_{ss}$, so that we then have for the discriminant:

$$\Delta = (\theta_{ss} - 1)^2 - 2\kappa(1 + \theta_{ss})e^{\theta_{ss}} + \kappa^2 e^{2\theta_{ss}}. \tag{4.45}$$

Thus, for any positive value of θ_{ss}, there are two values of κ, given by the roots of this quadratic, which satisfy eqn (4.44):

$$\kappa_{\pm} = (\theta_{ss}^{1/2} \pm 1)^2 e^{-\theta_{ss}}. \tag{4.46}$$

The corresponding values of μ are then given by

$$\mu_{\pm} = \kappa_{\pm}\theta_{ss}. \tag{4.47}$$

Some of the κ–μ pairs which are given by (4.46) and (4.47) as θ_{ss} is varied are given in Table 4.2 to illustrate the use of this parametric approach.

Table 4.2

Parametric location of conditions for change from nodal to focal responses, given by eqns (4.46) and (4.47)

θ_{ss}	κ_+	μ_+	κ_-	μ_-
0	1	0	1	0
0.1	1.568	0.1568	0.4231	0.0423
0.25	1.752	0.4380	0.1947	0.0487
0.5	1.768	0.8838	0.0520	0.0260
1.0	1.472	1.472	0	0
2.0	0.7888	1.578	0.0232	0.0464
3.0	0.3716	1.115	0.0267	0.0800
4.0	0.1648	0.6594	0.0183	0.0733
5.0	0.0707	0.3528	0.0103	0.0515
7.0	0.0121	0.0848	0.0025	0.0179
10.0	0.0008	0.0079	0.0002	0.0021

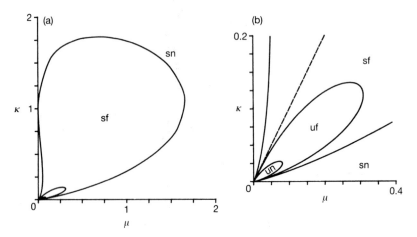

FIG. 4.3. The μ–κ parameter plane showing loci of changes in local stability or character for the model with the exponential approximation: (a) full plane; (b) enlargement of region near origin showing, in particular, the locus of Hopf bifurcation (change from stable to unstable focus) and the locus corresponding to the maximum in the $\alpha_{ss}(\mu)$ curves (broken line).

The full curves in the κ–μ parameter plane, expressed by (4.46) and (4.47), are shown in Fig. 4.3. There are two closed loops emerging from the origin. The outer loop, given by the upper root in (4.46) and the lower root with $0 < \theta_{ss} < 1$, also touches the κ axis at ($\mu = 0$, $\kappa = 1$). Outside this locus the stationary state is a stable node; inside this loop it is a focus (we will discuss stability within this region in the next subsection). The maximum in this curve occurs for $\kappa = \frac{1}{2}(3 + \sqrt{5}) \exp[-\frac{1}{2}(3 - \sqrt{5})] \approx 1.787$. For larger values of the dimensionless reaction rate constant, e.g. for high ambient temperatures, no damped oscillatory states will be found.

With $\theta_{ss} > 1$, the lower root in (4.46) describes the small lower loop, which corresponds to the conditions for which the stationary state regains nodal character. Inside this region, the eigenvalues $\lambda_{1,2}$ are real and are positive, so we have an unstable node. This curve has a maximum at $\kappa = \frac{1}{2}(3 - \sqrt{5}) \exp[-\frac{1}{2}(3 + \sqrt{5})] \approx 0.0279$, so this response is not to be found over a wide range of experimental conditions.

(b) Transition from stable to unstable states

The condition for a change in the local stability of the stationary state in this model is that the trace of the Jacobian matrix should be zero. We can also recognize this as the first requirement for Hopf bifurcation, about which we shall have more to say in the next section. The condition $\mathrm{tr}(\mathbf{J}) = 0$ is also most easily handled parametrically by replacing μ by $\kappa\theta_{ss}^*$ wherever possible in eqn (4.42). This leads to

$$\mathrm{tr}(\mathbf{J}) = \theta_{ss}^* - 1 - \kappa e^{\theta_{ss}^*} = 0. \tag{4.48}$$

Thus the conditions for loss of stability are given in terms of the parameters κ and μ as

$$\kappa^* = (\theta_{ss}^* - 1)e^{-\theta_{ss}^*} \tag{4.49}$$

$$\mu^* = \kappa^*\theta_{ss}^* \tag{4.50}$$

with θ_{ss}^* taking any value greater than or equal to unity.

The locus of these Hopf bifurcation points is also shown in Fig. 4.3 and can be seen to be another closed loop emanating from the origin. It lies in the region between the loci for changes between nodal and focal character, so the condition $\mathrm{tr}(\mathbf{J})$ separates stable focus from unstable focus. The curve has a maximum at

$$\kappa = (\kappa^*)_{\mathrm{max}} = e^{-2}. \tag{4.51}$$

Unstable stationary states can only be found therefore if the reaction rate constant $k_1(T_a) < e^{-2}/t_N$. For any κ less than this, eqn (4.49) has two solutions for the stationary-state temperature excess θ_{ss}^*. For example, with $\kappa = 0.05$ the roots can be found numerically and are $\theta_{ss}^* = 1.1594$ and 4.1399.

There are then, also, two values of the dimensionless concentration of the reactant, μ_1^* and μ_2^* with $\mu_1^* > \mu_2^*$ say, on this locus. For our example these are $\mu_2^* = 0.05797$ and $\mu_1^* = 0.2070$. In between these solutions, the stationary state is unstable. For any other particular system with a different value of κ, the appropriate Hopf bifurcation points can be calculated in a similar way, as given in Table 4.3, or read off Fig. 4.3. However, if κ is small, we can also estimate μ_1^* and μ_2^* directly by using an approximate, but quite accurate, solution to eqn (4.49):

lower root $\qquad \theta_{ss}^- = 1 + e\kappa + (e\kappa)^2 + \cdots$ $\qquad\qquad$ (4.52)

upper root $\qquad \theta_{ss}^+ \approx 1.378 \ln(\kappa^{-1})$. $\qquad\qquad$ (4.53)

These forms provide better than 1 per cent accuracy for θ_{ss}^- if $\kappa < 0.06$, and for θ_{ss}^+ if $0.01 < \kappa < 0.05$. μ_1^* and μ_2^* are then given by

$$\mu_1^* = 1.378\kappa \ln(\kappa^{-1}) \qquad\qquad (4.54)$$

$$\mu_2^* = \kappa[1 + e\kappa + (e\kappa)^2 + \cdots]. \qquad\qquad (4.55)$$

We may also note from this last result that the lower Hopf bifurcation point μ_2^* lies at a slightly higher value of μ than the maximum in the α_{ss} locus (which occurs at $\mu = \kappa$).

Because we have only a single stationary state, we can use the Poincaré–Bendixson theorem to recognize that sustained oscillatory responses will be found at least over the whole range of κ–μ parameter space corresponding to instability. (Although we must also check that the concentration α and

Table 4.3

Conditions for Hopf bifurcation with exponential approximation

κ	Lower bifurcation			Upper bifurcation		
	θ_2^*	μ_2^*	α_{ss}^*	θ_1^*	μ_1^*	α_{ss}^*
0.01	1.0280	0.0103	0.3677	6.2665	0.0627	0.0119
0.02	1.0576	0.0212	0.3673	5.3918	0.1078	0.0246
0.03	1.0892	0.0327	0.3665	4.8563	0.1457	0.0378
0.04	1.1230	0.0449	0.3653	4.4602	0.1784	0.0516
0.05	1.1594	0.0580	0.3637	4.1399	0.2070	0.0659
0.06	1.1990	0.0719	0.3615	3.8665	0.2320	0.0809
0.07	1.2425	0.0870	0.3587	3.6239	0.2537	0.0967
0.08	1.2909	0.1033	0.3550	3.4021	0.2722	0.1133
0.09	1.3457	0.1211	0.3504	3.1934	0.2874	0.1310
0.10	1.4093	0.1409	0.3443	2.9914	0.2991	0.1502
0.11	1.4863	0.1635	0.3362	2.7888	0.3068	0.1715
0.12	1.5862	0.1903	0.3247	2.5737	0.3088	0.1963
0.13	1.7425	0.2265	0.3051	2.3110	0.3004	0.2290
0.1353	2.0000	0.2707	0.2707	2.0000	0.2707	0.2707

temperature rise θ cannot become unbounded.) We may also note that the oscillations are favoured by small values of the dimensionless reaction rate constant, i.e. by conditions where the chemical heat release proceeds slowly compared with heat transfer.

To learn more about the details of the extent, the onset, and the growth of oscillations we are able to perform a complete Hopf analysis on the present model. We do this in detail, as an introduction to the full application of the Hopf technique, in the next chapter. The main results are, however, presented in the next section.

4.7. Oscillations in the model with the exponential approximation

We have already determined the following information about the behaviour of the pool chemical model with the exponential approximation. There is a unique stationary-state solution for α_{ss}, the concentration of the intermediate A, and θ_{ss}, the temperature rise, for any given combination of the experimental conditions μ and κ. If the dimensionless reaction rate constant κ is larger than the value e^{-2}, then the stationary state is always stable. If heat transfer is more efficient, so that $\kappa < e^{-2}$, then there will be two Hopf bifurcation points along the stationary-state locus as μ varies (Fig. 4.4). If these bifurcation points are μ_1^* and μ_2^* (with $\mu_1^* > \mu_2^*$), the stationary state will be stable for concentrations of the reactant such that $\mu > \mu_1^*$ and for $\mu < \mu_2^*$, but it will be unstable over the range $\mu_2^* < \mu < \mu_1^*$. We also know that the stationary-state value of the dimensionless temperature rise at the

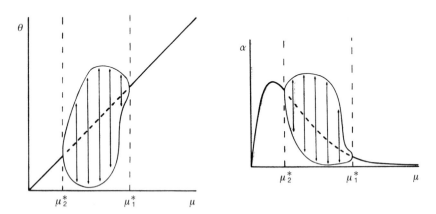

FIG. 4.4. The change in stability and growth of oscillatory solutions in the intermediate concentration and temperature excess as functions of the reactant concentration showing Hopf bifurcations at μ_1^* and μ_2^* (parameter details as given in Table 4.1 except for $\gamma = 0$): (a) $\alpha_{ss}(\mu)$; (b) $\theta_{ss}(\mu)$.

upper Hopf bifurcation point θ_1^* must be larger than 2, and at the lower point it must lie in the range $2 > \theta_2^* > 1$.

In the next chapter we will derive additional aspects about the birth and growth of oscillations in this system. The natural frequency ω_0 of the oscillations as they are born at one or other of the Hopf points can be evaluated explicitly and is related to the stationary-state dimensionless temperature rise:

$$\omega_0 = (\theta^* - 1)^{1/2}. \tag{4.56}$$

At a given Hopf bifurcation point a limit cycle emerges. If this limit cycle is stable, the corresponding motion around it gives rise to physically observable oscillations. The stability is determined by the sign of a parameter, customarily designated β_2, which acts for a limit cycle in the way the eigenvalues $\lambda_{1,2}$ act for a stationary state: if β_2 is negative, the limit cycle is stable; if β_2 is positive, the limit cycle is unstable. For the present model β_2 is given simply by

$$\beta_2 = (1 - \theta^*)/8. \tag{4.57}$$

The emerging limit cycle is born when the dimensionless reactant concentration has the value μ^*: the cycle grows as μ then varies away from μ^*. There are two possibilities: the limit cycle can grow as μ increases, i.e. for $\mu > \mu^*$, or as μ decreases, with $\mu < \mu^*$. Which of these two applies at any given bifurcation point is determined by the sign of a parameter μ_2 (we retain the conventional notation for this quantity at the slight risk of confusion between this and the value of the dimensionless reactant concentration at the lower Hopf bifurcation point, μ_2^*). The appropriate form for μ_2 for the present model is

$$\mu_2 = (\theta^* - 1)^2 e^{-\theta^*}/8(2 - \theta^*). \tag{4.58}$$

If μ_2 is positive, the limit cycle grows as μ increases beyond the Hopf point μ^*. The magnitude of μ_2, as well as its sign, is of significance, governing the growth of oscillatory amplitude which increases as

$$A \sim [(\mu - \mu^*)/\mu_2]^{1/2}. \tag{4.59}$$

This form only applies close to the Hopf bifurcation point, but it is here that numerical methods such as direct integration of the equations converge most slowly.

The natural frequency or period of the oscillations at the Hopf point can be calculated from eqn (4.56). The growth in period as we move away from μ^* then follows the leading-order form

$$T = (2\pi/\omega_0)[1 + (\tau_2/\mu_2)(\mu - \mu^*)] \tag{4.60}$$

where the factor τ_2/μ_2 is given for the present thermokinetic scheme by

$$\tau_2/\mu_2 = -(2\theta^{*3} - \theta^{*2} + 3\theta^* - 6)e^{\theta^*}/6(\theta^* - 1)^3. \tag{4.61}$$

4.8. Development of oscillations away from Hopf points

For the upper Hopf bifurcation point, with $\theta_1^* > 2$, the stability exponent β_2 is negative as is the term μ_2. The first fact ($\beta_2 < 0$) means that the limit cycle emerging from the bifurcation is stable. The particular sign of μ_2 means that the limit cycle grows as μ is decreased below μ_1^*, as shown in Fig. 4.4.

At the lower bifurcation point, where $1 < \theta_2^* < 2$, β_2 is again negative but μ_2 is now positive. Thus the limit cycle emerging here is stable, but now grows as μ is increased beyond μ_2^*.

The period of the oscillations at the bifurcation point is given by $2\pi/\omega_0$. Thus the period is shorter at the upper bifurcation than at the lower one.

Table 4.4 illustrates the application of the above formulae for systems with a range of κ values. The Hopf bifurcation points are located by solving eqns (4.49) and (4.50) for a given κ.

Using μ_2 and τ_2, we can see from eqns (4.59) and (4.60) how the oscillations associated with the limit cycle grow in amplitude and period. If the magnitude of μ_2 is small, the oscillatory amplitude grows quickly: the period changes quickly if $|\tau_2/\mu_2|$ is large. Note that the polynomial $2\theta^{*3} - \theta^{*2} + 3\theta^* - 6$, which occurs in the numerator of the group τ_2/μ_2, changes sign at $\theta^* = 1.2404$ (corresponding to the lower Hopf bifurcation point with $\kappa = 0.069\,55$). The sign of τ_2/μ_2 determines whether the period lengthens or decreases as we move away from the Hopf point. Thus, for $\kappa > 0.069\,55$ the period decreases across the whole oscillatory range as μ is increased. For smaller values of κ, however, the period at first increases as μ is increased above μ_2^*. We also know that for all values of κ that the period

Table 4.4

Example applications of the Hopf formulae for pool chemical model with exponential approximation. The two sets of data for each κ correspond to lower and upper Hopf bifurcation points respectively

κ	μ^*	θ^*	ω_0	β_2	μ_2	τ_2
0.01	0.010 28	1.028	0.1673	-0.0035	3.61×10^{-5}	1.503
	0.062 76	6.276	2.5052	-0.6595	-1.53×10^{-3}	0.557
0.025	0.026 85	1.074	0.2720	-0.0093	2.53×10^{-4}	0.567
	0.1282	5.128	2.0317	-0.5160	-4.04×10^{-3}	0.533
0.05	0.0580	1.160	0.4000	-0.0200	1.19×10^{-3}	0.240
	0.2070	4.140	1.7720	-0.3925	-9.17×10^{-3}	0.532
0.075	0.0950	1.266	0.5158	-0.0333	3.40×10^{-3}	0.098
	0.2633	3.511	1.5846	-0.3139	-1.56×10^{-2}	0.558
0.10	0.1410	1.410	0.6403	-0.0513	8.70×10^{-3}	-0.034
	0.2992	2.992	1.4114	-0.2490	-2.51×10^{-2}	0.627
e^{-2}	$2e^{-2}$	2	1	$-1/8$	$\pm\infty$	$\pm\infty$

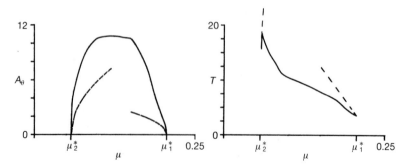

FIG. 4.5.   The development of oscillatory amplitude A_θ and period T across the range of instability, $\mu_2^* \leqslant \mu \leqslant \mu_1^*$, for the pool chemical model with $\kappa = 0.05$. The broken curves give the limiting forms predicted by eqns (4.59)–(4.61).

decreases as μ approaches the upper Hopf point μ_1^*, for the latter case there must be a maximum in the period somewhere in the oscillatory region.

The latter case is illustrated in Fig. 4.5, which compares the limiting results with numerical solutions over the whole range of instability for a particular example with $\kappa = 0.05$. Here the appropriate forms at the lower and upper bifurcation points respectively become

$$A = 29.0(\mu - \mu_2^*)^{1/2} \quad \text{and} \quad 10.44(\mu_1^* - \mu)^{1/2}$$

$$T = 15.71[1 + 201.7(\mu - \mu_2^*) + \cdots] \quad \text{and}$$

$$3.55[1 - 44.35(\mu_1^* - \mu) - \cdots].$$

Clearly higher-order terms soon become important, but the simple forms hold close to the bifurcation points (where accurate numerical integration is not always easy).

The behaviour exhibited by this model is relatively simple. There is only ever one limit cycle. This is born at one bifurcation point, grows as the system traverses the range of unstable stationary states, and then disappears at the second bifurcation point. Thus there is a qualitative similarity between the present model and the isothermal autocatalysis of the previous chapter. The limit cycle is always stable and no oscillatory solutions are found outside the region of instability.

Some typical oscillatory records are shown in Fig. 4.6. For conditions close to the Hopf bifurcation points the excursions are almost sinusoidal, but this simple shape becomes distorted as the oscillations grow. For all cases shown in Fig. 4.6, the oscillations will last indefinitely as we have ignored the effects of reactant consumption by holding μ constant. We can use these computations to construct the full envelope of the limit cycle in μ–α–θ phase space, which will have a similar form to that shown in Fig. 2.7 for the previous autocatalytic model. As in that chapter, we can think of the time-dependent

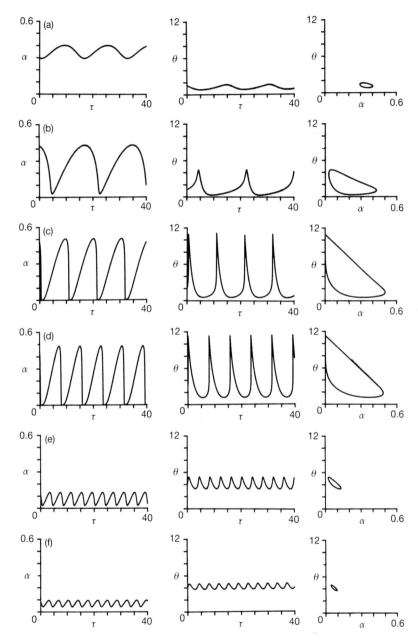

FIG. 4.6. The development of oscillatory amplitude, period, and the associated limit cycle with the reactant concentration for the pool chemical model. Parameter details as given in Table 4.1 except for $\gamma = 0$: (a) $\mu = 0.0581$, (b) $\mu = 0.06$, (c) $\mu = 0.1$, (d) $\mu = 0.15$, (e) $\mu = 0.2$, (f) $\mu = 0.205$.

solutions following the locus while it is stable, but then moving off it at a Hopf bifurcation point to wind around the limit cycle envelope.

4.9. Exact Arrhenius temperature dependence

The temperature dependence of the rate constant for the step A → B leads to the term $f(\theta)$ in the dimensionless mass- and heat-balance eqns (4.24) and (4.25). The exact representation of an Arrhenius rate law is $f(\theta) = \exp[\theta/(1 + \gamma\theta)]$, where γ is a dimensionless measure of the activation energy RT_a/E. As mentioned before, γ will typically be a small quantity, perhaps about 0.02. Provided the dimensionless temperature rise θ remains of order unity ($\theta < 10$, say) then the term $\gamma\theta$ may be neglected in the denominator of the exponent as a first simplification.

If the temperature rise becomes large at any time during the reaction then $\gamma\theta$ may become large enough to contribute significantly to $f(\theta)$. The full Arrhenius form may also show qualitative differences in response from those with the exponential approximation. We have already seen values of θ larger than 10, even under stationary-state conditions ($\theta_{ss} = \mu/\kappa$ and κ may take very small values), and even larger extents of self-heating may arise during non-stationary periods, e.g. at the oscillatory maxima.

In this section, therefore, we briefly investigate the stationary-state and Hopf bifurcation patterns that are found with the exact Arrhenius temperature dependence.

4.9.1. Stationary-state solutions

The governing equations have the form

$$d\alpha/d\tau = \mu - \kappa\alpha \exp[\theta/(1 + \gamma\theta)] \tag{4.62}$$

$$d\theta/d\tau = \alpha \exp[\theta/(1 + \gamma\theta)] - \theta. \tag{4.63}$$

The stationary states are thus given by

$$\theta_{ss} = \mu/\kappa \tag{4.64}$$

$$\alpha_{ss} = \theta_{ss} \exp[-\theta_{ss}/(1 + \gamma\theta_{ss})]. \tag{4.65}$$

The dependences of θ_{ss} and α_{ss} on the ratio of the two parameters μ/κ for a particular small value of γ are shown in Fig. 4.7(a).

For small γ, corresponding loosely to large values for the activation energy, the α_{ss} locus has both the maximum which was displayed with the exponential approximation and now a minimum which occurs at relatively distant values of μ/κ. This qualitative change in the locus means that the stationary-state concentration of A increases with high-enough reactant concentra-

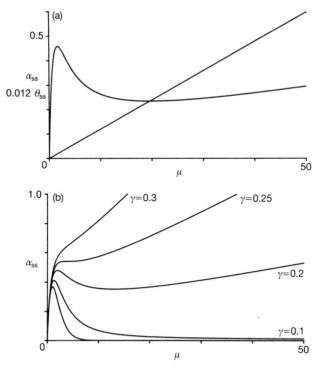

FIG. 4.7. Stationary-state loci for the intermediate concentration and temperature excess for model with full Arrhenius temperature dependence: (a) $\gamma = 0.175$, showing maximum and minimum in $\alpha_{ss}(\mu)$; (b) the disappearance of extrema in the $\alpha_{ss}(\mu)$ locus with increasing γ. $\gamma = 0$, 0.1, 0.2, 0.25, and 0.3.

tions. The location of these turning points is easily achieved by looking for the extrema in the right-hand side of eqn (4.65), i.e. for θ_{ss} given by

$$\mu/\kappa^{\neq} = \theta_{ss}^{\neq} = \tfrac{1}{2}\gamma^{-2}[1 - 2\gamma \pm (1 - 4\gamma)^{1/2}]. \tag{4.66}$$

For very small γ the limiting forms of (4.66) are

$$\mu/\kappa^{-} = \theta_{ss}^{-} \approx 1 + 2\gamma + \cdots \Rightarrow \alpha_{ss}^{-} \approx (1 + \gamma)e^{-1} \tag{4.67}$$

$$\mu/\kappa^{+} = \theta_{ss}^{+} \approx 1/(\gamma^{2}) \cdots \Rightarrow \alpha_{ss}^{+} \approx \gamma^{-2}e^{-1/\gamma}. \tag{4.68}$$

The lower root for small γ lies at μ/κ slightly greater than unity and α_{ss} slightly greater than e^{-1}, tending to these values as γ tends to zero. The upper root corresponds to the minimum in the curve and for small γ clearly lies at large μ/κ with α_{ss} exponentially small. In the limit of the exponential approximation, $\gamma \to 0$, the minimum goes off to infinity and $\alpha_{ss} = 0$.

As γ increases (high ambient temperature or low activation energy) the maximum and minimum move closer together: μ/κ^{-} increases and μ/κ^{+}

decreases as shown in Fig. 4.7(b). The two turning points merge when $\gamma = \frac{1}{4}$, when the discriminant in eqn (4.66) is zero, and there is then an inflexion point in the α_{ss} locus at $\mu/\kappa = 4$, $\alpha_{ss} = \frac{1}{4}e^{-2}$. For larger values of γ, the stationary-state concentration of A is a monotonically increasing function of the initial reactant concentration.

4.9.2. Local stability

The conditions for a change from *nodal to focal character* can be obtained in parametric form:

$$\kappa = \left(\frac{\theta^{1/2}}{1 + \gamma\theta} \pm 1 \right)^2 \exp\left(\frac{\theta}{1 + \gamma\theta} \right) \tag{4.69}$$

$$\mu = \kappa\theta. \tag{4.70}$$

The loci typically drawn out by these equations as θ varies are shown in Fig. 4.8(a). For $\gamma < \frac{1}{4}$ there is a small closed loop near the origin. Inside this loop, the stationary state is an unstable node. The larger outer loop separates stable focal character (inside curve) from stable nodal states (outside curve). As γ increases beyond $\frac{1}{4}$ the small inner loop shrinks to zero size; the outer loop still exists. Stable focal character exists over some values of the parameters μ and κ for any value of γ.

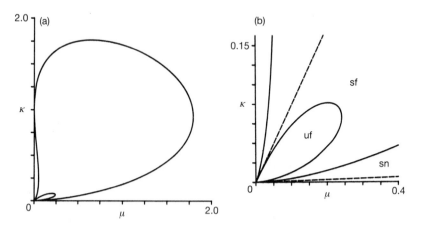

FIG. 4.8. The μ–κ parameter plane showing changes in local stability and character for $\gamma = 0.1$: (a) full parameter plane; (b) enlargement for small μ and κ showing locus of Hopf bifurcation (transition from stable to unstable focus and (as broken lines) the loci for the maximum and minimum in $\alpha_{ss}(\mu)$.

The condition for *Hopf bifurcation*, i.e. for a change from stable to unstable focus, is also shown in Fig. 4.8. This is given parametrically by

$$\kappa^* = \left(\frac{\theta^*_{ss}}{(1 + \gamma\theta^*_{ss})^2} - 1 \right) \exp\left(\frac{\theta^*_{ss}}{1 + \gamma\theta^*_{ss}} \right) \tag{4.71}$$

$$\mu^* = \kappa\theta^*_{ss} \tag{4.72}$$

$$\alpha^*_{ss} = \theta^*_{ss} \exp[- \theta^*_{ss}/(1 + \gamma\theta^*_{ss})]. \tag{4.73}$$

The values of κ^* predicted from these equations are only positive if $\theta^*_{ss} > (1 + \gamma\theta^*_{ss})^2$: $\kappa = 0$ for $\theta^*_{ss} = (\theta^*_{ss})_+$ and for $(\theta^*_{ss})_-$ where these are functions of γ given by

$$\theta^*_{\pm} = \frac{(1 - 2\gamma) \pm (1 - 4\gamma)^{1/2}}{2\gamma^2}. \tag{4.74}$$

Positive values for κ^* arise when θ^*_{ss} varies between these roots: $\theta^*_- \leqslant \theta^*_{ss} \leqslant \theta^*_+$.

Equation (4.74) has distinct real roots provided $\gamma < \frac{1}{4}$. Hopf bifurcation cannot occur if the activation energy E becomes too small compared with the thermal energy RT_a, i.e. if $E < 4RT_a$. This is the same condition on γ as that for the existence of the maximum and minimum in the α_{ss} locus. In fact, the Hopf bifurcation points always occur for μ values between the maximum and minimum, i.e. on the part of the locus where α_{ss} is decreasing, as shown in Fig. 4.8(b) where the loci of turning points are shown as broken lines.

For systems with $\gamma < \frac{1}{4}$, Hopf bifurcation occurs for some μ provided κ is small enough. The maximum value of κ for Hopf bifurcation is

$$\kappa^*_{max} = (1 - 4\gamma)e^{-2} \tag{4.75}$$

which again emphasizes the importance of requiring $\gamma < \frac{1}{4}$.

We may view eqns (4.71)–(4.73) in another way. Choose a system with $\gamma < \frac{1}{4}$. Next choose the dimensionless rate constant κ. If κ is less than $(1 - 4\gamma)e^{-2}$, eqn (4.71) can be solved to yield two positive roots θ^*_1 and θ^*_2. From these values for the stationary-state temperature excess we calculate the reactant concentration required for Hopf bifurcation from eqn (4.72) whilst (4.73) gives the stationary-state concentration of the intermediate A.

As an example, let us consider $\gamma = 0.02$: this gives $\kappa^*_{max} = 0.92e^{-2} \approx 0.1245$. Now let us choose $\kappa = 0.05 < \kappa^*_{max}$. Then we find that for the lower root $\theta^*_2 = 1.2226$, which then gives $\mu^*_2 = 0.061\ 13$ and $\alpha^*_2 = 0.3707$, and for the upper root $\theta^*_1 = 4.3160$, so that $\mu^*_1 = 0.2158$ and $\alpha^*_1 = 0.0812$.

These results may be compared with those found in § 4.6.2(b) with the exponential approximation (which gave $\theta^*_2 = 1.1594$, $\mu^*_2 = 0.057\ 97$, $\alpha^*_2 = 0.3637$ and $\theta^*_1 = 4.1399$, $\mu^*_1 = 0.2070$, $\alpha^*_1 = 0.0659$) to see the effect of

Table 4.5

Conditions for Hopf bifurcation, $\gamma = 0.02$

	Lower bifurcation			Upper bifurcation		
κ	θ_2^*	μ_2^*	α_{ss}^*	θ_1^*	μ_1^*	α_{ss}^*
0.01	1.0732	0.0107	0.3753	6.9110	0.0691	0.0159
0.02	1.1064	0.0221	0.3748	5.8183	0.1164	0.0317
0.03	1.1421	0.0343	0.3739	5.1677	0.1550	0.0478
0.04	1.1806	0.0472	0.3726	4.6945	0.1878	0.0642
0.05	1.2226	0.0611	0.3707	4.3160	0.2158	0.0812
0.06	1.2688	0.0761	0.3681	3.9952	0.2397	0.0988
0.07	1.3205	0.0924	0.3648	3.7116	0.2598	0.1172
0.08	1.3793	0.1103	0.3603	3.4520	0.2762	0.1367
0.09	1.4480	0.1303	0.3545	3.2063	0.2886	0.1576
0.10	1.5317	0.1532	0.3465	2.9642	0.2964	0.1806
0.11	1.6422	0.1806	0.3349	2.7106	0.2982	0.2072
0.12	1.8217	0.2186	0.3145	2.4004	0.2880	0.2430
0.1245	2.0833	0.2594	0.2819	2.0833	0.2594	0.2819

using the full Arrhenius form. Table 4.5 lists the Hopf bifurcation points for other values of κ with γ still equal to 0.02, obtained in the same way.

The various Hopf bifurcation parameters β_2, μ_2, and τ_2 can again be determined explicitly but have much more complex forms. We will discuss the details in the next chapter and only consider here the stability of the emerging limit cycle through β_2. With the full Arrhenius form, this parameter is given by

$$\beta_2 = \frac{1 - \theta^* - 2\gamma(3 - 2\theta^*) - 3\gamma^2(4 - \theta^*) - 6\gamma^3\theta^{*2}}{8(1 + \gamma\theta^*)^6}. \qquad (4.76)$$

The most significant aspect of this new expression is that β_2 can now change sign. There are some combinations of the dimensionless activation energy and the stationary-state temperature rise for which β_2 becomes positive. Under such conditions, which if they occur do so at the upper Hopf bifurcation μ_1^*, the emerging limit cycle is unstable. Furthermore, the unstable cycle then grows as the reactant concentration μ increases above μ_1^*, as shown in Fig. 4.9. This unstable limit cycle grows in amplitude until it merges with the stable limit cycle which was born at the lower Hopf point μ_2^*. The two cycles collide and disappear together in a stable–unstable limit cycle bifurcation at μ_{su}. Thus, over the range of reactant concentrations $\mu_1^* < \mu < \mu_{su}$, the system has two possible stable modes of behaviour (sitting at the stationary state or moving on the larger limit cycle) from which to choose, these modes being separated by the unstable limit cycle. The full implications of this added complexity, and the role played by unstable limit cycles, will be discussed in the next chapter.

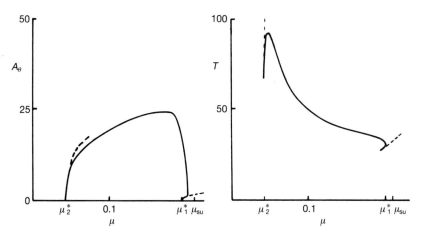

FIG. 4.9. The development of oscillatory amplitude A_θ and period T across the range of instability, $4.2 \times 10^{-3} = \mu_2^* \leqslant \mu \leqslant \mu_1^* = 0.0195$, for the pool chemical model with $\kappa = 2 \times 10^{-3}$ and $\gamma = 0.21$, typical of a system with a subcritical Hopf bifurcation at which an unstable limit cycle emerges at μ_1^*. The broken curves give the limiting forms predicted by eqns (4.59)–(4.61).

4.10. Time-dependent behaviour

We now relax our implicit approximation in which the consumption of the reactant has been ignored. The full time-dependent behaviour of the dimensionless equations will be considered. The situation is not greatly affected by use of either the exponential approximation or the full Arrhenius form, so we return to the former for simplicity. We will take the example data from Table 4.1: $\kappa = 0.05$, $\mu_0 = 0.5$, and $\varepsilon = 10^{-2}$. As we are employing the exponential approximation the value of γ is not important.

First, can we expect any oscillatory behaviour? Instability is possible only if $\kappa < e^{-2}$. This requirement is satisfied here. From the data in Table 4.4, the Hopf bifurcation points for this system occur for $\mu_1^* = 0.207$ and $\mu_2^* = 0.058$. For our example, the initial value $\mu_0 = 0.5$ exceeds the upper bifurcation point, so the system at first has a stable pseudo-stationary state to approach, with $\theta_{ss} \approx 10$ and $\alpha_{ss} \approx 4.54 \times 10^{-4}$. From Fig. 4.3 we may also estimate that the approach to this state will be monotonic since the initial conditions lie outside the region of damped oscillations.

The pre-oscillatory period, during which the temperature excess decreases and the concentration of A increases, will last at least until μ has fallen from μ_0 to the value μ_1^*, i.e. until the time τ_1^* given by

$$\tau_1^* = \varepsilon^{-1} \ln (\mu_0/\mu_1^*) = 88.2. \tag{4.77}$$

Following this the pseudo-stationary state becomes unstable and the concentration and temperature *histories* are expected to move away into oscillatory behaviour. The frequency ω_0 at this point is 1.772, so the oscillatory period should be 3.55. The stationary-state concentration and temperature excess have the values 6.59×10^{-2} and 4.14 at the bifurcation point respectively.

As seen in the previous chapter, the growth of observable excursions is not immediate and may take a considerable time. If oscillations do develop, we expect them to last until the pseudo-stationary state regains stability at time τ_2^*:

$$\tau_2^* = \varepsilon^{-1} \ln (\mu_0/\mu_2^*) = 215.4. \qquad (4.78)$$

After this time, there should be a damped oscillatory return to the pseudo-stationary locus with $\theta_{ss} \approx 1.16$ and $\alpha_{ss} \approx 0.313$. The concentration of

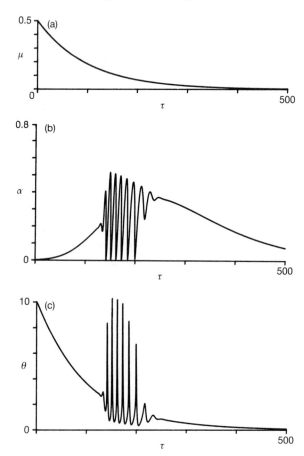

FIG. 4.10. Computed dimensionless concentration and temperature histories for the thermokinetic model with parameters given in Table 4.1 showing monotonic decay of precursor reactant μ and oscillations in the concentration α of intermediate A and the temperature excess θ: (a) $\mu(\tau)$, (b) $\alpha(\tau)$, and (c) $\theta(\tau)$.

A would then be close to its maximum value, so the post-oscillatory period should see both θ and α decreasing, eventually to zero. The actual computed behaviour is shown in Fig. 4.10 which is the dimensionless analogue of Fig. 4.1.

References

Gray, B. F. and Roberts, M. J. (1988). Analysis of chemical kinetic systems over the entire parameter space. *Proc. R. Soc.*, **A416**, 391–402.

Gray, P., Kay, S. R., and Scott, S. K. (1988). Oscillations of an exothermic reaction in a closed system I. *Proc. R. Soc.*, **A416**, 321–41.

Kay, S. R. and Scott, S. K. (1988). Oscillations of an exothermic reaction in a closed system II. *Proc. R. Soc.*, **A416**, 343–59.

Salnikov, I. Ye. (1948). Thermokinetic model of a homogeneous periodic reaction. *Dokl. Akad. Nauk SSSR*, **60**, 405–8.

Salnikov, I. Ye. (1949). Thermokinetic model of a homogeneous periodic reaction. *Zh. Fiz. Khim.*, **23**, 258–60.

HOPF BIFURCATIONS, THE GROWTH OF SMALL OSCILLATIONS, RELAXATION OSCILLATIONS, AND EXCITABILITY

Hopf bifurcation analysis commonly signals the onset of oscillatory behaviour. This chapter uses a particular two-variable example to illustrate the essential features of the approach and to explore the relationship to relaxation oscillations. After a careful study of this chapter the reader should be able to:

(1) devise and use dimensionless forms of the conservation equations to locate stationary states;

(2) set up the appropriate Jacobian to establish local stability of stationary states, particularly the points at which stability is lost as $\text{tr}(\mathbf{J})$ changes sign;

(3) map the possible modes of behaviour in parameter space;

(4) establish that at these points the necessary requirements for bifurcation to limit cycles are satisfied;

(5) establish the stability of the limit cycle by evaluating the sign of the Floquet multiplier β_2;

(6) determine the initial growth in oscillatory amplitude from zero by determining the coefficient μ_2;

(7) determine the shift in the oscillatory period and frequency away from their values at birth of oscillations by determining the coefficient τ_2;

(8) analyse the large-amplitude relaxation oscillations in a simple fashion.

In this chapter we give an introduction and recipe for the full Hopf bifurcation analysis for chemical systems. Rather than work in completely general and abstract terms, we will illustrate the various stages by using the thermokinetic model of the previous chapter, with the exponential approximation for simplicity. We can draw many quantitative conclusions about the oscillatory solutions in that model. In particular we will be able to show: (i) that the parameter values given by eqns (4.49) and (4.50) for $\text{tr}(\mathbf{J}) = 0$ satisfy all the requirements of the Hopf theorem; (ii) that oscillatory behaviour is completely confined to the conditions for which the stationary state is unstable; and (iii) that the growth in oscillatory amplitude and period can be expressed in very simple terms involving κ and μ. Despite our best intentions, this is perhaps the most mathematical section of the book. Those readers not interested in the details of the method can, however, happily omit this chapter on a first reading.

5.1. Introduction to Hopf bifurcation analysis

We present the analysis in fairly general terms, as well as giving the appropriate expressions at each stage for the present model, so that we can refer back when applying the method to slightly more complex examples later on. We will also follow closely the notation used in the book by Hassard *et al.* (1980) which we have found particularly helpful (the 'recipe' given in chapter 2 of their book is especially recommended and essentially a generalization of that given here).

To save continual referencing back to the previous chapter we collect here the important equations for the scheme $P \rightarrow A \rightarrow B + heat$. The governing rate equations are

$$d\alpha/d\tau = \mu - \kappa\alpha e^{\theta} \tag{5.1}$$

$$d\theta/d\tau = \alpha e^{\theta} - \theta. \tag{5.2}$$

The stationary states are then

$$\theta_{ss} = \mu/\kappa \qquad \alpha_{ss} = (\mu/\kappa)e^{-(\mu/\kappa)}. \tag{5.3}$$

The eigenvalues λ_1 and λ_2 which determine the local stability of the stationary state are given by the roots of the equation

$$\lambda^2 - (tr(\mathbf{J}))\lambda + \det(\mathbf{J}) = 0 \tag{5.4}$$

where the trace and determinant of the Jacobian matrix are

$$tr(\mathbf{J}) = (\mu/\kappa) - 1 - \kappa e^{\mu/\kappa} = \theta_{ss} - 1 - \kappa e^{\theta_{ss}} \tag{5.5}$$

$$\det(\mathbf{J}) = \kappa e^{\mu/\kappa} = \kappa e^{\theta_{ss}}. \tag{5.6}$$

The conditions for Hopf bifurcation require the trace to become zero and can be expressed parametrically as

$$\kappa^* = (\theta_{ss}^* - 1)e^{-\theta_{ss}^*} \tag{5.7}$$

$$\mu^* = \kappa^*\theta_{ss}^*. \tag{5.8}$$

The locus described by these equations in the κ–μ parameter plane is reproduced in Fig. 5.1, which also shows a typical stationary-state bifurcation diagram for fixed κ. Hopf bifurcations occur at two values of the precursor reactant concentration $\mu_{1,2}^*$, with $\mu_2^* < \mu_1^*$, for any given κ less than a

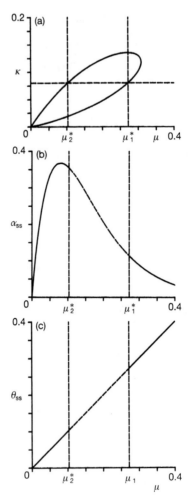

FIG. 5.1.   (a) A typical 'parameter plane' showing a locus of Hopf bifurcation points. For any given value of the parameter κ on the ordinate we may construct a horizontal (broken line): the Hopf bifurcation points, μ_1^* and μ_2^*, are then located as shown. The corresponding station-ary-state loci, shown in (b) and (c), have unstable solutions between μ_1^* and μ_2^*.

maximum $\kappa_{max} = e^{-2}$. This maximum κ corresponds to $\theta_{ss}^* = 2$ in the above equations.

5.1.1. Behaviour of eigenvalues at Hopf bifurcation

Close to the conditions at which the trace is becoming zero, the eigenvalues $\lambda_{1,2}$ will have the form of a complex conjugate pair:

$$\lambda_{1,2} = v(\mu) \pm i\omega(\mu) \tag{5.9}$$

where $v(\mu)$ is the real part of the eigenvalues given simply by

$$v(\mu) = \tfrac{1}{2}\operatorname{tr}(\mathbf{J}). \tag{5.10}$$

The imaginary part $\omega(\mu)$ is related to the discriminant $\Delta = (\operatorname{tr}(\mathbf{J}))^2 - 4\det(\mathbf{J})$ by

$$\omega(\mu) = \tfrac{1}{2}(-\Delta)^{1/2} \tag{5.11}$$

(recalling that as $\operatorname{tr}(\mathbf{J})$ tends to zero, Δ remains negative).

We have stressed that both the real and imaginary parts depend on the parameter μ because we are imagining experiments where the reactant concentration will be varied whilst κ is held constant. If we were doing the experiments another way so that μ was held fixed and the dimensionless reaction rate constant varied in the vicinity of the Hopf bifurcation point we would then wish to consider $v(\kappa)$ and $\omega(\kappa)$.

At the conditions specified by eqns (5.7) and (5.8) we have $\operatorname{tr}(\mathbf{J}) = 0$, and so

$$v_0 = v(\mu^*) = 0. \tag{5.12}$$

We must now check that the imaginary part of $\lambda_{1,2}$ is not zero under these conditions. With $\operatorname{tr}(\mathbf{J}) = 0$ we have, quite generally,

$$\omega_0 = \omega(\mu^*) = (\det(\mathbf{J}))^{1/2} \tag{5.13}$$

and for the particular model here

$$\omega_0 = (\kappa e^{\mu/\kappa})^{1/2} > 0. \tag{5.14}$$

The quantity ω_0 is of considerable importance. It gives the natural frequency of the oscillations (corresponding to a period of $2\pi/\omega_0$) as they first appear at the point of Hopf bifurcation.

Next we must check that the real part of the eigenvalues actually passes through zero, as required in §3.4.1. This condition will be satisfied if the derivative of v, or equivalently the trace of the Jacobian, with respect to the parameter being varied (in this case μ) is non-zero when evaluated at μ^*. For the present model

$$(\mathrm{d}v/\mathrm{d}\mu)_{\mu^*} = \tfrac{1}{2}(\mathrm{d}\operatorname{tr}(\mathbf{J})/\mathrm{d}\mu)_{\mu^*} = (2 - \theta_{\mathrm{ss}})/2\kappa. \tag{5.15}$$

This will be non-zero, except when $\theta_{\mathrm{ss}} = 2$. From the parametric solutions (5.7) and (5.8) we can see that when $\theta_{\mathrm{ss}} = 2$ we have $\kappa^* = e^{-2}$, $\mu^* = 2e^{-2}$. This corresponds to the maximum in the locus of Hopf bifurcations in Fig. 4.2 and to conditions for which μ_1^* and μ_2^* merge. For all $\kappa < e^{-2}$, the two bifurcation points have $(\mathrm{d}v/\mathrm{d}\mu)_{\mu^*} \neq 0$ and hence satisfy this degeneracy condition.

Another derivative evaluated at the Hopf bifurcation point of interest, which we will need later on, is that of the imaginary part of frequency $(\mathrm{d}\omega/\mathrm{d}\mu)_{\mu^*}$. For our model here,

$$(\mathrm{d}\omega/\mathrm{d}\mu)_{\mu^*} = 2e^{\theta_{\mathrm{ss}}}/\omega_0. \tag{5.16}$$

When we come to look at the stability of the limit cycle which is born at the Hopf bifurcation point, we shall meet a quantity known as the 'Floquet multiplier', conventionally denoted β_2, which plays a role similar to that played for the stationary state by the eigenvalues λ_1 and λ_2. If β_2 is negative, the limit cycle will be stable and should correspond to observable oscillations; if β_2 is positive the limit cycle will be unstable.

5.1.2. Evaluation of limit cycle stability by means of β_2

In order to evaluate β_2 we need to consider how the governing equations for mass and energy balance themselves vary with changes in the variables. In the case of the present model this means evaluating various partial derivatives of (5.1) and (5.2) with respect to α and θ. Before proceeding, however, we should take a look at the elements of the Jacobian matrix evaluated for Hopf bifurcation conditions:

$$\mathbf{J}_{\mu*} = \begin{pmatrix} -\kappa e^{\theta_{ss}} & -\kappa \theta_{ss} \\ e^{\theta_{ss}} & \theta_{ss} - 1 \end{pmatrix}. \tag{5.17}$$

All four terms are non-zero. It will actually make the analysis much easier if we can transform the matrix so that it has the form

$$\mathbf{J}_{\mu*} = \begin{pmatrix} 0 & -\omega_0 \\ \omega_0 & 0 \end{pmatrix}. \tag{5.18}$$

This can be achieved without too much difficulty by introducing two new variables, x and y, which will be simple linear combinations of α and θ. In other words we want to use

$$x = a\alpha + b\theta \tag{5.19}$$

$$y = c\alpha + h\theta \tag{5.20}$$

where we will choose a, b, c, and h so that when we evaluate the Jacobian of dx/dt and dy/dt it will have the form of (5.18). This may still sound rather vague, but the present model gives us an excellent example on which to practice.

(a) Transformation of variables

We can start, quite generally, by taking $a = 0$ and $b = 1$, so that

$$x = \theta. \tag{5.21}$$

This choice has been made as the functional dependence of the mass- and energy-balance equations on θ is slightly more complex than that on α. In terms of the new variables, we can express α and θ as

$$\alpha = (y - hx)/c \tag{5.22}$$

$$\theta = x. \tag{5.23}$$

We also have that

$$dx/d\tau = d\theta/d\tau \tag{5.24}$$

$$dy/d\tau = c(d\alpha/d\tau) + h(d\theta/d\tau). \tag{5.25}$$

If we substitute into (5.24) and (5.25) from the original governing equations (5.1) and (5.2), and replace α and θ by (5.22) and (5.23), we obtain the new system of differential equations:

$$dx/d\tau = g_1(x, y) = c^{-1}(y - hx)e^x - x \tag{5.26}$$

$$dy/d\tau = g_2(x, y) = c\mu + c^{-1}(h - c\kappa)(y - hx)e^x - hx. \tag{5.27}$$

We are still free to choose the constants c and h, and the middle term of (5.27) suggests that we should take $h = c\kappa$ so that $dy/d\tau$ does not then depend on y. The equations then become:

$$dx/d\tau = g_1(x, y) = (y/c - \kappa x)e^x - x \tag{5.28}$$

$$dy/d\tau = g_2(x, y) = c(\mu - \kappa x). \tag{5.29}$$

The Jacobian for this system is

$$J = \begin{pmatrix} (y/c - \kappa x - \kappa)e^x - 1 & e^x/c \\ -c\kappa & 0 \end{pmatrix}_{ss}. \tag{5.30}$$

The trace of this new Jacobian becomes zero when the first element vanishes, i.e. when

$$(y/c - \kappa x - \kappa)e^x - 1 = 0 \tag{5.31}$$

subject to the stationary-state conditions $g_1(x, y) = g_2(x, y) = 0$. These three are all satisfied for the same μ^* and κ^* as found with the original variables α and θ, given by eqns (5.7) and (5.8) with the corresponding values of x^*_{ss} and y^*_{ss} and then given by (5.21) and (5.20). At such a point the Jacobian becomes

$$J = \begin{pmatrix} 0 & e^{x^*_{ss}}/c \\ -c\kappa & 0 \end{pmatrix}. \tag{5.32}$$

This will have the same form as the desired matrix (5.18) if we choose

$$c = -\omega_0/\kappa \qquad \text{so that} \qquad h = -\omega_0. \tag{5.33}$$

We can now finally write down our new transformed variables and governing equations:

$$x = \theta \tag{5.34}$$

$$y = -\omega_0(\alpha/\kappa + \theta) \tag{5.35}$$

and

$$dx/d\tau = g_1(x, y) = -(\kappa/\omega_0)(y + \omega_0 x)e^x - x \qquad (5.36)$$

$$dy/d\tau = g_2(x, y) = -(\omega_0/\kappa)(\mu - \kappa x). \qquad (5.37)$$

We can now go on to determine the necessary partial derivatives of $g_1(x, y)$ and $g_2(x, y)$, all of which will need to be evaluated at the point of Hopf bifurcation (which was located in terms of the original variables).

(b) Evaluation of partial derivatives

The list of partial derivatives that we require is quoted here without derivation or comment. The specific forms correspond to the model under consideration, but again could be evaluated in general terms. We require

$$\partial^2 g_1/\partial x^2 \quad = -(\kappa/\omega_0)(y + \omega_0 x + 2\omega_0)e^x$$

$$\partial^2 g_1/\partial x \partial y \quad = -(\kappa/\omega_0)e^x$$

$$\partial^2 g_1/\partial y^2 \quad = 0$$

$$\partial^3 g_1/\partial x^3 \quad = -(\kappa/\omega_0)(y + \omega_0 x + 3\omega_0)e^x$$

$$\partial^3 g_1/\partial x^2 \partial y \quad = -(\kappa/\omega_0)e^x$$

$$\partial^3 g_1/\partial x \partial y^2 \quad = 0$$

$$\partial^3 g_1/\partial y^3 \quad = 0$$

$$\partial^2 g_2/\partial x^2 \quad = 0$$

$$\partial^2 g_2/\partial x \partial y \quad = 0$$

$$\partial^2 g_2/\partial y^2 \quad = 0$$

$$\partial^3 g_2/\partial x^3 \quad = 0$$

$$\partial^3 g_2/\partial x^2 \partial y \quad = 0$$

$$\partial^3 g_2/\partial x \partial y^2 \quad = 0$$

$$\partial^3 g_2/\partial y^3 \quad = 0.$$

Only four of these partial derivatives are not identically zero. For these we may use the following relationships, which apply at the Hopf bifurcation point, to simplify

$$\kappa^* = (x - 1)e^{-x} \qquad \mu^* = x(x - 1)e^{-x} \qquad \omega_0 = (x - 1)^{1/2}$$

$$\kappa^*/\omega_0 = (x - 1)^{1/2}e^{-x} \qquad y = -x^2/(x - 1)^{1/2}.$$

These give:

$$\partial^2 g_1/\partial x^2 = (2 - x)$$
$$\partial^2 g_1/\partial x \partial y = -(x - 1)^{1/2}$$
$$\partial^3 g_1/\partial x^3 = (3 - 2x)$$
$$\partial^3 g_1/\partial x^2 \partial y = -(x - 1)^{1/2}.$$

(We must remember that these forms apply only at the bifurcation point, but clearly once we know x, which is the value of the dimensionless temperature rise θ_{ss}, at the Hopf bifurcation, all the above terms can be evaluated easily.)

The fact that so many of these terms are identically zero comes from the appropriateness of the linear transformation chosen and the especially simple form of g_2.

(c) Evaluation of four complex numbers

Next, there are four complex numbers g_{11}, g_{02}, g_{20}, and g_{21} which we need to evaluate. These are defined in general terms as

$$g_{11} = \tfrac{1}{4}[\partial^2 g_1/\partial x^2 + \partial^2 g_1/\partial y^2 + i(\partial^2 g_2 \partial x^2 + \partial^2 g_2/\partial y^2)] \qquad (5.38)$$

$$g_{02} = \tfrac{1}{4}[\partial^2 g_1/\partial x^2 - \partial^2 g_1/\partial y^2 - 2\partial^2 g_2/\partial x \partial y$$
$$+ i(\partial^2 g_2/\partial x^2 - \partial^2 g_1/\partial y^2 + 2\partial^2 g_1/\partial x \partial y)] \qquad (5.39)$$

$$g_{20} = \tfrac{1}{4}[\partial^2 g_1/\partial x^2 - \partial^2 g_1/\partial y^2 - 2\partial^2 g_2/\partial x \partial y$$
$$+ i(\partial^2 g_2/\partial x^2 - \partial^2 g_2/\partial y^2 - 2\partial^2 g_1/\partial x \partial y)] \qquad (5.40)$$

$$g_{21} = \tfrac{1}{8}[\partial^3 g_1/\partial x^3 + \partial^3 g_1/\partial x \partial y^2 + \partial^3 g_2/\partial x^2 \partial y$$
$$+ \partial^3 g_2/\partial y^3 + i(\partial^3 g_2/\partial x^3 + \partial^3 g_2/\partial x \partial y^2$$
$$- \partial^3 g_1/\partial x^2 \partial y - \partial^3 g_1/\partial y^3)]. \qquad (5.41)$$

For our specific example, these numbers can be written in terms of x in very much simpler forms:

$$g_{11} = \tfrac{1}{4}(2 - x)$$
$$g_{02} = \tfrac{1}{4}(2 - x) - \tfrac{1}{2}i(x - 1)^{1/2}$$
$$g_{20} = \tfrac{1}{4}(2 - x) + \tfrac{1}{2}i(x - 1)^{1/2}$$
$$g_{21} = \tfrac{1}{8}(3 - 2x) + \tfrac{1}{8}i(x - 1)^{1/2}.$$

(d) Evaluation of limit cycle stability

As mentioned previously the stability of the emerging limit cycle is determined by the sign of a quantity β_2. This term is defined generally by

$$\beta_2 = \mathrm{Re}(g_{21}) - (1/\omega_0)[\mathrm{Re}(g_{11})\mathrm{Im}(g_{20}) + \mathrm{Re}(g_{20})\mathrm{Im}(g_{11})]. \quad (5.42)$$

For our present non-isothermal model the specific forms above reduce to the simple result

$$\beta_2 = (1 - x)/8. \quad (5.43)$$

Recalling that x is identically θ_{ss} and hence always greater than or equal to unity, we see that the Floquet multiplier is always negative with this scheme; only a stable limit cycle emerges.

5.1.3. Growth in size and period of limit cycle

At the point of Hopf bifurcation, the emerging limit cycle has zero amplitude and an oscillatory period given by $2\pi/\omega_0$. As we begin to move away from the bifurcation point the amplitude A and period T grow in a form we can calculate according to the formulae

$$A = [(\mu - \mu^*)/\mu_2]^{1/2} + \cdots \quad (5.44)$$

$$T = (2\pi/\omega_0)[1 + (\tau_2/\mu_2)(\mu - \mu^*) + \cdots]. \quad (5.45)$$

In these equations th term $(\mu - \mu^*)$ represents how far we have moved away from the bifurcation point, in terms of the dimensionless concentration of reactant. There are two new quantities μ_2 and τ_2 which tell us a number of things. The amplitude A grows as the square root of the distance from the bifurcation point $(\mu - \mu^*)$, and so the term $(\mu - \mu^*)/\mu_2$ must be positive. If μ_2 turns out to be positive, then the limit cycle must grow as μ is increased beyond μ^*; if μ_2 is negative, the limit cycle grows as μ decreases below μ^*. The growth (or decrease) in oscillatory period is linear in $(\mu - \mu^*)$ and depends on the ratio τ_2/μ_2.

First, let us consider μ_2. This is defined as

$$\mu_2 = -\beta_2/(\mathrm{d}\,\mathrm{tr}(\mathbf{J})/\mathrm{d}\mu)_{\mu^*} = -\tfrac{1}{2}\beta_2/(\mathrm{d}v/\mathrm{d}\mu)_{\mu^*}. \quad (5.46)$$

Note that this involves the derivative of the real parts of the eigenvalues $\lambda_{1,2}$ or, equivalently, of the trace of $\mathbf{J}$ evaluated at the Hopf bifurcation point. We know that $\mathrm{Re}(\lambda_{1,2})$ is passing through zero at this point.

If $\mathrm{d}\,\mathrm{tr}(\mathbf{J})/\mathrm{d}\mu$ is *positive* at μ^*, then $\mathrm{tr}(\mathbf{J})$ will be negative (corresponding to a stable stationary state) for $\mu < \mu^*$ and positive (corresponding to an unstable stationary state) for $\mu > \mu^*$. In these circumstances μ_2 will have the opposite sign to β_2: thus if β_2 is negative, μ_2 will be positive. The limit cycle then grows as μ increases beyond μ^* (because μ_2 is positive) and is stable (because β_2 is negative). Conversely, if β_2 is positive an unstable limit cycle emerges as μ decreases below μ^*. These possibilities, and those arising when $d\,\mathrm{tr}(\mathbf{J})/d\mu$ is *negative*, are illustrated in Fig. 5.2.

The growth of the period involves τ_2/μ_2, where τ_2 is defined by

$$\tau_2 = - [C + \mu_2(d\omega/d\mu)_{\mu^*}]/\omega_0 \qquad (5.47)$$

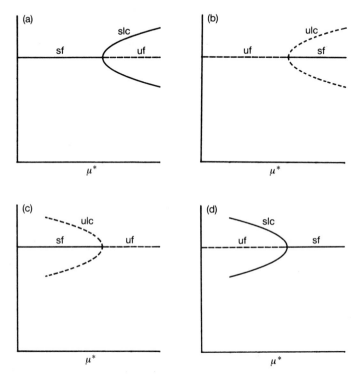

FIG. 5.2.   The development of limit cycles at Hopf bifurcation points. The figure can be interpreted as follows. The straight line represents the locus of the (constant) stationary-state solution; a solid line indicates stability (sf = stable focus); a broken line indicates an unstable solution (uf = unstable focus). The point at which the stationary state loses stability, μ^*, corresponds to the Hopf bifurcation. At that point a limit cycle, represented by the parabola, emerges from the stationary state: again, a solid curve indicates stability and a broken curve instability. (a) $d(\mathrm{tr}(\mathbf{J})/d\mu > 0$ and $\beta_2 < 0$, the stationary state loses stability and a stable limit cycle emerges as μ is increased beyond μ^*; (b) $d(\mathrm{tr}(\mathbf{J})/d\mu > 0$ and $\beta_2 > 0$, the stationary state is again stable for $\mu < \mu^*$ and unstable for $\mu > \mu^*$, but now an unstable limit cycle emerges from the bifurcation point, growing around the unstable state as μ is decreased; (c) $d(\mathrm{tr}(\mathbf{J}))/d\mu < 0$ and $\beta_2 > 0$, now the stationary state is unstable for $\mu < \mu^*$ and gains stability as μ is increased, an unstable limit cycle emerges and grows for $\mu > \mu^*$; (d) $d(\mathrm{tr}(\mathbf{J}))/d\mu < 0$ and $\beta_2 < 0$, the stationary state loses stability as μ is decreased below μ^* and a stable limit cycle emerges in the same direction. The locus of limit cycles represented in Fig. 4.4, say, then has case (a) at the lower Hopf point μ_2^* and case (d) at the upper Hopf bifurcation μ_1^*.

with

$$C = \tfrac{1}{2}\mathrm{Im}(g_{21}) + (1/2\omega_0)[\mathrm{Re}(g_{11})\mathrm{Re}(g_{20}) - \mathrm{Im}(g_{11})\mathrm{Im}(g_{20})$$
$$- 2|g_{11}|^2 - \tfrac{1}{3}|g_{02}|^2].$$

For τ_2, therefore, we need the derivative of the imaginary part of the eigenvalues evaluated at the Hopf bifurcation point. We may also note that the sign of the quotient τ_2/μ_2 is of less immediate significance than those of β_2 and μ_2.

Equations (5.44) and (5.45) give only the leading-order growth of A and T. The higher-order terms become more important as the magnitude of the departure from the bifurcation point $(\mu - \mu^*)$ increases and then, in general, we need to compute the size of the limit cycles.

For the non-isothermal pool chemical model we can again find simple expressions with

$$d\,\mathrm{tr}(\mathbf{J})/d\mu = (2 - x)/\kappa^* = (2 - x)e^x/(x - 1) \qquad (5.48)$$

so that

$$\mu_2 = (x - 1)^2 e^{-x}/8(2 - x) \qquad (5.49)$$

$$\tau_2/\mu_2 = -(2x^3 - x^2 + 3x - 6)e^x/6(x - 1)^3. \qquad (5.50)$$

From eqn (5.49) we see that μ_2 changes sign as x (or θ_{ss}) passes through the value 2. Thus μ_2 is positive at the lower Hopf bifurcation (for which $1 < \theta^* < 2$). This in turn means that the emerging limit cycle always grows, as μ is increased above μ_2^*, into the region of the unstable stationary state. At the upper Hopf bifurcation point (with $\theta^* > 2$), μ_2 is negative. Here, then, the limit cycle grows as the reactant concentration decreases past the bifurcation point μ_1^*, again as we move into the region of stationary-state instability.

This completes the general recipe as far as we shall require it in this book. In some special cases, e.g. when β_2 is exactly zero, the bifurcation point becomes degenerate. Higher-order terms are important from the moment we move away from μ^*. We shall not consider these cases in fine detail here, although we will be interested in locating any such points in later chapters.

5.2. Hopf bifurcation analysis with Arrhenius model: birth and growth of oscillations

We have already discussed the expressions resulting from a full Hopf bifurcation analysis of the thermokinetic model with the exponential approximation ($\gamma = 0$). We may do the same for the exact. Arrhenius temperature dependence ($\gamma \neq 0$). Although the algebra is somewhat more onerous, we still arrive at analytical expressions for the stability of the emerging or vanishing limit cycle and the rate of growth of the amplitude and period at the bifurcation points. With the exponential approximation it was found that the limit cycle was always stable (the parameter β_2 was always negative). In the present case, however, we will see a more flexible response and unstable limit cycles may be found under some conditions.

First, we recall the conditions for Hopf bifurcation derived in §4.9.2(b). These are

$$\kappa^* = \left(\frac{\theta^*}{(1 + \gamma\theta^*)^2} - 1\right)\exp\left(-\frac{\theta^*}{1 + \gamma\theta^*}\right) \tag{5.51}$$

$$\mu^* = \kappa\theta^*. \tag{5.52}$$

5.2.1. Stability of limit cycle by means of β_2

The evaluation of β_2 follows the same recipe as that given above through the transformation to new variables x and y, the various partial derivatives $\partial^2 g_1/\partial x^2$ etc., and then to g_{11}. This manipulation requires patience or a computer algebra package (preferably both), but really we are only interested in the resulting expression which is

$$\beta_2 = \frac{1 - \theta^*_{ss} - 2\gamma(3 - 2\theta^*_{ss}) - 3\gamma^2\theta^*_{ss}(4 - \theta^*_{ss}) - 6\gamma^3\theta^{*2}_{ss}}{8(1 + \gamma\theta^*_{ss})^6}. \tag{5.53}$$

This rather cluttered equation at least reduces very simply to the correct form for the exponential approximation (eqn (5.43)) in the limit as $\gamma \to 0$. The numerator is a quadratic in θ^*_{ss}. We will be interested to see if there are any conditions under which β_2 becomes zero. If β_2 is negative, the emerging limit cycle is stable as before. If β_2 becomes positive, the emerging limit cycle will lose its stability and become unstable.

The numerator of eqn (5.53) can be rewritten as

$$(1 - 6\gamma) - (1 - 4\gamma + 12\gamma^2)\theta^*_{ss} + 3\gamma^2(1 - 2\gamma)\theta^{*2}_{ss} = 0 \tag{5.54}$$

and so is satisfied when θ^*_{ss} is given by the (upper) root of

$$\theta^*_{ss} = \frac{(1 - 4\gamma + 12\gamma^2) \pm (1 - 8\gamma + 28\gamma^2)^{1/2}}{6\gamma^2(1 - 2\gamma)}. \tag{5.55}$$

It turns out that for systems with $\gamma \leqslant 2/9$ there is always a point along the Hopf bifurcation locus at which $\beta_2 = 0$, as indicated in Fig. 5.3. The change in stability of limit cycle always occurs on the upper branch of Hopf bifurcation points. The corresponding values of μ^*_1 and κ are obtained by substituting from eqn (5.5) into (5.51) and (5.52). So for $\gamma = 1/50$, the location is $\theta^*_{ss} = 801.8$, $\kappa = 6.40 \times 10^{-21}$, $\mu = 5.13 \times 10^{-18}$. For κ smaller than this value, β_2 is positive and the emerging limit cycle is unstable; for larger κ, β_2 is negative, leading to the familiar stable limit cycle.

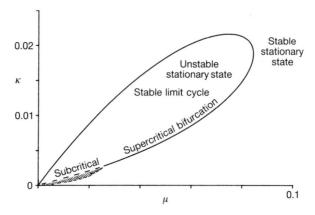

FIG. 5.3.   Locus of Hopf bifurcation points in κ–μ parameter plane for thermokinetic model with the full Arrhenius temperature dependence and $\gamma = 0.21$. The nature of the Hopf bifurcation point and, hence, the stability of the emerging limit cycle changes along this locus at $\kappa = 2.77 \times 10^{-3}$. Supercritical bifurcations are denoted by the solid curve, subcritical bifurcations occur along the broken segment, i.e. at the upper bifurcation point for the lowest κ. The stationary-state solution is unstable and surrounded by a stable limit cycle for all parameter values within the enclosed region. Oscillatory behaviour also occurs in the small shaded region below the Hopf curve, where the stable stationary state is surrounded by both an unstable and a stable limit cycle (see text).

Table 5.1 gives the location of the $\beta_2 = 0$ point for various values of γ. Points such as this, where the nature of the limit cycle changes qualitatively, are known as 'degenerate Hopf bifurcations'.

5.2.2. Growth of stable and unstable limit cycles

The expression μ_2, which determines the direction in which the limit cycle grows, is

$$\mu_2 = \kappa \frac{- 3\gamma^2(1 - 2\gamma)\theta_{ss}^{*2} + (1 - 4\gamma + 12\gamma^2)\theta_{ss}^* - (1 - 6\gamma)}{8[2 - (1 - 2\gamma)\theta_{ss}^*](1 + \gamma\theta_{ss}^*)^2}. \qquad (5.56)$$

At the upper Hopf bifurcation point, μ_2 and β_2 have the same sign. For values of κ above the degenerate Hopf point, β_2 and μ_2 are negative. A stable limit cycle emerges as μ is reduced below μ_1^* (Fig. 5.4(a)) in the same way as found with the exponential approximation earlier; this is known as a supercritical Hopf bifurcation. If κ is less than the degenerate value, then β_2 and μ_2 are positive. Now an unstable limit cycle emerges, and it exists and grows for values of μ greater than the upper Hopf bifurcation point (Fig. 5.4(b)); this is a subcritical bifurcation.

At the lower Hopf bifurcation, β_2 is always negative and μ_2 is positive. Thus a stable limit cycle emerges from μ_2^*, growing as the reactant concentration μ is increased.

Table 5.1

Location of degenerate Hopf bifurcation points from eqn (5.55)

γ	κ	θ_{ss}^*	μ
0	0	∞	0
0.05	4.88×10^{-8}	122.11	5.96×10^{-6}
0.10	5.12×10^{-4}	29.43	1.51×10^{-2}
0.15	4.97×10^{-3}	14.03	6.97×10^{-2}
0.20	4.46×10^{-3}	9.73	4.33×10^{-2}
0.21	2.77×10^{-3}	9.34	2.58×10^{-2}
0.22	5.57×10^{-4}	9.05	5.04×10^{-3}
2/9	0	9	0

The variation in oscillatory amplitude across the whole range of instability must be completed numerically (a suitable method for this is described in the appendix to this chapter). With the full Arrhenius form there are two possible scenarios, corresponding to the two different types of Hopf bifurcation at μ_1^*.

When the Hopf bifurcation at μ_1^* is supercritical ($\beta_2 < 0$) the system has just a single stable limit cycle. This emerges at μ_1^* and exists across the range $\mu_2^* < \mu < \mu_1^*$, within which it surrounds the unstable stationary-state solution. The limit cycle shrinks back to zero amplitude at the lower bifurcation point μ_2^*. This behaviour is qualitatively the same as that shown with the simplifying exponential approximation and is illustrated in Fig. 5.4(a).

For a subcritical Hopf bifurcation at μ_1^*, the situation is significantly different. We now have a stable limit cycle emerging at the lower bifurcation and an unstable cycle from the upper Hopf point. Both of these grow as μ increases. For μ in the range $\mu_2^* < \mu < \mu_1^*$ there is an unstable stationary-state solution surrounded by a stable limit cycle. Above μ_1^*, however, the stationary state becomes stable again and is now surrounded by two limit cycles: one stable and the other unstable. A typical phase plane representation is shown in Fig. 5.5. The unstable limit cycle lies between the two stable responses, separating the phase plane into two regions or 'basins of attraction'. Any trajectory starting within the unstable cycle eventually winds on to the stable stationary state; those starting outside the unstable cycle move towards the stable limit cycle. Thus, for these values of the reactant concentration, the reaction may be either steady or oscillatory, depending on the initial conditions.

As μ increases further, the unstable limit cycle grows within the stable cycle. Eventually the two coalesce (at μ_{su} in Fig. 5.4(b)), whereupon limit cycle behaviour disappears.

Often the range $\mu_1^* < \mu < \mu_{su}$ over which stable stationary state and oscillatory behaviour coexist is extremely small, but it allows the possibility of oscillations outside the region enclosed by the Hopf locus in Fig. 5.3, as

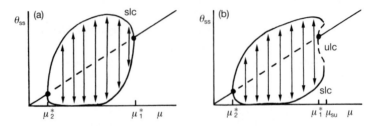

FIG. 5.4. The birth and growth of oscillatory solutions for the thermokinetic model with the full Arrhenius temperature dependence. (a) The Hopf bifurcations μ_1^* and μ_2^* are both supercritical, with $\beta_2 < 0$, and the stable limit cycle born at one dies at the other. (b) The upper Hopf bifurcation is subcritical, with $\beta_2 > 0$. An unstable limit cycle emerges and grows as the dimensionless reactant concentration μ increases—at μ_{su} this merges with the stable limit cycle born at the lower supercritical Hopf bifurcation point μ_2^*.

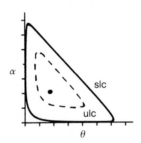

FIG. 5.5. A typical phase portrait for a system with $\mu_1^* < \mu < \mu_{su}$, showing a stable stationary-state solution (singular point) surrounded first by an unstable limit cycle (broken curve) and then by a stable limit cycle (solid curve). The unstable limit cycle separates those initial conditions, corresponding to points in the parameter plane lying within the ulc, which are attracted to the stationary state from those outside the ulc, which are attracted on to the stable limit cycle and hence which lead to oscillations.

indicated by the shaded region below the degenerate Hopf point. Unstable limit cycles also arise quite commonly in open systems such as the well-stirred flow reactor which we consider in the next three chapters. Again they play an important role, separating different 'basins of attraction' in the phase plane.

5.3. Relaxation oscillations

The Hopf bifurcation approach is a mathematically rigorous technique for locating and analysing the onset of oscillatory behaviour in general dynamical systems. Another approach which has been particularly well exploited for chemical systems is that of looking for 'relaxation' oscillations. Typically, the wave form for such a response can be broken down into distinct periods, some where the concentrations are varying slowly and others where there are sharp changes.

We can illustrate this latter technique with the simple thermokinetic model with the Arrhenius temperature dependence discussed above. This will also allow us to see that the two approaches are not separate, but that oscillations change smoothly from the basically sinusoidal waveform at the Hopf bifurcation to the 'relaxation' form in other parts of the parameter plane.

5.3.1. Thermokinetic oscillator with small parameters

The realm in which relaxation oscillations arise for equations such as the present scheme is that in which the different participants vary on quite different timescales. If we take the rate equations for the concentration of A and the temperature rise in their dimensionless form we have

$$d\alpha/d\tau = \mu - \kappa\alpha\exp(\theta/1 + \gamma\theta) \tag{5.57}$$

$$d\theta/d\tau = \alpha\exp(\theta/1 + \gamma\theta) - \theta. \tag{5.58}$$

The two parameters μ and κ appear in the first of these equations, one in each of the two terms on the right-hand side, but not in the second. If we are operating the reaction under conditions such that these are both small quantities, then the two terms in $d\alpha/d\tau$ will both be small. In such a case, the concentration of the intermediate will vary only slowly compared to the rate of change of temperature.

If we look at the parameter plane, Fig. 5.6, we can see that small values of these parameters correspond to the region close to the origin, but that this region still includes conditions for which our Hopf analysis has shown that

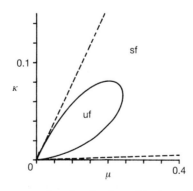

FIG. 5.6. Typical $\kappa-\mu$ parameter plane for the thermokinetic model with the full Arrhenius temperature depedence: a region of stationary-state instability lies within the locus of Hopf bifurcation points (solid curve). Also shown, as broken lines, are the loci corresponding to the maximum and minimum in the $g(\alpha, \theta) = 0$ nullcline (see text).

the reaction is oscillatory. In fact as we get closer to the origin, so the oscillatory range occupies more and more of the region between the two straight lines, which correspond to the locations of the maximum and minimum in the $\alpha_{ss}(\mu)$ locus (see §4.9).

If μ and κ are particularly small, we can usefully introduce new dimensionless groups M and K defined by

$$\mu = \varepsilon M \qquad \kappa = \varepsilon K \tag{5.59}$$

where ε is a small parameter and M and K are of order unity. We can also use a new timescale T, which is related to the previous dimensionless time τ by

$$\tau = \varepsilon^{-1} T. \tag{5.60}$$

This new timescale T is a 'slow time' and is the timescale on which the slow variable α is changing.

Equations (5.57) and (5.58) can now be rewritten as

$$d\alpha/dT = M - K\alpha \exp(\theta/1 + \gamma\theta) = f(\alpha, \theta) \tag{5.61}$$

$$\varepsilon d\theta/dT = \alpha \exp(\theta/1 + \gamma\theta) - \theta \quad = g(\alpha, \theta) \tag{5.62}$$

where we use the functions f and g as a shorthand representation of the right-hand sides.

5.3.2. 'Fast' and 'slow' motions

The rate at which the temperature excess θ changes on the new timescale is given by $d\theta/dT = g(\alpha, \theta)/\varepsilon$. Because ε has been defined as a small parameter, this will be a large quantity unless the function $g(\alpha, \theta)$ is close to zero. Thus we can expect that the temperature excess will try to adjust very quickly (i.e. while α changes very little) until.

$$g(\alpha, \theta) = \alpha \exp(\theta/1 + \gamma\theta) - \theta = 0. \tag{5.63}$$

This is the 'fast motion' of the system. We can draw this 'nullcline', so called because one of the rates of change is zero along this curve, in the α–θ plane in Fig. 5.7. It has a maximum and a minimum whose coordinates are relatively easily located as a function of γ, as we will see below (they also correspond to the maximum and minimum in the stationary-state locus).

As α varies according to its own rate eqn (5.61), the 'slow motion' of the system, we can expect the trajectory in the phase plane to try to stay as close as possible to the $g(\alpha, \theta) = 0$ nullcline. We can also draw the $f(\alpha, \theta) = 0$ nullcline on to the phase plane. This is a curve on which $d\alpha/dT$ is zero, so the concentration of A passes through a maximum or a minimum in time whenever a trajectory crosses it. There is one point where the two nullclines intersect. Here both time derivatives vanish simultaneously; this is the sta-

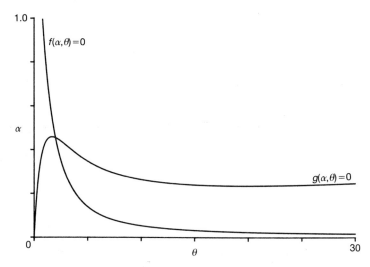

FIG. 5.7. The nullclines $f(\alpha, \theta) = 0$ and $g(\alpha, \theta) = 0$ in the phase plane for the thermokinetic oscillator, specifically with $\gamma = 0.175$ and $M/K = 2.0$, although the qualitative form of these loci is general for all parameter values (with $\gamma < \frac{1}{4}$).

tionary-state solution. The location of the intersection point depends on the values of M and K: $\theta_{ss} = M/K$, $\alpha_{ss} = (M/K)\exp\{-(M/K)/[1 + \gamma(M/K)]\}$. Figure 5.8(a) corresponds to a particular set of the parameters such that the stationary state occurs at low θ, below the maximum in the $g(\alpha, \theta) = 0$ nullcline. The trajectory described by the concentration and temperature rise as they evolve from an arbitrary initial condition is also shown. As described above, the temperature excess adjusts rapidly (on the T timescale) moving the system on to the $g(\alpha, \theta) = 0$ nullcline. Over a longer period, α and θ then move along this curve until they reach the stationary-state intersection, where they come to rest. A typical time-depedent trace is shown in Fig. 5.8(b).

If, however, the intersection of the two nullclines is positioned as in Fig. 5.8(c)—between the maximum and minimum in the $g(\alpha, \theta) = 0$ null-cline—the stationary state is not attained. Again, there is an initial jump on to the $g(\alpha, \theta) = 0$ nullcline, e.g. to point E. The concentration and temperature excess then evolve along this curve: $f(\alpha, \theta)$ is positive along this branch of the nullcline, since it lies below the $f(\alpha, \theta) = 0$ curve, so α increases and we move towards the maximum at point A. When we reach A, $f(\alpha, \theta)$ is still positive, so we must move off the $g(\alpha, \theta) = 0$ nullcline. Now we have $d\theta/dT = g(\alpha, \theta)/\varepsilon$, with $g(\alpha, \theta)$ not close to zero. The rate of temperature change will thus be very large. The system now jumps rapidly to the other branch of the $g(\alpha, \theta) = 0$ nullcline, whilst α changes very little, to point B. There we again have $g(\alpha, \theta) = 0$, but now with $f(\alpha, \theta) < 0$. Thus α decreases along the nullcline and we move to point C. There the system undergoes another fast jump, to D.

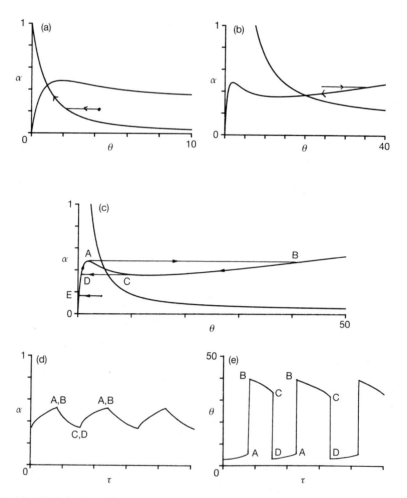

FIG. 5.8. Variation in stationary-state intersection relative to the maximum and minimum in the $g(\alpha, \theta) = 0$ nullcline with the quotient M/K for a system with $\gamma = 0.2$. (a) Intersection below maximum, $M/K = 1.6$: a given trajectory moves quickly to the $g(\alpha, \theta) = 0$ nullcline which it then moves along to the stationary-state solution. (b) Intersection above the minimum, $M/K = 20$: again a given trajectory will approach the stationary state along the $g(\alpha, \theta) = 0$ nullcline. (c) Intersection lying between the extrema, $M/K = 5$: now the stationary state is not approached and the time-dependent solutions cycle around the phase plane on the $g(\alpha, \theta) = 0$ nullcline (slow motion) with rapid jumps from one branch to the other (fast motion) at the turning points. (d), (e) Schematic representation of the relaxation oscillations for the conditions in (c).

The continuous motion around the circuit ABCD which thus ensues gives the relaxation oscillation shown schematically in Fig. 5.8(d).

This qualitative description only holds strictly in the limit $\varepsilon \to 0$. Given this qualification, however, we can use the condition for a unique stationary-state intersection lying between the extrema in the $g(\alpha, \theta) = 0$ nullcline to obtain quantitative estimates of the existence, amplitude, and period of oscillations.

5.3.3. Quantitative analysis of relaxation oscillator

In order to make quantitative predictions, we must locate the four 'corners' of the oscillation: A, B, C, and D. Two of these, A and C, correspond to the turning points in the $g(\alpha, \theta) = 0$ nullcline. For the present model these are also the turning points in the $\alpha_{ss}(\mu)$ locus, whose coordinates have been determined in eqns (4.66)–(4.68). We can content ourselves here with the leading-order forms with small γ:

$$\text{point A:} \quad \theta_A \approx 1 + 2\gamma \qquad \alpha_A \approx (1 + \gamma)e^{-1} \tag{5.64}$$

$$\text{point C:} \quad \theta_C \approx (1 - 2\gamma)/\gamma^2 \qquad \alpha_C \approx (1 - 2\gamma)\gamma^{-2}e^{(\gamma - 1)/\gamma}. \tag{5.65}$$

The α ordinates of points B and D will be the same as those of A and C respectively, as the jumps are represented by vertical lines in the phase plane. Thus we seek the other solutions to $g(\alpha, \theta) = 0$ for these values of α. At point B, we must satisfy

$$(1 + \gamma)e^{-1} = \theta_B \exp[-\theta_B/(1 + \gamma\theta_B)]. \tag{5.66}$$

However, the temperature excess is large here, certainly $\theta_B > \theta_C \approx \gamma^{-2}$, so the exponent on the right-hand side reduces to $e^{-1/\gamma}$. Using this we then have for point B

$$\text{point B:} \quad \theta_B \approx (1 + \gamma)\exp((1/\gamma) - 1) \qquad \alpha_B \approx (1 + \gamma)e^{-1}. \tag{5.6.7}$$

For point D we must solve

$$(1 - 2\gamma)\gamma^{-2}\exp(1 - 1/\gamma) = \theta_D \exp[-\theta_D/(1 + \gamma\theta_D)]. \tag{5.68}$$

Here $\theta_D \ll 1$, so we can neglect the departure of the exponential term from unity, giving

$$\text{point D:} \quad \theta_D \approx \alpha_D \approx (1 - 2\gamma)\gamma^{-2}\exp(1 - 1/\gamma). \tag{5.69}$$

As a typical example, if $\gamma = 0.1$ we obtain the following (θ, α) locations:

$$A = (1.2, 0.405) \qquad\qquad B = (8100, 0.405)$$

$$C = (80, 9.87 \times 10^{-3}) \qquad D = (9.87 \times 10^{-3}, 9.87 \times 10^{-3}).$$

5.3.4. Existence of oscillations

We require that the stationary state should lie on the section of the $g(\alpha, \theta)$ nullcline along which α is decreasing. Thus θ_{ss}, which is simply the quotient

M/K, must lie between θ_A and θ_C. The requirement for relaxation oscillations is, therefore,

$$1 + 2\gamma < M/K < (1 - 2\gamma)/\gamma^2. \qquad (5.70)$$

5.3.5. Oscillatory amplitude

The amplitudes of the oscillation in terms of the temperature rise and the concentration of the intermediate are given by

$$A_\theta = \theta_B - \theta_D \approx \theta_D \qquad A_\alpha = \alpha_A - \alpha_C. \qquad (5.71)$$

For our example above, then $A_\theta \approx 8100$ and $A_\alpha \approx 0.395$. Table 5.2 compares the predicted amplitudes and oscillatory periods for a system with $M = 0.5$, $K = 0.1$, and $\gamma = 0.1$ with decreasing values of the small parameter ε. Although the various leading-order formulae also strictly require $\gamma \ll 1$, which is not completely satisfied here, we can clearly see that eqn (5.71) provides a very good order of magnitude prediction for the smallest values of ε. Thus the relaxation oscillator analysis is successful in describing the behaviour of the thermokinetic model under these conditions (although with the very large temperature excursions predicted and found, we may begin to worry about the physical reality of the model itself—especially the continued neglect of precursor consumption—in these circumstances).

5.3.6. Oscillatory period

The fast motion jumps from A to B and from C to D are virtually instantaneous on the slow timescale T (for sufficiently small ε). Thus the period of the oscillation is determined by the sum of times for motion along BC and DA. In fact, for relatively small γ, only the latter will make any

Table 5.2

Computed oscillatory amplitudes and period as functions for the small parameter ε for the thermokinetic model with $M = 0.5$, $K = 0.1$, and $\gamma = 0.1$ (the original dimensional parameters are given by $\mu = \varepsilon M$ and $\kappa = \varepsilon K$)

ε	A_θ	A_α	T
0.1	53.59	0.5729	1.690
0.01	257.6	0.4536	1.045
0.001	2138	0.4215	0.9194
0.0001	5855	0.4069	0.8801

significant contribution, because the value of θ is so high along the upper branch that reaction is very fast there.

In general terms we wish to calculate the time for motion along the $g(\alpha, \theta) = 0$ nullcline. Along this curve we have

$$\alpha = \theta \exp[-\theta/(1 + \gamma\theta)] \tag{5.72}$$

and hence

$$d\alpha = \left(1 - \frac{\theta}{(1 + \gamma\theta)^2}\right) \exp[-\theta/(1 + \gamma\theta)] d\theta. \tag{5.73}$$

The time ΔT taken to traverse a particular section of the nullcline can be evaluated by integrating $d\alpha/dT$ between the appropriate limits. Using the relationship (5.72) in the rate eqn (5.61) we have

$$\Delta T = \int_{\alpha_1}^{\alpha_u} \frac{d\alpha}{M - K\theta(\alpha)} \tag{5.74}$$

or, substituting from (5.73),

$$\Delta T = \int_{\alpha_1}^{\alpha_u} \frac{[1 - \theta(1 + \gamma\theta)^{-2}]}{M - K\theta} \exp\left(\frac{-\theta}{1 + \gamma\theta}\right) d\theta \tag{5.75}$$

We wish to evaluate this integral along the two sections BC and DA.

Along BC we have $\gamma\theta \gg 1$ and $\theta \gg \theta_{ss} = M/K$. We may use these to simplify the integrand, giving

$$\Delta T = K^{-1}e^{-1/\lambda} \int_{\theta_1}^{\theta_u} \frac{1 - \gamma^2\theta}{\gamma^2\theta^2} d\theta. \tag{5.76}$$

The limits of integration are $\theta_u = \theta_B \approx (1 + \gamma)\exp((1/\gamma) - 1)$ and $\theta_1 = \theta_C \approx (1 - 2\gamma)/\gamma^2$, yielding ultimately

$$\Delta T_{BC} \approx K^{-1}e^{-1/\gamma}[\gamma^{-1} + \gamma - 2 + 2\ln\gamma]. \tag{5.77}$$

As the dimensionless rate constant K is of order unity, this time will be very short for small γ. For our example, $\gamma = 0.1$ and $\Delta T_{BC} \approx 1.6 \times 10^{-4} \, \text{K}^{-1}$.

The slowest motion of all, and hence that which dominates the total oscillatory period, is that along DA. On this section the temperature rise is relatively low, so $\gamma\theta \ll 1$ and we may make the exponential approximation to the Arrhenius term. The resulting integral to be evaluated is

$$\Delta T = K^{-1} \int_{\theta_1}^{\theta_u} \frac{1 - \theta}{M/K - \theta} e^{-\theta} d\theta. \tag{5.78}$$

If, additionally, we assume that the quotient M/K is large compared with unity (and hence also always large compared with θ in the denominator of the integrand), we can use the approximate form

$$\Delta T = M^{-1} \int_{\theta_1}^{\theta_u} (1 - \theta) e^{-\theta} \, d\theta.$$ (5.79)

The limits are $\theta_u = \theta_A \approx 1 + 2\gamma$ and $\theta_1 = \theta_D \approx (1 - 2\gamma)\gamma^{-2} \exp(1 - 1/\gamma)$, so that the time becomes

$$\Delta T_{DA} \approx M^{-1}(1 + 2\gamma)\exp[(1 + 2\gamma)].$$ (5.80)

This is of order unity. For our example, $\Delta T_{DA} \approx 0.36/M$, and this is a leading-order estimate of the total period

$$\Delta T_p \approx M^{-1}(1 + 2\gamma)\exp[-(1 + 2\gamma)]$$ (5.81)

(A better estimate of the time ΔT_{DA}, and hence of the period ΔT_p, can be obtained using an integration of eqn (5.78). This involves the exponential integral, $\mathrm{Ei}(x_i)$ where $x_i = (M/K) - \theta_i$, a function freely available in standard tabulations. We then find

$$\Delta T_{DA} \approx K^{-1}\exp(-M/K)\{e^{x_u} - e^{x_1} + (M/K - 1)[\mathrm{Ei}(x_u) - \mathrm{Ei}(x_1)]\}.$$

If $M/K = 10$, eqn (5.81) can be used with only 13 per cent error; this error reduces to 5 per cent and 1 per cent for $M/K = 20$ and 100 respectively.)

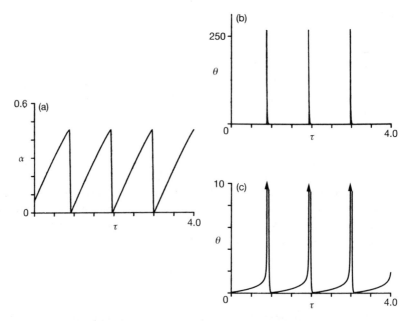

FIG. 5.9. The actual form of the relaxation oscillations for the thermokinetic model with $\varepsilon = 0.01$, $M = 0.5$, $K = 0.1$, and $\gamma = 0.1$.

5.3.7. Oscillatory waveform

An idealized relaxation waveform, based on slow and fast motion arguments, has been presented in Fig. 5.8(d). We can now give a more accurate representation, in Fig. 5.9, from the information derived in the previous subsection. Because the 'slow' motion along BC is in fact rather fast compared with the total period (determined almost completely by the motion along DA), we do not see a 'plateau' at high temperature rises. Instead the $\theta(T)$ trace is one of sharp spikes of high self-heating separated by periods in which the temperature rise builds up slowly from virtually zero to unity. The corresponding variation in the intermediate concentration shows a sawtooth form, with a sharp decrease followed by a gradual build-up. We have already seen such sawtooth and spiky waveforms for the models in chapters 2–4, and will do so again when we study open systems, so these clearly are a representative pattern for chemical oscillations in general.

5.4. Comparison between relaxation and Hopf bifurcation analyses

The results of the previous sections allow us to compare and unify these two apparently different approaches to investigating oscillatory behaviour.

First, we may discuss the respective predictions for the conditions for oscillations in terms of the parameters μ and κ. The relaxation oscillation approach suggests that, in the limit $\varepsilon \to 0$, any stationary state lying on the section of the $\alpha_{ss}(\mu)$ locus between the maximum and the minimum will be unstable. This situation corresponds to any μ–κ parameter value lying between the two straight lines in Fig. 5.6, so oscillatory behaviour is predicted throughout this region. The Hopf analysis, on the other hand, predicts that the onset and existence of oscillations will be confined to the petal-shaped region described by eqns (5.51) and (5.52), and to a small shaded region outside this, close to points of subcritical bifurcations and which cannot be determined from the local analysis.

To all intents and purposes, the Hopf prediction is the exact result. It is not difficult to construct an intersection between the maximum and minimum which is a stable stationary state. For instance, with $\gamma = 0.02$, $\kappa = 0.12$, and $\mu = 0.2$ we have $\theta_A = 1.042$, $\theta_C = 2400$, and $\theta_{ss} = \mu/\kappa = 5/3$. The corresponding phase plane nullclines are shown in Fig. 5.10, together with a trajectory spiralling in to the stationary-state intersection. The trace of the Jacobian matrix is negative for this solution $(\mathrm{tr}(\mathbf{J}) = -4.1 \times 10^{-2})$ indicating its local stability. This is not, however, a particularly fair test of the relaxation analysis because the parameters μ and κ are not especially small. In the vicinity of the origin (where ε is small) both approaches converge.

Close to the Hopf boundary, the oscillations will be almost sinusoidal and are described by the various formulae for β_2, μ_2, etc. of §§5.1 and 5.2. As we

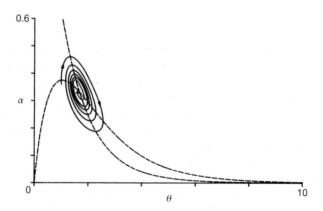

FIG. 5.10. The possibility of a stable stationary-state intersection on the middle branch of the $g(\alpha, \theta) = 0$ nullcline: the nullclines are shown as broken curves, the solid curve gives the evolution of a typical trajectory towards the stable focal state.

move into the interior of the oscillatory region, however, higher-order non-linear effects become important and the Hopf predictions cannot be used accurately. It is in just these conditions that the relaxation analysis comes into its own—so we have two complementary techniques for studying the same oscillations in different parameter ranges.

5.5. Excitability

Another form of behaviour exhibited by a number of chemical reactions, including the Belousov–Zhabotinskii system, is that of excitability. This concerns a mixture which is prepared under conditions outside the oscillatory range. The system sits at the stationary state, which is stable. Infinitesimal perturbations decay back to the stationary state, perhaps in a damped oscillatory manner. The effect of finite, but possibly still quite small, perturbations can, however, be markedly different. The system ultimately returns to the same state, but only after a large excursion, resembling a single oscillatory pulse. Excitable B–Z systems are well known for this propensity for supporting spiral waves (see chapter 1).

The driving force for this excitable behaviour can be revealed from the model studied above, again assuming ε is small and looking at the nullclines in the phase plane. A suitable orientation of the curves $f(\alpha, \theta) = 0$ and $g(\alpha, \theta) = 0$ is shown in Fig. 5.11. The stationary-state intersection lies outside the range of 'instability', i.e. just before the maximum in the $g(\alpha, \theta) = 0$ nullcline, reflecting stability. If a small perturbation momentarily decreases α, or induces either a decrease or a very small increase in θ, the system merely jumps back on to that nullcline and then moves along it to the intersection

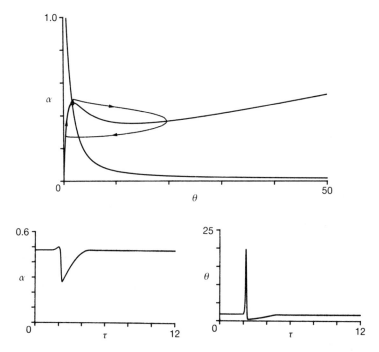

FIG. 5.11. Excitability in a chemical system. (a) The nullclines $f(\alpha, \theta) = 0$ and $g(\alpha, \theta) = 0$ intersect just to the left of the maximum. A suitable perturbation must make a full circuit, as shown by a typical trajectory, before returning to the stable stationary state. (b), (c) The corresponding evolution of the concentration of intermediate A and the temperature excess in time showing the large-amplitude excursion.

point. Should the perturbation increase α above the value corresponding to the maximum in the $g(\alpha, \theta) = 0$ nullcline, however, the return to the stationary state takes the scenic route. There is a fast jump to the upper branch BC, along which the system then evolves before jumping back to D and returning to the stationary state along the lower branch. Figure 5.11 also shows how the concentration and temperature excess would typically vary in time during such an excursion.

Appendix. Shooting methods for limit cycles

We will frequently wish to determine the amplitude and period of motion around a limit cycle, corresponding to oscillatory behaviour. The simplest approach is just to integrate the governing equations for long enough, with time running forwards if we want to look at stable limit cycles and backwards if we have an unstable cycle. This is not always particularly efficient, and for systems with more than two variables it cannot always be used to obtain unstable responses. An alternative approach is to employ a shooting method, which we describe in a rough form below. This will converge faster than the simple time integration and can be used without any information about the stability of the cycle.

We describe the application to two-variable systems first. Here, the system is moving on a two-dimensional phase plane. Let the variables be x and y. The phase plane is shown in Fig. A5.1, and we wish to find the limit cycle surrounding the stationary state. To do this we can make an initial guess of a point which might lie on the limit cycle. A convenient choice is to take y equal to its stationary-state value y_{ss} and x as some x_0 which must differ from x_{ss}: we will take $x_0 > x_{ss}$. We now integrate the governing equations for one circuit around the plane, i.e. until $y = y_{ss}$ again and x is again larger than x_{ss} (there will of course be a point 'half-way' round at which $y = y_{ss}$ with $x < x_{ss}$, but we want a full circuit). Let the new value of x be x_1 and the difference $\Delta x = x_1 - x_0$.

If out first guess was very lucky, and we started from a point exactly on the limit cycle, Δx will be zero—we return exactly to the starting point, irrespective of whether the cycle is stable or unstable. Otherwise, Δx will be non-zero, and its value will depend on the starting point x_0. Thus we have some functional dependence

$$\Delta x = \Delta x(x_0). \tag{A5.1}$$

We can treat the problem of locating the limit cycle as that of finding a zero of this new function, i.e. of finding x_0 such that $\Delta x(x_0) = 0$. Of the various algorithms which can be used for locating zeros of functions, a Newton–Raphson approach is quite suitable. Thus, having taken an estimate of the appropriate initial value x_0^i, which gives rise to a non-zero difference Δx^i, we may obtain an improved value x_0^{i+1} from the formula

$$x_0^{i+1} = x_0^i - \frac{\Delta x^i}{\partial \Delta x / \partial x_0}. \tag{A5.2}$$

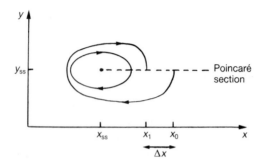

FIG. A5.1.   Construction of a Poincaré section in the phase plane illustrating the shooting method of locating limit cycle solutions.

The derivative $\partial\Delta x/\partial x_0$ must also be computed numerically. A simple method is to carry out two integrations, one from x_0^i and one from $x_0^i + \varepsilon$, where $\varepsilon = 10^{-3}x_0^i$ say, resulting in two differences Δx^i and $\Delta x(x_0^i + \varepsilon) = \Delta x^i + \delta$. The derivative is then approximated by $\partial\Delta x/\partial x_0 \approx \delta/\varepsilon$. We can then iterate through eqn (A5.2) until the difference Δx^i is smaller than some specified tolerance.

This process is easily generalized to systems with more variables and hence higher-dimension phase spaces. In general we specify some $(N - 1)$-dimensional plane in the N-dimensional phase space and perform a similar single-cycle integration. We then have $N - 1$ differences, which form a vector $\Delta\mathbf{x}$ which is a function of the $N - 1$ initial concentration $\mathbf{x}_0^i$. We wish to make all the components of $\Delta\mathbf{x}$ zero, and the appropriate Newton–Raphson form is then

$$\mathbf{x}_0^{i+1} = \mathbf{x}_0^i - \mathbf{J}^{-1}\Delta\mathbf{x}^i \tag{A5.3}$$

where $\mathbf{J}$ is the matrix of derivatives of the differences Δx_j with respect to all the initial points $x_{0,k}$ ($j, k = 1, N - 1$).

References

The following are additonal to the references in chapter 3:

Guckenheimer, J. and Holmes, P. (1986). *Nonlinear oscillations, dynamical systems and bifurcations of vector fields*. Springer, New York.

Hassard, B. D., Kazarinoff, N. D., and Wan, Y.-H. (1980). *Theory and applications of the Hopf bifurcation*. Cambridge University Press.

Hopf, E. (1942). Abzweigung einer periodischen Losung von einer stationaren Losung eines differential system. *Ber. Math.-Phys. Kl. Sachs. Acad. Wiss. Leipzig*, **94**, 1, 22.

Iooss, G. and Joseph, D. D. (1980). *Elementary stability and bifurcation theory*. Springer, New York.

Marsden, J. E. and McCracken, M. (1976). *The Hopf bifurcation and its applications*. Springer, New York.

AUTOCATALYSIS IN WELL-STIRRED OPEN SYSTEMS: THE ISOTHERMAL CSTR

This chapter and chapter 7 introduce open systems. Though of the simplest kind, they permit multiple stationary states. These are truly stationary and may be quite different from the state of final equilibrium. After a study of this chapter, which deals with stationary states in isothermal systems, the reader should be able to:

(1) appreciate more fully the versatility and value of the simple autocatalysis model system;

(2) recognize and quantify the rich variety of stationary states—steadily more varied as catalyst decay B → C and an uncatalysed reaction A → B are added to the model;

(3) understand the diminution in opportunity for exotic behaviour as the reverse reactions are introduced and increase in importance;

(4) see the value of flow diagrams in representing the location of unique or multiple stationary states and especially the influences of reversibility and of initially seeding the system with catalyst;

(5) understand the significance of the parameter space diagrams.

The sort of experiment which we have been considering so far involves the classic 'closed vessel'. The term 'closed' is used here in its thermodynamic sense, i.e. there is no exchange of matter and, in particular, no inflow of fresh reactants, rather than implying that the vessel has a lid on it. Thus we may have been talking about a bulb of reacting gas closed by a tap or an ordinary beaker containing various solutions mixed together. Because such experiments are easiest to set up, most of the early work in oscillatory reaction kinetics was done in this way. The remarkable longevity of oscillations in the Belousov–Zhabotinskii system has perhaps conspired to help continue this approach. It is not difficult to obtain hundreds of colour changes over several hours. The record from a Pt–calomel electrode pair (Fig. 1.1(a)) often shows very little variation in amplitude, suggesting that there is very little change in the system between successive excursions.

From a theoretical point of view, however, this arrangement is not so satisfactory. Even for the B–Z system, the bromide electrode record

(Fig. 1.1(b)) reveals quite clearly that each oscillation is different from its predecessor and its successor. As the concentrations of the major reactants, bromate ion and malonic acid, are also falling throughout we must recognize that the observed responses arise in a system which has a continually varying background composition. Our analyses of the previous chapters have sometimes hidden this momentarily, working with the pseudo-steady-state equations. Even when the time dependence is recognized explicitly, the resulting expressions, such as eqns (2.15) and (2.16), are really only leading-order approximations to the true evolution. For local stability analyses, we needed somehow to 'stop' the decay in the reactant concentration during the test, so its time dependence did not interfere. This latter attitude can, in some cases, be justified, often *a posteriori*, but care needs to be taken to check the influence of this approximation.

A much better experimental arrangement is one in which true stationary states can be realized. These are conditions under which the net rates of production or removal of all species become zero simultaneously and hence the composition within the reactor ceases to change. The reaction can then be studied against a constant and, in principle, fully characterized background,

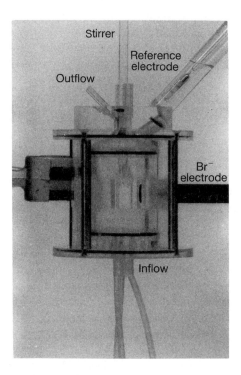

FIG. 6.1.   A continuously fed well-stirred tank reactor (CSTR) for solution-phase reactions.

unaffected by transients. Such states can also be maintained for as long as is desired by continuing the inflow.

There are two particularly common, and technically very important, manifestations of open systems and which have become increasingly used in the study of oscillatory processes. These are (i) the chemical engineers' continuously fed well-stirred tank reactor (CSTR) and (ii) cells in which there is transport across a boundary or membrane from an external reservoir and in which diffusional processes are usually important.

The CSTR is, in many ways, the easier to set up and operate, and to analyse theoretically. Figure 6.1 shows a typical CSTR, appropriate for solution-phase reactions. In the next three chapters we will look at the wide range of behaviour which chemical systems can show when operated in this type of reactor. In this chapter we concentrate on stationary-state aspects of isothermal autocatalytic reactions similar to those introduced in chapter 2. In chapter 7, we turn to non-isothermal systems similar to the model of chapter 4. There we also draw on a mathematical technique known as singularity theory to explain the many similarities (and some differences) between chemical autocatalysis and thermal feedback. Non-stationary aspects such as oscillations appear in chapter 8.

To introduce the appropriate features and concepts of reaction in flow systems, we start by considering the simplest irreversible examples—a single first-order step and then two consecutive first-order steps.

6.1. Continuous-flow reactors—simple deceleratory kinetics

Many of the important, but perhaps unfamiliar, concepts relevant to the CSTR can be revealed by examining the behaviour of a simple first-order reaction:

$$A \rightarrow B \qquad \text{rate} = k_1 a. \tag{6.1}$$

Our aim is to determine the concentration of A in the reactor as a function of time and in terms of the experimental conditions (inflow concentrations, pumping rates, etc.). We need to obtain the equation which governs the rate at which the concentration of A is changing within the reactor. This 'mass-balance equation' will have contributions from the reaction kinetics (the rate equation) *and* from the inflow and outflow terms. In the simplest case the reactor is fed by a stream of liquid with a volume flow rate of q dm^3 s^{-1} in which the concentration of A is a_0. If the volume of the reactor is V dm^3, then the average time spent by a molecule in the reactor is V/q s. This is called the mean residence time, t_{res}. The inverse of t_{res} has units of s^{-1}; we will call this the flow rate k_f, and see that it plays the role of a pseudo-first-order rate constant. We denote the concentration of A in the reactor itself by a.

6.1.1. Mass-balance equation

The rate at which A flows into the reactor is qa_0 mol s^{-1}, the rate of outflow being qa mol s^{-1}. The rate at which A is reacting to give B is $Vk_1 a$ mol s^{-1}. Thus the rate of change of a at any given time is given by

$$V\, da/dt \quad = \quad qa_0 \quad - \quad qa \quad - \quad Vk_1 a. \tag{6.2}$$

| net rate of | rate of | rate of | rate of |
| change | inflow | outflow | reaction |

Dividing throughout by V gives

$$da/dt = k_f a_0 - (k_f + k_1)a. \tag{6.3}$$

6.1.2. Stationary-state solutions

If the concentration in the reactor is initially a_0, then eqn (6.3) integrates to give

$$a = \frac{k_f + k_1 \exp[-(k_1 + k_f)t]}{k_f + k_1} a_0. \tag{6.4}$$

This tells us how the concentration of A varies with time in the reactor, as shown in Fig. 6.2.

At $t = 0$, eqn (6.4) gives us our assumed initial condition $a = a_0$. More interesting is the behaviour at long times, when the exponential term in the numerator tends to zero. Then we find that the concentration of A tends to

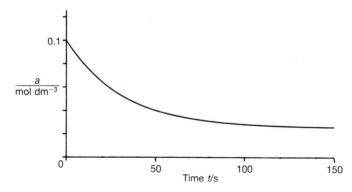

FIG. 6.2. The evolution of the reactant concentration towards its stationary-state value for a simple first-order reaction in a CSTR: $a_0 = 0.1$ mol dm^{-3}, $k_1 = 0.024$ s^{-1}, $k_f = 8 \times 10^{-3}$ s^{-1}.

a limiting value

$$a \to a_{ss} = \frac{k_f a_0}{k_f + k_1} = \frac{a_0}{1 + k_1 t_{res}}. \tag{6.5}$$

If this value for a is substituted into the mass-balance eqn (6.3), it gives $da/dt = 0$. This is thus the stationary-state solution. Once it has been achieved the concentration remains constant. Note that at the stationary state the chemical reaction rate is not zero; rather it is given by $k_1 a_{ss}$, and this rate of conversion of A to B just balances the net rate of mechanical inflow of A to the reactor.

The concept of the stationary state is vitally important, and we can learn many significant aspects about more complex systems where direct analytical integration of the mass-balance equations is not possible by locating the conditions under which the various derivatives become zero.

The numerical value of the stationary-state concentration of A depends on the reaction rate constant k_1, on the inflow concentration a_0, and on the flow rate k_f. In a given experiment, it is k_f which is most easily varied—by altering the pumping speed, for instance. Figure 6.3(a) shows how a_{ss} and the stationary-state reaction rate $k_1 a_{ss}$ depend on the flow rate; Fig. 6.3(b) shows the equivalent forms in terms of the residence time, t_{res}. At low flows (long residence times) there can be virtually complete conversion of A to B as the system comes close to its chemical equilibrium state (complete conversion in the present irreversible model). Thus, $a_{ss} \approx 0$ for $k_f \approx 0$ and the rate also goes to zero. At very high flow rate, the mean residence time is very short, so molecules do not on average have much time to react. Only low extents of reaction can be achieved, with a_{ss} being close to the inflow concentration a_0. Here the chemical reaction rate is close to its maximum value $k_1 a_0$. In between these extremes, the stationary-state concentration varies smoothly: if we choose any flow rate, there will be one corresponding stationary state (and only one).

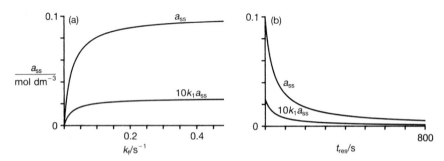

FIG. 6.3.   The dependence of the reactant concentration a_{ss} and the reaction rate $k_1 a_{ss}$ at the stationary state on (a) flow rate and (b) residence time for a first-order reaction in a CSTR. Rate data as in Fig. 6.2.

6.1.3. Extent of reaction and bifurcation diagrams

Diagrams such as Fig. 6.3, which show the dependence of the stationary-state composition on a particular parameter, are known as bifurcation diagrams. It is customary, when trying to judge the efficiency of various processes for instance, to discuss the *extent of reaction* rather than the concentration of the reactant. The former is the fractional conversion of A into products and is given by the difference between the inflow concentration of A and its stationary-state value, i.e. how much A has reacted, divided by the original (inflow) concentration. For the extent of reaction we use the symbol γ, and under stationary-state conditions

$$\gamma_{ss} = \frac{a_0 - a_{ss}}{a_0}. \tag{6.6}$$

This is a dimensionless group: $\gamma = 0$ corresponds to no conversion; $\gamma = 1$ to complete conversion.

Figure 6.4 shows how the extent of reaction depends on the mean residence time $(1/k_f)$ for the simple first-order reaction model, with

$$\gamma_{ss} = \frac{k_1 t_{res}}{1 + k_1 t_{res}}.$$

Thus, at short residence times, where there is little conversion, the extent of conversion increases almost linearly with t_{res}; at long residence times the system tends to complete reaction $\gamma_{ss} \to 1$.

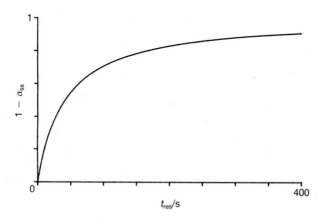

FIG. 6.4. The dependence of the stationary-state extent of reaction, γ_{ss}, on residence time for a first-order reaction in a CSTR.

6.1.4. Flow diagrams

When proceeding further, with more complex models, an extra construction will prove useful—the flow diagram. In such a diagram, we plot the net rate of inflow L and the chemical reaction rate R as functions of the extent of reaction γ. Equation (6.2) can be rewritten as

$$\frac{d\gamma}{dt} = \frac{\gamma}{t_{res}} - k_1(1 - \gamma). \tag{6.7}$$

Figure 6.5 shows the flow diagram for the present model and those appropriate to a number of other deceleratory systems. In each case, the reaction rate curve R falls as the extent of conversion increases (this is, of course, the meaning of the term deceleratory); for the first-order reaction there is a linear

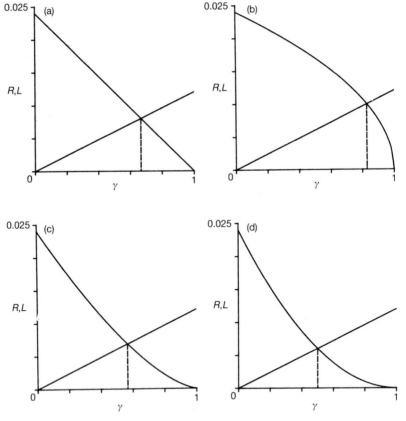

FIG. 6.5.   Flow diagrams for deceleratory reactions in a CSTR: (a) first-order reaction, $n = 1$; (b) $n = \frac{1}{2}$; (c) $n = \frac{3}{2}$; (d) $n = 2$. In each case, $k_1 = 0.024 \text{ s}^{-1}$ and $t_{res} = 20.83 \text{ s}$.

decrease with slope $-k_1$. The flow curve L is linear, with a positive slope given by k_f or $1/t_{res}$. Any point of intersection of R and L corresponds to a composition for which these terms are equal and hence for which $d\gamma/dt = 0$ (a stationary state). The intersection will clearly depend on the slopes of the two lines and thus on k_1 and t_{res}, in exactly the way described in the equations above.

6.1.5. Preliminary stability analysis

The flow diagram also allows a quick, qualitative stability analysis. Consider a system sitting at the point of intersection in Fig. 6.5(a). If the extent of conversion γ in the reactor now increases slightly (corresponding to a decrease in the concentration of A) the net rate of inflow increases whilst the rate of reaction decreases. Thus, there will be a compensating effect: the concentration of A will tend to increase and, hence, the extent of reaction decrease again, back to the stationary-state value. Similarly, a momentary decrease in γ results in the reaction rate increasing beyond the inflow rate, so additional A is converted to B, increasing γ back to its intersection value.

This argument shows that for the first-order reaction model the stationary state always has some sort of stability to perturbations. In fact, this is only a first step and will not reveal Hopf bifurcations or oscillatory solutions, should they occur. A full stability analysis of typical flow–reaction schemes will appear in the next chapter.

6.2. Cubic autocatalysis in a CSTR

As well as deceleratory reactions, kineticists often find that some chemical systems show a rate which increases as the extent of reaction increases (at least over some ranges of composition). Such acceleratory, or autocatalytic, behaviour may arise from a complex coupling of more than one elementary kinetic step, and may be manifest as an empirically determined rate law. Typical dependences of R on γ for such systems are shown in Figs 6.6(a) and (b). In the former, the curve has a basic parabolic character which can be approximated at its simplest by a quadratic autocatalysis, rate $\propto \gamma(1 - \gamma)$.

Figure 6.6(b) is better approximated by a cubic form, rate $\propto \gamma^2(1 - \gamma)$. Cubic autocatalysis has already provided us with behaviour of interest in chapter 2. In the remainder of this chapter we consider the stationary-state responses of schemes with this feedback mechanism in flow reactors. We will consider three models, with increasingly varied possible behaviour: first an autocatalytic step on its own; next we allow the autocatalytic species to undergo a subsequent reaction; finally we add an uncatalysed reaction in competition with the autocatalysis. The local stability of such systems is

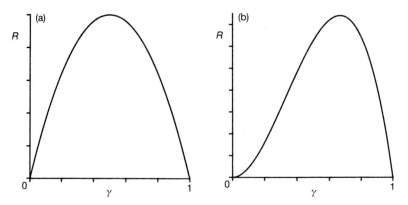

Fig. 6.6.   Dependences of reaction rate R on extent of reaction γ typical of self-accelerating (autocatalytic) systems: (a) prototype quadratic autocatalysis; (b) prototype cubic autocatalysis.

touched upon in the relevant sections and will be discussed in some detail for the second of these three cases in chapter 8.

6.2.1. Cubic autocatalysis: mass-balance equations

The application of the flow diagram to a single cubic autocatalytic step

$$A + 2B \rightarrow 3B \qquad \text{rate} = k_1 ab^2 \qquad (6.8)$$

has been discussed in § 1.7, so we may proceed relatively quickly here.

The mass-balance equation is obtained in the same way as that in § 6.1.1 above, giving

$$\frac{da}{dt} = \frac{a_0 - a}{t_{res}} - k_1 ab^2. \qquad (6.9)$$

Because of the autocatalysis, this equation involves the concentrations of both A and B. A similar mass-balance equation could be written for the autocatalyst in the form

$$\frac{db}{dt} = \frac{b_0 - b}{t_{res}} + k_1 ab^2. \qquad (6.10)$$

In the present case, however, we can use the stoichiometry of the reaction so the concentrations of A and B are linked to the (constant) inflow concentrations:

$$a + b = a_0 + b_0 \quad \text{so that} \quad b = (a_0 + b_0) - a. \qquad (6.11)$$

Substituting this into (6.9) we have

$$\frac{da}{dt} = \underbrace{\frac{a_0 - a}{t_{res}}}_{L} - \underbrace{k_1 a(a_0 + b_0 - a)^2}_{R}. \tag{6.12}$$

Because the behaviour of this model is determined by just a single differential equation, it is known as a 'one-variable' system.

Although this cubic ordinary differential equation can be integrated in closed form, we can obtain full information more easily by looking at the stationary states. We will use this opportunity of revision to introduce the appropriate dimensionless forms, but of course the various results will be equivalent to those from chapter 1.

6.2.2. Dimensionless groups

The process of obtaining suitable dimensionless forms for the present model is very similar to those followed in the previous chapters, §§ 2.8 and 3.4: we need first to choose a reference concentration and a reference timescale. For the first of these, the concentration, there is a natural choice, as we have seen already in eqn (6.6), i.e. the inflow concentration of the reactant a_0. Again, using Greek letters for dimensionless quantities, we have

$$\alpha = a/a_0 \qquad \beta = b/a_0 \qquad \text{and} \qquad \beta_0 = b_0/a_0. \tag{6.13}$$

We may immediately put limits on the values that these terms may take: α and β lie in the ranges

$$0 \leqslant \alpha \leqslant 1 \qquad 0 \leqslant \beta \leqslant 1 + \beta_0. \tag{6.14}$$

Although there are no *a priori* limits on β_0, which is simply the ratio of inflow concentrations, (other than $\beta_0 > 0$) we might expect this to be less than unity under most circumstances.

We may also note the relationship between these dimensionless concentrations and the extent of reaction γ. In particular, α and γ will always be related simply by

$$\gamma = 1 - \alpha. \tag{6.15}$$

For the present model, for which eqn (6.11) holds, we also have

$$\gamma = \beta - \beta_0. \tag{6.16}$$

In choosing a timescale, we could take the residence time t_{res} as our basis. However, if the experiments envisaged involve the variation of t_{res} as discussed in § 6.1.3 above, this is not so convenient. It is better that we should choose a constant. In the present case we are rather limited for other possibilities. However, the rate constant k_1 can be recruited if treated properly.

We must note that k_1 is effectively a third-order rate constant, with units of (concentration)$^{-2}$ (time)$^{-1}$. Thus multiplying by a_0^2 and then taking the inverse gives a chemical timescale

$$t_{ch} = 1/k_1 a_0^2 \qquad (6.17)$$

which we use to give a dimensionless time τ and a dimensionless residence time τ_{res}:

$$\tau = k_1 a_0^2 t \qquad = t/t_{ch} \qquad (6.18)$$

$$\tau_{res} = k_1 a_0^2 t_{res} \qquad = t_{res}/t_{ch}. \qquad (6.19)$$

The mass-balance equation can now be written as

$$\frac{d\alpha}{d\tau} = \frac{1-\alpha}{\tau_{res}} - \alpha(1 + \beta_0 - \alpha)^2. \qquad (6.20)$$

$$L \qquad\qquad\qquad R$$

6.2.3. Flow diagram and stationary states

In the spirit of plotting the rates of reaction and net inflow against extent of conversion, Figs 6.7(a) and (b) show R and L as functions of the extent of conversion, $1 - \alpha$, from eqn (6.20) for a typical non-zero value of β_0.

The reaction rate curve R is zero at complete conversion and also has low (but non-zero) values close to $1 - \alpha = 0$, with a maximum close to two-thirds conversion (actually at $1 - \alpha = \frac{2}{3} - \frac{1}{3}\beta_0$). Importantly, R does not depend on the residence time τ_{res}, although it does vary if β_0 is changed. The flow line L is zero when $1 - \alpha = 0$ since the inflow and outflow have the same composition (no conversion of A to B). The gradient of the flow line (Fig. 6.7(b)) is given by $1/\tau_{res}$, so it is steep for short residence times (fast flow rates) and relatively flat for long τ_{res}. (Note how τ_{res} actually compares t_{res} and t_{ch}, so 'short' residence times are those that are much less than the chemical timescale etc.) The flow line is, however, unaffected by the inflow concentration of the autocatalyst β_0.

Stationary states, corresponding to $d\alpha/d\tau = 0$, are again denoted by intersections on the flow diagram, as shown in Figs 6.7(c) and (d). For short residence times, L is steep and there is only one intersection, close to the origin. If the residence time is increased, two additional intersections, corresponding to significant extents of reaction, appear as R and L become tangential. Further flattening of the flow line sees the lower two intersections move together. These merge and disappear at a second tangency at long residence times, leaving only one stationary state with almost complete reaction.

The full $1 - \alpha_{ss}$ versus τ_{res} bifurcation diagram is shown in Fig. 6.7(e) and reveals an S-shaped curve. Between the two turning points is a region of

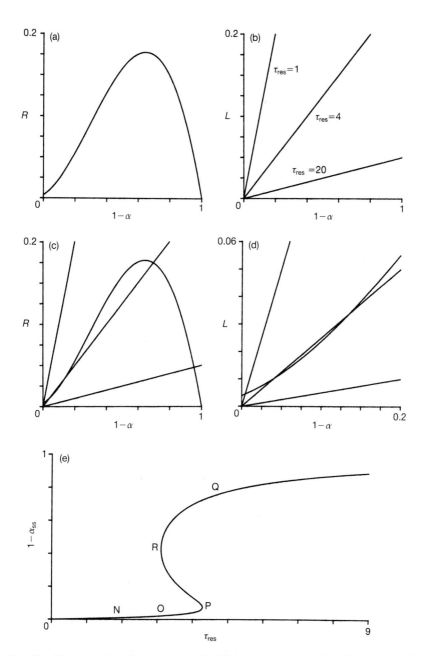

FIG. 6.7. The onset of stationary-state multiplicity represented on a flow diagram for cubic autocatalysis with $\beta_0 = \frac{1}{16}$: (a) the reaction rate curve R; (b) three typical flow lines L, with $\tau_{res} = 1$, 4, and 20; (c) intersections of R and L on the flow diagram, giving stationary-state solutions: (d) enlargement of flow diagram in vicinity of origin showing the unique intersection for the steepest flow line ($\tau_{res} = 1$), the two intersections for $\tau_{res} = 4$, and none at low extents of reaction with $\tau_{res} = 20$; (e) the stationary-state locus appropriate to the present scheme, showing ignition, extinction, and the hysteresis loop OPQR (see text).

multiple stationary states. Suppose an experiment is started with a relatively short residence time. The system has only one stationary state, of low extent of reaction, to which it moves. Thus we then sit on the stationary-state curve at a point such as N in Fig. 6.7(e). If the residence time is now increased, by reducing the pumping speed, we move along the stationary-state locus, with the extent of reaction increasing slightly. This branch of stationary-state solutions is sometimes known as the 'flow branch'. At the point O, two new solutions appear above the flow branch. However, in the absence of any dramatic changes in our experiment, there is no reason for the system to move away from its low reaction state. Throughout the range between O and P, the system has this property of sitting at one stationary state, but also having other states available to it. On the other hand, if the residence time is increased even infinitesimally beyond P, then our low extent of reaction state disappears. The system must now move to the uppermost branch at point Q. This corresponds to high degrees of conversion of A to B and to a higher chemical reaction rate, and is known as the 'thermodynamic branch' because it approaches the equilibrium state (complete reaction) in the limit τ_{res} tending to infinity.

There is thus a discontinuity in the variation of the stationary-state extent of conversion with the residence time. In other circumstances, this jump from low to high reaction rate is called an ignition. The term is less common and less appropriate for the isothermal systems we are envisaging here, although it does convey the spirit of a sharp change in the behaviour of the system. Typically a solution rich in B may differ from one rich in A, say in colour or the potential of a suitable electrode couple, so the transition from one branch to the other may be readily detected. Further increase in the residence time only produces a smooth increase in extent of reaction. If the residence time is decreased again, the system can stay on the upper branch, even as τ_{res} is reduced below the 'ignition' point Q through the region of multiple stationary states. Only at point R does the thermodynamic branch disappear, and the extent of reaction decreases abruptly to point O. This is known as an 'extinction' of the reaction or 'washout'. The system thus shows 'hysteresis', the conditions for ignition and extinction points being different.

6.2.4. Influence of autocatalyst inflow on multiple stationary states

The above scenario only holds if there are turning points, corresponding to ignition and extinction, in the stationary-state locus. These, in turn, arise when the reaction rate and flow rate curves are tangential in the flow diagram. Thus multistability and discontinuous jumps rely on tangency. This allows us to locate the conditions for ignition and extinction relatively easily. For tangency we require that at some extent of reaction the curves R and

L should be equal *and* they should have equal slope, i.e. we require simultaneously that

$$R = L \quad \text{and} \quad dR/d(1 - \alpha) = dL/d(1 - \alpha). \tag{6.21}$$

Following the procedure of § 1.7.2 we obtain the dimensionless residence times for ignition and extinction in terms of β_0:

$$\frac{8}{\tau_{res}} = 1 + 20\beta_0 - 8\beta_0^2 \pm (1 - 4\beta_0)(1 - 8\beta_0)^{1/2} \tag{6.22}$$

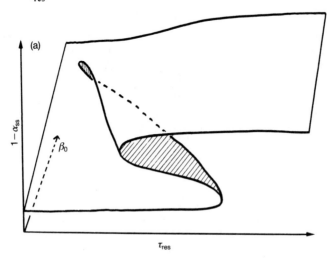

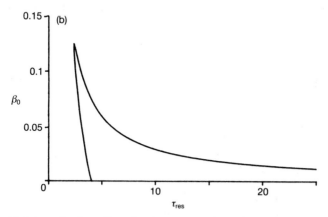

FIG. 6.8. (a) Schematic three-dimensional representation of the stationary-state surface $(1 - \alpha_{ss})$–τ_{res}–β_0 showing the folding at low autocatalyst inflow concentrations which gives rise to ignition, extinction, and multiplicity. (b) The projection onto the β_0–τ_{res} parameter plane of the two lines of fold points in the stationary-state surface, forming a typical cusp at $\beta_0 = \frac{1}{8}$, $\tau_{res} = \frac{64}{27}$: inside this cusp region the system has multiple stationary states; outside, there is only a unique solution.

with the upper root giving the lower residence time and corresponding to the extinction point. The two roots move closer together as β_0 increases. Figure 6.8 shows a three-dimensional representation of the stationary-state surface (rather than curve) plotted versus both τ_{res} and β_0. As the inflow concentration of the autocatalyst is increased, the S-shape 'unfolds'. The loci of the ignition and extinction points, as given by eqn (6.22), are shown as a projection from this surface on to the τ_{res}–β_0 plane at the bottom of the figure, where it describes a cusp. Thus the model provides a classic example of a cusp catastrophe.

The final term in eqn (6.22) involves the discriminant $(1 - 8\beta_0)^{1/2}$. For real roots, we therefore require that the dimensionless inflow concentration of the autocatalyst should be less than $\frac{1}{8}$, so $b_0 < \frac{1}{8}a_0$ in real terms.

When $\beta_0 = \frac{1}{8}$, the two roots of eqn (6.22) are exactly equal. The ignition and extinction points are coincident at $\tau_{res} = 64/27$: multistability is lost. For larger inflow concentrations of B the stationary-state extent of reaction increases smoothly with the residence time and the distinction between the flow and thermodynamic branches is lost.

6.2.5. Local stability

Using the qualitative stability analysis from the flow diagram as outlined in §§ 1.7.3 and 6.1.5, we find the following pattern. For conditions where there is a single stationary state, it is always stable to perturbations: in regions of multiple solutions, the uppermost and lowest are stable whilst the middle

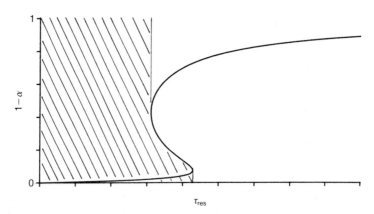

FIG. 6.9. Domains of attraction for the two stable branches of stationary-state solution: systems which have initial extents of conversion lying within the shaded region evolve to the lower branch; those in the unshaded region approach the highest extent of reaction. In the region of multiple stationary states, the middle branch acts as a separatrix.

branch is unstable. This alternation of stable–unstable–stable is typical of one-variable kinetic systems.

It may also seem sensible, if there are multiple solutions, to ask which of the states is 'the most stable'? In fact, however, this is not a valid question, partly because we have only been asking about very small disturbances. Each of the two stable states has a 'domain of attraction'. If we start with a particular initial concentration of A the system will move to one or other. Some initial conditions go to the low extent of reaction state (generally those for which $1 - \alpha$ is low initially), the remainder go to the upper stationary state. The shading in Fig. 6.9 shows which initial states go to which final stationary state. It is clear from the figure that the middle branch of (unstable) solutions plays the role of a boundary between the two stable states, and so is sometimes known as a separatrix (in one-dimensional systems only, though).

Systems like the present model, which have only one independent concentration, cannot show oscillations.

6.3. Reversibility

The flow diagram technique can also be used to illustrate clearly the rather limited effects on autocatalytic systems which arise from the reversibility of chemical reactions. For this we replace step (6.8) by its reversible counterpart

$$A + 2B \rightarrow 3B \qquad k_1, k_{-1}. \qquad (6.23)$$

We denote the equilibrium constant by $K_e = k_1/k_{-1} = b_{eq}/a_{eq}$. If this reaction is carried out in a closed vessel with initial concentrations a_0 and b_0 then at chemical equilibrium we have, using eqn (6.11),

$$a_{eq} = \frac{a_0 + b_0}{1 + K_e} \qquad b_{eq} = \frac{K_e(a_0 + b_0)}{1 + K_e}. \qquad (6.24)$$

Returning to open systems, the mass-balance equation can be written as

$$\frac{da}{dt} = \frac{a_0 - a}{t_{res}} - k_1 a(a_0 + b_0 - a)^2 + k_{-1}(a_0 + b_0 - a)^3$$

$$= \frac{a_0 - a}{t_{res}} - k_1(a_0 + b_0 - a)^2 [a - K_e^{-1}(a_0 + b_0 - a)]. \qquad (6.25)$$

In dimensionless terms this becomes

$$\frac{d\alpha}{d\tau} = \frac{1 - \alpha}{\tau_{res}} - (1 + \beta_0 - \alpha)^2 [(1 + K_e^{-1})\alpha - K_e^{-1}(1 + \beta_0)]$$

$$= \frac{1 - \alpha}{\tau_{res}} - (1 + K_e^{-1})(1 + \beta_0 - \alpha)^2 (\alpha - \alpha_{eq}). \qquad (6.26)$$

$$L \qquad\qquad\qquad R$$

Here α_{eq} is simply the dimensionless concentration of A at equilibrium:

$$\alpha_{eq} = \frac{1 + \beta_0}{1 + K_e}. \tag{6.27}$$

In the limit of irreversibility $k_{-1} \to 0$, so $K_e^{-1} \to 0$ and $\alpha_{eq} \to 0$, and eqn (6.26) reduces smoothly to the form of (6.20), as we would hope.

6.3.1. Flow diagram

The flow rate curve L is unaffected by reversibility: it still shows a linear dependence on the extent of reaction $1 - \alpha_{ss}$ with a gradient determined by the inverse of the residence time. The reaction rate curve R now depends upon two parameters, β_0 and K_e (the third apparent parameter α_{eq} is uniquely determined by these). We will examine the effects of these separately below, but may note some general features here.

R again shows a cubic dependence on the extent of reaction (Fig. 6.10 is typical of small β_0). Near to the origin there is a double root, where R is both zero and has a minimum at $1 - \alpha_{ss} = -\beta_0$ (in theory negative values for the extent of reaction may have physical meaning if the B entering the reactor is converted to A, but it is also of general interest to pay some attention to this part of the diagram). For zero extent of reaction, the reaction rate $R = \beta_0^2(1 - K_e^{-1}\beta_0)$ and this decreases as the system becomes reversible (K_e

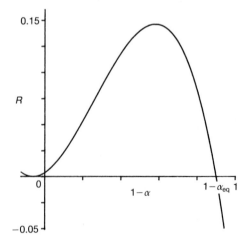

FIG. 6.10.   The influence of reaction reversibility on the rate curve R: the position of the maximum and point $R = 0$ are shifted to lower extents of reaction. Exact numerical values correspond to $\beta_0 = \frac{1}{16}$ and $K_e = 9.625$.

decreases). The curve has a maximum, R_{max}, given by

$$R_{max} = \frac{4(1 + \beta_0)^3}{27(1 + K_e^{-1})^2} \quad \text{at} \quad (1 - \alpha) = \frac{2 - (1 + K_e^{-1})\beta_0}{3(1 + K_e^{-1})}. \tag{6.28}$$

Comparing this with the maximum rate in the absence of reversibility found in § 6.2.3, we see that reversibility reduces R_{max} by a factor $(1 + K_e^{-1})^2$. Perhaps the most significant effect of reversibility on the reaction curve concerns the extent of conversion at which R falls to zero after the maximum. This no longer occurs at complete conversion of A but at

$$1 - \alpha = 1 - \alpha_{eq} = 1 - \frac{(1 + \beta_0)}{(1 + K_e)} \tag{6.29}$$

which corresponds to the equilibrium state. For larger extents of reaction ($\alpha < \alpha_{eq}$) the rate becomes negative, corresponding to a net production of A from B.

6.3.2. Special case of no catalyst inflow

If the reactor is fed by an inflow of pure A, so $\beta_0 = 0$, matters are especially simple. The stationary-state condition is given by

$$\underset{L}{\underbrace{\frac{1 - \alpha_{ss}}{\tau_{res}}}} = \underset{R}{\underbrace{(1 + K_e^{-1})(1 - \alpha_{ss})^2 (\alpha_{ss} - \alpha_{eq})}} \tag{6.30}$$

with

$$\alpha_{eq} = 1/(1 + K_e). \tag{6.31}$$

Figure 6.11 shows the reaction rate curve R for a number of decreasing values of the equilibrium constant K_e. These curves all have a double root at the origin. As K_e decreases the curves are squeezed more and more to the left corner of the flow diagram as α_{eq} increases and the maximum rate $(4/27)$ $(1 + K_e^{-1})^{-2}$ decreases.

The condition for tangency is independent of K_e in this system, requiring

$$1 - \alpha = \tfrac{1}{4}. \tag{6.32}$$

Multiple intersections occur at long residence times, when the flow line L has a low gradient, and require

$$\tau_{res} > \frac{(1 - \alpha_{eq})^2}{4(1 + K_e^{-1})}. \tag{6.33}$$

This latter result compares with the condition $\tau_{res} > \tfrac{1}{4}$, appropriate to the irreversible case.

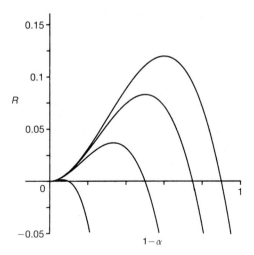

FIG. 6.11.   The influence on the reaction rate curve R of increasing the degree of reversibility of the cubic autocatalytic step for a system with $\beta_0 = 0$: (a) $K_e = 9$, (b) $K_e = 4$, (c) $K_e = 1$, (d) $K_e = \frac{1}{9}$. The slope of the extinction tangent does not vary, so multistability is possible for all K_e.

One solution of eqn (6.30) is simply the zero conversion state

$$1 - \alpha_{ss} = 0. \tag{6.34}$$

Two other real solutions exist if eqn (6.33) is satisfied, given by the roots of a quadratic

$$1 - \alpha_{ss} = \frac{1}{2}\left[1 - \alpha_{eq} \pm \left((1 - \alpha_{eq})^2 - \frac{4}{\tau_{res}(1 + K_e^{-1})}\right)^{1/2}\right]. \tag{6.35}$$

As the residence time tends to infinity, the upper root of this equation tends to the equilibrium state $1 - \alpha_{ss} = 1 - \alpha_{eq}$. Multistability exists over all residence times satisfying inequality (6.33), so there is no upper limit corresponding to an ignition point because there is no inflow of autocatalyst.

6.3.3. General tangency condition with inflow of autocatalyst

The condition for tangency for a system with autocatalyst in the inflow is obtained from eqns (6.21) and (6.26):

$$1 - \alpha = \frac{1 - K_e^{-1}\beta_0 \pm \{(1 - K_e^{-1}\beta_0)[1 - (8 + 9K_e^{-1})\beta_0]\}^{1/2}}{4(1 + K_e^{-1})}. \tag{6.36}$$

In the limit of irreversibility $K_e \to \infty$, so $K_e^{-1} \to 0$, and this reduces to

$$1 - \alpha = \tfrac{1}{4}[1 \pm (1 - 8\beta_0)^{1/2}] \tag{6.37}$$

which is equivalent to eqn (1.47).

The tangency condition can only have real roots so long as $(8 + 9K_e^{-1})\beta_0 < 1$. This provides limits on either the autocatalyst inflow concentration

$$\beta_0 \leqslant \frac{1}{8 + 9K_e^{-1}} \tag{6.38}$$

or, equivalently, on the equilibrium constant

$$K_e \geqslant \frac{8\beta_0}{1 - 8\beta_0}. \tag{6.39}$$

Thus, reversibility decreases the range of inflow concentrations over which multiple stationary states can exist. If the reactor has no autocatalyst in the inflow, multistability exists over some range of residence times, no matter how small the equilibrium constant becomes. Otherwise, increasing the inflow concentration decreases the extent of reversibility (i.e. raises the minimum value for K_e) over which multistability can be found.

6.3.4. Influence of autocatalyst inflow

We can illustrate quite clearly the effect of increasing the autocatalyst inflow concentration on the possibility of multiple stationary states using a flow diagram. Figures 6.12(a)–(e) show the relative orientations of R and L for increasing β_0. The reaction rate curve R moves across the diagram in the following sequence. In Fig. 6.12(a) the inflow concentration of B satisfies inequality (6.38), so R and L can have multiple intersections as shown, allowing multiple stationary states. Figure 6.12(b) shows the behaviour when (6.38) becomes an equality: R and L have a tangency, but at one point only, so there is no multiplicity. As β_0 increases further, the location of the maximum in the reaction rate curve $R_{\max}$ moves to lower extents of reaction, as described by eqn (6.28). When $\beta_0 = 2/(1 + 3K_e^{-1})$, the maximum rate occurs at zero conversion, as shown in Fig. 4.12(c): the reaction is now deceleratory in the sense that the rate falls monotonically as the extent of reaction increases from zero. In all these cases, the curve R falls to zero after the maximum, as the extent of reaction increases. The location of this zero is given by eqn (6.29) and this location also moves to lower extents of reaction as the autocatalyst inflow concentration increases. When $\beta_0 = K_e$ the inflow corresponds to the equilibrium composition, so $R = 0$ at zero extent of reaction, $1 - \alpha_{eq} = 0$. The only intersection (Fig. 6.12d) of the reaction and

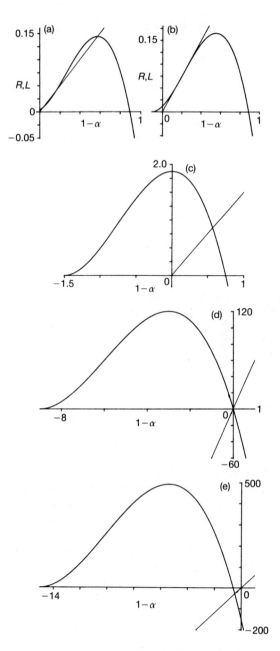

FIG. 6.12. The influence of autocatalyst inflow on the reaction rate curve R for a system with reversibility, $K_e = 9$: (a) $\beta_0 = \frac{1}{16}$, showing multiple stationary-state intersections; (b) $\beta_0 = 1/(8 + 9K_e^{-1}) = \frac{1}{9}$, for which multiplicity just vanishes; (c) $\beta_0 = 2/(1 + 3K_e^{-1}) = 1.5$, with the maximum in R at zero extent of reaction, so the process is now deceleratory; (d) $\beta_0 = K_e$, inflow has equilibrium composition so no net reaction occurs—the only intersection occurs at the origin for all residence times; (e) $\beta_0 > K_e$, intersections lie at negative extents of reaction indicating a conversion of B to A.

flow rate curves occurs at the origin, so there is no net reaction under stationary-state operation for any flow rate with this inflow. If the inflow concentration is even higher, $\beta_0 > K_e$, then the reaction rate curve has a negative value at the origin and for all positive extents of reaction as shown in Fig. 6.12(e): R and L do intersect, but this occurs for negative extents of reaction, so the reaction goes 'backwards' converting B to A and moving in the direction of the equilibrium composition.

6.4. Stationary-state behaviour for systems with catalyst decay

In the previous sections, the autocatalytic species B is only removed from the reactor by means of outflow or by conversion back to the original reactant A. We now consider the case where B can undergo an additional decay process, $B \rightarrow C$, similar to that considered in the model of chapter 2. Returning to irreversible steps, the kinetic scheme becomes, for autocatalysis,

$$A + 2B \rightarrow 3B \qquad \text{rate} = k_1 ab^2 \qquad (6.40)$$

and for decay

$$B \rightarrow C \qquad \text{rate} = k_2 b. \qquad (6.41)$$

We will now need to consider two mass-balance equations, one for A and one for B, so we now have a 'two-variable' system.

This decoupling of the concentrations of A and B may appear initially to be only a small modification, but it really has far-reaching effects. We will see below that even the stationary-state behaviour can be much more complex, and there is a much greater flexibility in the pattern of local stabilities: the uppermost and lowest states are no longer always stable, nor are unique states, and oscillatory responses are now possible.

6.4.1. Mass-balance equations and dimensionless groups

The equations appropriate to this augmented scheme, equivalent to eqns (6.9) and (6.10) of the previous section, are

$$\frac{da}{dt} = \frac{a_0 - a}{t_{res}} - k_1 ab^2 \qquad (6.42)$$

$$\frac{db}{dt} = \frac{b_0 - b}{t_{res}} + k_1 ab^2 - k_2 b. \qquad (6.43)$$

One significant difference between this pair of equations and (6.9) and (6.10) is that the net chemical rate of formation of B is not simply equal to the net chemical rate of removal of A. If (6.9) and (6.10) are added together all the reaction rate terms cancel, but here the term $-k_2 b$ remains.

The concentration of the final product C can be obtained by conservation of mass from the inflow composition once a and b are known. Assuming no inflow of C, then

$$c = (a_0 + b_0) - (a + b). \tag{6.44}$$

We cannot, however, use any relationship such as eqn (6.11) to relate a and b, in general, and so these are said to be 'decoupled'.

We can use our previous groups α, β, τ_{res}, etc. to make eqns (6.42) and (6.43) dimensionless. One extra term is needed, to incorporate k_2. As before, we use $k_1 a_0^2$ as our basis and define the dimensionless decay rate constant κ_2 as

$$\kappa_2 = k_2/k_1 a_0^2 = k_2 t_{\text{ch}}. \tag{6.45}$$

The mass-balance equations then become

$$\frac{d\alpha}{d\tau} = \frac{1 - \alpha}{\tau_{\text{res}}} - \alpha\beta^2 \tag{6.46}$$

$$\frac{d\beta}{d\tau} = \frac{\beta_0 - \beta}{\tau_{\text{res}}} + \alpha\beta^2 - \kappa_2\beta. \tag{6.47}$$

6.4.2. Stationary-state relationship and flow diagram

The stationary-state composition has dimensionless concentration α_{ss} and β_{ss} satisfying the simultaneous equations

$$0 = \frac{1 - \alpha_{\text{ss}}}{\tau_{\text{res}}} - \alpha_{\text{ss}}\beta_{\text{ss}}^2 \tag{6.48}$$

$$0 = \frac{\beta_0 - \beta_{\text{ss}}}{\tau_{\text{res}}} + \alpha_{\text{ss}}\beta_{\text{ss}}^2 - \kappa_2\beta_{\text{ss}}. \tag{6.49}$$

Adding these gives a simple linear relationship between the two concentrations:

$$\beta_{\text{ss}} = \frac{1 + \beta_0 - \alpha_{\text{ss}}}{1 + \kappa_2\tau_{\text{res}}}. \tag{6.50}$$

A number of points need to be made about this result. First, unlike the similar relationship (6.11), eqn (6.50) *only applies at the stationary state*. Secondly, the numerator really contains two contributions: the inflow of B, as β_0, and the amount of A that has been converted to B, as $1 - \alpha_{\text{ss}}$. The denominator then shows that the stationary-state concentration of the autocatalyst is always less than this. Of course this shortfall between the amount of B present in the reactor and that which has *flowed* in or been produced merely reflects the number of such molecules which have then reacted further to produce C. Thus, the denominator increases as the rate constant k_2 increases or as the mean residence time lengthens, indicating that more B has reacted during its stay in the reactor. Note that in our original dimensional quantities, the term $\kappa_2 \tau_{res}$ is simply $k_2 t_{res}$ and does not involve $k_1 a_0^2$ at all.

To obtain a single stationary-state equation, we now substitute for β_{ss}, from (6.50) into (6.48), giving

$$\underbrace{\frac{(1 + \kappa_2 \tau_{res})^2}{\tau_{res}}(1 - \alpha_{ss})}_{L} \quad - \quad \underbrace{\alpha_{ss}(1 + \beta_0 - \alpha_{ss})^2}_{R} = 0. \tag{6.51}$$

Again we will be particularly interested in the way the stationary-state extent of reaction $1 - \alpha_{ss}$ depends on the residence time, and how the locus drawn out in this way changes if β_0 and κ_2 are varied between successive experiments. We will make use of the flow diagram technique again, but first note one or two points. Equation (6.51) is a cubic in the stationary-state concentration of A: for given values of τ_{res} etc. there will be either one or three real positive roots, as found in the previous section. However, (6.51) is now also a quadratic equation in the residence time, so for a given concentration there will either be two or no solutions for τ_{res}. This is quite different from the behaviour in § 6.2: when $\kappa_2 = 0$ the stationary-state condition becomes linear in τ_{res}, so that for every α_{ss} there is only a single corresponding residence time.

6.4.3. Flow diagram

In the flow diagram, Fig. 6.13, the reaction rate curve R is still given by the simple cubic

$$R = \alpha_{ss}(1 + \beta_0 - \alpha_{ss})^2 \tag{6.52}$$

and so involves the inflow concentration of the autocatalyst but is independent of the residence time and of the decay rate constant κ_2. The flow rate L is slightly more involved, although it is still a linear function of the extent of conversion $1 - \alpha_{ss}$:

$$L = \frac{(1 + \kappa_2 \tau_{res})^2}{\tau_{res}}(1 - \alpha_{ss}). \tag{6.53}$$

The gradient of L is given by $(1 + \kappa_2 \tau_{res})^2 / \tau_{res}$ and so now depends on the mean residence time and on the new parameter κ_2. Figure 6.13(b) shows how the flow line varies in slope with the residence time: initially the gradient decreases as τ_{res} increases, as seen in the previous section. Now, however, the gradient of the flow line does not decrease indefinitely but passes through a minimum. The minimum value of the gradient is achieved when the rates of

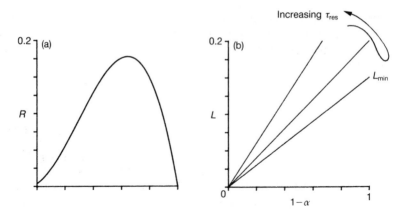

FIG. 6.13.   Flow diagram for reaction with autocatalyst decay. (a) The reaction rate curve R is unaffected by the decay step and has the same cubic form. (b) The variation of the gradient of the flow line L with the residence time is altered by the decay step: there is now a line of minimum gradient L_{min} corresponding to $\tau_{res} = \kappa_2^{-1}$ for which the gradient is $4\kappa_2$.

removal of B by flow and by reaction are equal, i.e. when $\kappa_2 \tau_{res} = 1$ and so

$$dL/d(1 - a_{ss})_{min} = 4\kappa_2 \quad \text{when} \quad \tau_{res} = 1/\kappa_2. \qquad (6.54)$$

For longer residence times, the flow line becomes steeper again.

We are again concerned with intersections of R and L on the flow diagram. The larger the value of κ_2, the steeper the minimum gradient of the flow line and hence L will not cut as far into R as τ_{res} varies. In particular we may lose the possibility of one or even both tangencies between the curves, and hence lose points of ignition and extinction. To illustrate the effect of the auto-catalyst decay through κ_2 on the stationary-state response we can consider a CSTR which is fed only by the reactant A, so $\beta_0 = 0$.

6.4.4. Systems with no autocatalyst inflow

We will proceed with the flow diagram analysis in a moment. First, however, we note that with $\beta_0 = 0$, the stationary-state condition (6.51) becomes significantly simpler:

$$\underbrace{\frac{(1 + \kappa_2 \tau_{res})^2}{\tau_{res}}(1 - \alpha_{ss})}_{L} - \underbrace{\alpha_{ss}(1 - \alpha_{ss})^2}_{R} = 0.$$
$$(6.55)$$

The term $1 - \alpha_{ss}$ appears directly in both L and R. Thus one solution of this equation is simply

$$1 - \alpha_{ss} = 0 \qquad (6.56)$$

corresponding to no reaction, for reasons we have seen previously.

Other stationary-state solutions must then satisfy the reduced equation

$$\frac{(1 + \kappa_2 \tau_{res})^2}{\tau_{res}} - \alpha_{ss}(1 - \alpha_{ss}) = 0. \qquad (6.57)$$

In terms of the extent of conversion, this quadratic has the roots

$$1 - \alpha_{ss} = \frac{1}{2}\left[1 \pm \left(1 - \frac{4(1 + \kappa_2 \tau_{res})^2}{\tau_{res}}\right)^{1/2}\right]. \qquad (6.58)$$

Figure 6.14 shows these stationary-state solutions as a function of residence time for various small values of κ_2. The non-zero states exist over a limited range of τ_{res}; they lie on the upper and lower shores of a closed curve, known as an 'isola'. The size of the isola decreases as κ_2 increases. At each end of the isola there is a turning point in the locus, corresponding to extinction or washout. There are no ignition points in these curves.

The ends of the isola are easily located. They correspond to residence times at which the upper and lower roots of (6.58) become equal, and hence to values of τ_{res} for which the discriminant becomes zero. Thus, we require

$$4(1 + \kappa_2 \tau_{res})^2 = \tau_{res} \qquad (6.59)$$

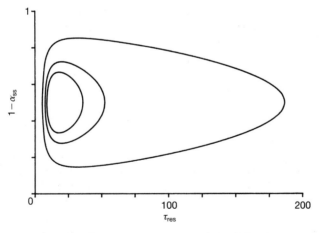

FIG. 6.14.   Stationary-state loci for reaction with no autocatalyst inflow but autocatalyst decay and $\kappa_2 < \frac{1}{16}$. The zero-reaction state $1 - \alpha_{ss} = 0$ exists as a solution for all conditions; the non-zero solutions form a closed curve (isola) which grows as κ_2 is decreased. The isola patterns shown are for $\kappa_2 = \frac{1}{18}, \frac{1}{20}$, and $\frac{1}{32}$ in order of increasing size.

which is a quadratic in the residence time for a given κ_2. The roots of this equation, and hence the extinction points, are given by

$$\tau_{res}^{\pm} = \frac{1 - 8\kappa_2 \pm (1 - 16\kappa_2)^{1/2}}{8\kappa_2^2}. \qquad (6.60)$$

Put another way, for a given κ_2 there is a range of residence times over which the stationary-state eqn (6.58) has real solutions given by

$$\tau_{res}^- \leqslant \tau_{res} \leqslant \tau_{res}^+. \qquad (6.61)$$

This range decreases in size as κ_2 increases. In fact the form of the numerator in eqn (6.60) shows that there is an upper limit on the dimensionless rate constant, i.e. for an isola to exist at all we require that

$$\kappa_2 \leqslant \tfrac{1}{16} \quad \text{i.e.} \quad k_2 \leqslant \tfrac{1}{16} k_1 a_0^2. \qquad (6.62)$$

The variation of τ_{res}^+ and τ_{res}^- with κ_2 given by (6.60) can be drawn out as a projection in the parameter plane, in a similar way to that portrayed for the residence time–inflow concentration plane in Fig. 6.8 for the model without decay. Again we see the characteristic cusp shape as $\kappa_2 \to \tfrac{1}{16}$.

We can see quite persuasively how an isola pattern arises, by considering the flow diagram. The reaction rate curve R is shown in Fig. 6.15. For short residence times, the flow line L is steep and so only intersects with R at the origin: there is thus only one stationary state, corresponding to zero extent of reaction.

As τ_{res} is increased, the gradient of the flow line decreases, and L may eventually touch R away from the origin, at a point of tangency. Using the tangency condition (6.21), this occurs at the point $1 - \alpha_{ss} = \tfrac{1}{2}$, irrespective of the value of κ_2. The gradient of the flow line then has the value $\tfrac{1}{4}$. At this tangency, two new stationary states appear as an isolated point corresponding to 50 per cent conversion of A. These move apart as τ_{res} increases and the flow line cuts further into R.

Eventually, L attains its minimum slope, given by eqn (6.52). The two non-zero stationary states are now separated by their maximum distance. Any further increase in the residence time causes the non-zero states to move closer together as the flow line becomes steeper again. Eventually, R and L become tangential again, and the two solutions merge. The isola closes up at this point, where again $1 - \alpha_{ss} = \tfrac{1}{2}$ and the gradient is $\tfrac{1}{4}$. For yet longer residence times, only the intersection at the origin persists.

Tangency is clearly only possible if the flow line can attain a gradient of $\tfrac{1}{4}$. We know that the minimum gradient which L can have is $4\kappa_2$, so we have simply the requirement that this should be less than $\tfrac{1}{4}$, i.e. we need

$$4\kappa_2 < \tfrac{1}{4} \Rightarrow \kappa_2 < \tfrac{1}{16}$$

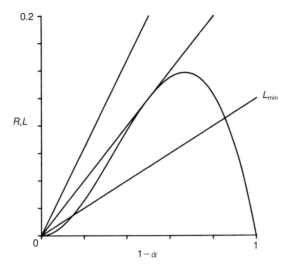

FIG. 6.15.   Flow diagram representation of the origin of isola solutions for cubic autocatalysis with decay. The gradient of the flow line L at first decreases with increasing residence time: non-zero intersections appear as R and L become tangential and these move apart as L becomes less steep. After L has attained its minimum slope the non-zero intersections move closer together again, and merge (and disappear) as R and L attain tangency for the second time. The intersection at zero extent of reaction exists for all residence times.

which, of course, is the same result as that derived above as eqn (6.62). The latter made use of the special feature of the stationary-state equation which allowed easy factorization to a quadratic. The flow diagram tangency condition is more generally applicable. In fact, we shall also use a third route to this same result in chapter 7, which is yet more general.

If κ_2 is larger than $\frac{1}{16}$, L and R do not have any intersections, except that at the origin, for any residence time, because the flow line can never have a low-enough slope. The only stationary-state solution is that corresponding to zero extent of reaction, so the $(1 - \alpha_{ss})$–τ_{res} bifurcation diagram is now almost completely featureless, although there is, in fact, a unique solution for all residence times.

Returning to the flow diagram approach, the extinction points are given by the tangencies and, hence, by the gradient of L taking the value $\frac{1}{4}$. The gradient of the flow line is $(1 + \kappa_2 \tau_{res})^2 / \tau_{res}$, and so extinction is located by the condition

$$\frac{(1 + \kappa_2 \tau_{res})^2}{\tau_{res}} = \frac{1}{4} \tag{6.63}$$

which is the same as eqn (6.59) and so again yields (6.60).

6.4.5. Inflow of autocatalyst

We can now consider how the relationship between isolas and unique stationary states, and indeed any other new patterns of behaviour, is affected by the inflow of some autocatalyst. In such a case β_0 will be non-zero.

The full expressions for R and L have been given in eqns (6.52) and (6.53). The only difference between these forms and those of the previous subsection is that the reaction rate curve R now involves the inflow concentration of the autocatalyst. The flow line L, its dependence on the residence time, and its minimum gradient are all independent of β_0.

We saw in the case of a perfectly stable autocatalyst, considered in § 6.2.3, that when β_0 is non-zero there are two different tangencies: one for extinction, the other for ignition. Now that the autocatalyst can react further, we can have up to four tangencies, one each for extinction and ignition as the slope of the flow line is decreasing, and one more each as L is on its way up again. Whether or not any or all of these possibilities are realized depends on the gradients required for each of the tangencies (which we shall see depend on β_0) and on the minimum gradient (which depends on κ_2).

The conditions for tangency are again located by applying eqn (6.21). This yields for the extent of reaction

$$(1 - \alpha_{ss})^{\pm} = \tfrac{1}{4}[1 \pm (1 - 8\beta_0)^{1/2}] \qquad (6.64)$$

which is, in fact, the same as eqn (1.47) derived in § 1.7.2. The upper root gives the condition for both of the extinction tangencies, at which the system moves from a high extent of reaction; the lower root corresponds to the ignitions. This expression is independent of the rate constant κ_2.

For small values of the autocatalyst inflow concentration, such that $\beta_0 \ll 1$, the roots of eqn (6.64) are given approximately by

$$(1 - \alpha_{ss})^{+} \approx \tfrac{1}{2}(1 - 2\beta_0 \ldots) \qquad (1 - \alpha_{ss})^{-} \approx \beta_0 + \cdots \qquad (6.65)$$

These expressions make it clear how the locations of the tangencies move as β_0 increases: the extinction tangent moves from $(1 - \alpha_{ss})^{+} = \tfrac{1}{2}$ to lower extents of reaction. In the absence of autocatalyst inflow, the ignition tangent occurs at the origin (when the flow line L would have to be completely flat), but it moves away from there as β_0 increases. Thus the two tangencies move towards each other as the inflow concentration of B increases relative to that of A. When $\beta_0 = \tfrac{1}{8}$, the two tangents merge and ignition and extinction become impossible.

The gradient of the flow line L at the tangencies is obtained from (6.64) and the condition for equal slopes:

$$\frac{(1 + \kappa_2 \tau_{res})^2}{\tau_{res}} = (1 + \beta_0 - \alpha_{ss})(3\alpha_{ss} - \beta_0 - 1) \qquad (6.66)$$

so that

$$\frac{(1 + \kappa_2 \tau_{res})^2}{\tau_{res}} = \tfrac{1}{8}[1 + 20\beta_0 - 8\beta_0^2 \pm (1 - 8\beta_0)^{3/2}]. \tag{6.67}$$

For any given value of β_0, each root of this equation gives a quadratic for the residence time at which extinction (upper root) or ignition (lower root) occurs. Clearly we again require $\beta_0 < \tfrac{1}{8}$, otherwise the right-hand side of (6.67) is not real. Whether or not the resulting quadratics for the extinction and ignition residence times have real roots then depends on the value of the dimensionless rate constant κ_2.

If we denote the right-hand side of (6.67) as S^+ for the upper root and S^- for the lower root, the residence times for extinction (S^+) and ignition (S^-) are given (two for each) by

$$\tau_{res} = \frac{S^{\pm} - 2\kappa_2 \pm [S^{\pm}(S^{\pm} - 4\kappa_2)]^{1/2}}{2\kappa_2^2}. \tag{6.68}$$

Let us consider a few specific examples. We will take a system with the inflow concentrations fixed so that $\beta_0 = \tfrac{1}{32}$. The tangencies for this case occur for

ignition $\quad (1 - \alpha_{ss})^- = 0.033\,49 \quad$ gradient of $L = 0.120\,96$

extinction $\quad (1 - \alpha_{ss})^+ = 0.4665 \quad$ gradient of $L = 0.283\,34$

and are shown as lines L_{ign} and L_{ext} on Fig. 6.16(a).

If the rate constant $\kappa_2 = 0.1$, then the minimum gradient which L can have is $4\kappa_2 = 0.4$. This is steeper than both the tangents for this system and might correspond to the flow line L_1 in Fig. 6.16(a). There are no ignition or extinction points in the stationary-state locus (Fig. 16.6(b)) which has a unique solution for all residence times.

We can locate the value of κ_2 at which isola responses first become possible in this example by finding the condition for the minimum gradient to have exactly the same slope as the extinction tangent, i.e. we require $4\kappa_2 = 0.283\,34$ so that $\kappa_2 = 0.0708$. With this value of κ_2, the stationary-state locus shows the birth of an isola as an isolated point lying above the lower branch, as shown in Fig. 6.16(c).

If we now take $\kappa_2 = 0.05$, the flow line L_2 has a minimum gradient of 0.2 and lies below the extinction tangency but above that for ignition. As the residence time is varied, therefore, the flow line can pass the extinction tangent twice: once on the way down and once on the way back up again. This behaviour is similar to that described in the previous subsection and draws out an isola in the stationary-state locus (Fig. 6.16(d)). The residence times corresponding to the turning points at the end of the isola can be obtained from eqn (6.68). Taking the upper root from (6.67) for extinction so

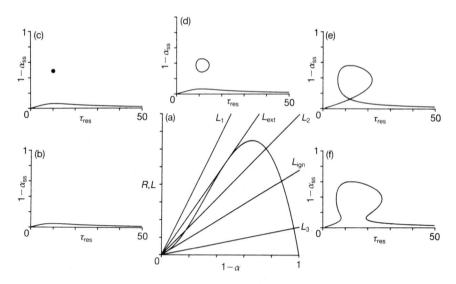

FIG. 6.16. (a) Flow diagram representation of the different stationary-state loci for cubic autocatalysis with inflow and decay of B: (b) unique response, minimum slope L_1; (c) birth of isola as isolated point, minimum slope L_{ext}; (d) isola, minimum slope L_2; (e) transition from isola to mushroom, minimum slope L_{ign}; (f) mushroom, minimum slope L_3.

S^+ has the value 0.283 34 gives $\tau_{res} = 5.93$ and 67.4. If the lower root, $S^- = 0.120\,96$, is taken from (6.67), there are no real roots of (6.68) for the corresponding ignition points.

If the flow line is to have a minimum gradient which coincides exactly with the ignition tangent T_i, then we require $4\kappa_2 = 0.120\,96$ so that $\kappa_2 = 0.030\,24$. For this value of the dimensionless rate constant, the stationary-state isola has grown so large that it just touches the lowest branch of solutions, as shown in Fig. 6.16(e).

Finally in this sequence, we can take $\kappa_2 = 0.01$. The minimum gradient of the flow line is 0.04, corresponding to line L_3, and this lies below the ignition tangent. The stationary-state locus, shown in Fig. 6.16(f), now shows two extinction points and two ignition points. The pattern is known as a mushroom. It has two regions of multiple solutions and hence of hysteresis: one S-shaped at low residence times, the other Z-shaped at longer residence times. We may use eqn (6.68) to calculate the locations of these turning points. For the extinctions we take $S^+ = 0.283\,34$, which yields $\tau_{res} = 3.80$ and 999.6; for the ignitions, the required slope S^- has the value 0.120 96, leading to $\tau_{res} = 10.0$ and 2629.8. (In fact, the stationary-state loci in Figs 6.16(b)–(f) are computed for a higher autocatalyst inflow concentration than that described above, $\beta_0 = 3/32$: the same qualitative features arise, but the residence

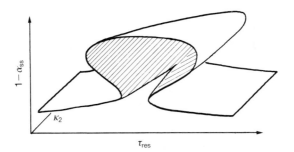

FIG. 6.17. Three-dimensional representation of the folding of the stationary-state surface $(1 - \alpha_{ss})$–τ_{res}–κ_2: slices through this surface with constant κ_2 give mushroom, isola, or unique patterns. A similar surface arises in the $(1 - \alpha_{ss})$–τ_{res}–β_0 space.

timescale is much more compact and hence allows all the aspects to be visualized simultaneously.)

The changes and overall trend just described in this example are summarized in Fig. 6.17, which shows a three-dimensional representation of the stationary-state extent of conversion as a function of both residence time and catalyst decay rate κ_2. This shows how the sequence unique $\rightarrow$ isola $\rightarrow$ mushroom, which comes about as κ_2 is decreased, corresponds to a 'folding' of a surface of solutions. Again we see cusps if the ignition and extinction points, given by eqn (6.68), are projected down on to the τ_{res}–κ_2 plane. A similar pattern is seen if κ_2 is held constant and the autocatalyst inflow varied, except that the changes take place as β_0 increases rather than as κ_2 decreases, and there is an extra change at relatively large β_0 as the mushroom opens and multiple stationary-state behaviour disappears (as β_0 passes through $\frac{1}{8}$).

In the above example we calculated the 'critical' values of the dimensionless decay rate constant κ_2, at which an isola first appears and at which it changes to a mushroom, for the particular values of the autocatalyst inflow concentration $\beta_0 = \frac{1}{32}$. Our approach was to locate the conditions when the extinction and ignition tangents have the same slope, given by the right-hand side of eqn (6.67), as the minimum gradient $4\kappa_2$. We can use this recipe to find a general condition for these changes in the stationary-state pattern: thus we require

$$\kappa_2 = \tfrac{1}{32}[1 + 20\beta_0 - 8\beta_0^2 \pm (1 - 8\beta_0)^{3/2}]. \tag{6.69}$$

The upper root of this equation gives the value of κ_2 at which an isola is born as a function of β_0; the lower root gives the change from isola to mushroom. For $\beta_0 = 0$, these roots tend to $\frac{1}{16}$ and 0 respectively, so the mushroom pattern is not found in the absence of autocatalyst inflow, as seen in the previous subsection. As β_0 increases the roots move closer together: they merge at the special point $\beta_0 = \frac{1}{8}$, $\kappa_2 = 27/256 = 3^3/4^4$.

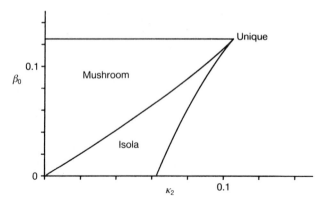

FIG. 6.18.   Regions of unique, isola, and mushroom patterns in β_0–κ_2 parameter plane. The loci are given by eqn (6.69) and the condition $\beta_0 = \frac{1}{8}$ and meet at the point $\kappa_2 = 3^3/4^4$, $\beta_0 = \frac{1}{8}$.

Equation (6.69) describes a pair of boundaries in the κ_2–β_0 'parameter plane', as shown in Fig. 6.18. Yet again, the influence of the square root term is to produce a cusp. Inside this cusp, we have combinations of κ_2 and β_0 for which the stationary-state locus $1 - \alpha_{ss}$ versus τ_{res} shows an isola. Also drawn on the figure is the line $\beta_0 = \frac{1}{8}$, which we have seen corresponds to the dimensionless autocatalyst inflow concentration at which the reaction rate curve R can no longer have any tangencies with the flow line L. This line meets the point of the cusp ($\beta_0 = \frac{1}{8}$, $\kappa_2 = 27/256$) and provides an upper boundary to the region of the plane for which the corresponding stationary-state locus has a mushroom shape. If the inflow concentration of B is increased, so we move across this line, then the two regions of multiple stationary states open out, at the same time, leaving a unique locus.

6.4.6. Behaviour at long residence times

One significant difference between the stationary-state loci observed in § 6.2, when the autocatalyst was infinitely stable, and those found above is the behaviour at long residence times. In the previous case, the reaction tends to completion, with $1 - \alpha_{ss}$ approaching unity as τ_{res} tends to infinity. This is equivalent to the reaction attaining its chemical equilibrium state as the system approaches a closed vessel situation. In terms used elsewhere, such behaviour may be described as 'a good thing'. The loci described in the previous subsections show a different response. Now, as the residence time lengthens, the stationary-state extent of conversion tends back to zero, i.e. to a state of no conversion of A. This is just about as far away from chemical equilibrium as the system can get and therefore 'a bad thing'. It arises despite the inflow of autocatalyst because flow is so slow that any B arriving in the reactor or being formed has time to decay to C, no matter how low κ_2 is: once the autocatalyst has decayed, there is no conversion of A.

This situation stems from the peculiarities of the simplest representations of autocatalytic reactions, as discussed previously. In chapter 2, we considered a kinetic scheme in which the autocatalysis proceeded in parallel with an uncatalysed step $A \to B$. We can expect that the inclusion of such a step here will have a strong influence on the behaviour at long residence times, as there will then always be a route from A to B. We consider this in the next section.

6.5. Influence of uncatalysed conversion of A to B

When the model of the previous section is augmented by a direct step from A to B, the scheme becomes

$$A \to B \qquad \text{rate} = k_u a$$

$$A + 2B \to 3B \qquad \text{rate} = k_1 ab^2$$

$$B \to C \qquad \text{rate} = k_2 b.$$

The dimensionless mass-balance equations are now

$$\frac{d\alpha}{d\tau} = \frac{1 - \alpha}{\tau_{res}} - \alpha\beta^2 - \kappa_u \alpha \tag{6.70}$$

$$\frac{d\beta}{d\tau} = \frac{\beta_0 - \beta}{\tau_{res}} + \alpha\beta^2 + \kappa_u \alpha - \kappa_2 \beta \tag{6.71}$$

where the only new parameter is the dimensionless uncatalysed rate constant κ_u defined by

$$\kappa_u = k_u / k_1 a_0^2. \tag{6.72}$$

6.5.1. Stationary-state condition

The relationship between the two dimensionless concentrations at the stationary state is not affected by the uncatalysed reaction. Thus setting $d\alpha/d\tau = d\beta/d\tau = 0$, and adding the resulting equations, yields

$$\beta_{ss} = \frac{1 + \beta_0 - \alpha_{ss}}{1 + \kappa_2 \tau_{res}}. \tag{6.73}$$

Substituting this into $d\alpha/d\tau = 0$ leads to the stationary-state condition in terms of α_{ss}:

$$\frac{(1 + \kappa_2 \tau_{res})^2}{\tau_{res}}(1 - \alpha_{ss}) - \alpha_{ss}[(1 + \beta_0 - \alpha_{ss})^2 + \kappa_u(1 + \kappa_2 \tau_{res})^2] = 0. \quad (6.74)$$

$$\qquad\qquad L \qquad\qquad\qquad\qquad\qquad\qquad R$$

Comparing this with eqn (6.55), we see that the flow rate term L is unaffected by the extra reaction, as might have been expected. The reaction rate term R, not surprisingly, does involve the uncatalysed reaction rate constant. An important change, however, is that R now also depends on κ_2 and, more significantly, on the residence time. Thus as τ_{res} varies both curves on the flow diagram will alter. There is even no simplification in setting $\beta_0 = 0$, since the stationary-state equation is irreducibly cubic. Proceeding by means of the tangency condition, some progress can be made. For instance, the tangency occurs when the stationary-state extent of conversion satisfies

$$2(1 - \alpha_{ss})^3 - (1 - 2\beta_0)(1 - \alpha_{ss})^2 + \beta_0^2 + (1 + \kappa_2 \tau_{res})^2 = 0 \quad (6.75)$$

and the slope of the flow line L will then have the value

$$\text{slope of } L = \frac{(1 + \kappa_2 \tau_{res})^2}{\tau_{res}} = - 3(1 - \alpha_{ss})^2 + 2(1 - 2\beta_0)(1 - \alpha_{ss})$$

$$+ \beta_0(2 - \beta_0) - (1 + \kappa_2 \tau_{res})^2. \quad (6.76)$$

Numerical solutions, and even exact equations, for the stationary state can be obtained in a quite straightforward manner if the values of all the other parameters are specified. It is, however, very difficult to unravel these two equations analytically and hence we cannot obtain general relationships of the form of eqns (6.68) and (6.69) for the points of ignition and extinction or the parameter values at which isolas are born or give way to mushrooms, etc. In the next section we will introduce a mathematical technique known as singularity theory which is specifically designed to locate changes in the stationary-state locus such as the isola to mushroom transition. For now, we just present the different qualitative responses, which this augmented model can exhibit, and some specific quantitative aspects.

6.5.2. Stationary-state patterns

Figures 6.19(a–d) show the four different types of 'bifurcation diagram' where the stationary-state extent of reaction is plotted as a function of residence time for the model with the uncatalysed reaction included *for the special case* of no catalyst inflow, $\beta_0 = 0$. Three patterns have been seen before: (a) unique, (b) isola, and (c) mushroom, in the absence of the uncatalysed step. Even with these, however, there is the significant difference at long residence times that the reaction approaches complete conversion, $1 - \alpha_{ss} \to 1$, as discussed above. The fourth pattern (d) is that of a single hysteresis loop, or 'breaking wave'. This response was not found in the previous section without the uncatalysed reaction. We have seen it before in

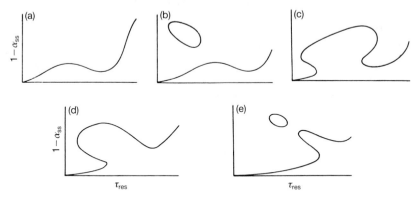

FIG. 6.19.   The five stationary-state loci patterns found when the uncatalysed conversion of A to B is included in the model: (a) unique, (b) isola; (c) mushroom; (d) single hysteresis loop (breaking wave); (e) hysteresis loop + isola.

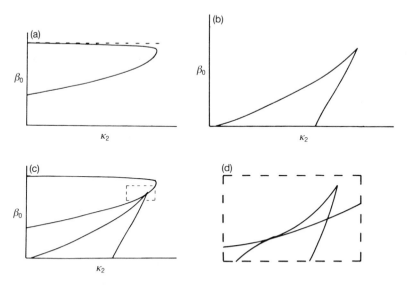

FIG. 6.20.   Typical forms for the boundaries in the β_0–κ_2 parameter plane separating regions of different stationary-state patterns for the system with uncatalysed reaction: (a) the hysteresis boundaries; (b) the isola boundaries; (c) relative positions of boundaries in (a) and (b), with four regions visible; (d) enlargement of area close to cusp in isola boundaries, showing existence of fifth region.

the case of an infinitely stable catalyst ($\kappa_2 = 0$, § 6.2.4), but then an inflow of the autocatalyst was required.

If the CSTR is fed with both A and B, so $\beta_0 > 0$, then a fifth pattern of response can also be found over a narrow range of experimental conditions. This is shown in Fig. 6.19(e) and has both a breaking wave and an isola. In total such a bifurcation diagram shows three extinction points and only one ignition.

The present model has three parameters other than the residence time. These are κ_2, β_0, and κ_u. Figure 6.18 showed the division of the κ_2–β_0 plane into different regions corresponding to unique, isola, and mushroom patterns for the case $\kappa_u = 0$. Mapping out the full κ_2–β_0–κ_u would require a three-dimensional diagram, which would (i) be difficult to draw accurately and (ii) not be easy to read quantitative results from. Instead, Figs 6.20 and 6.21

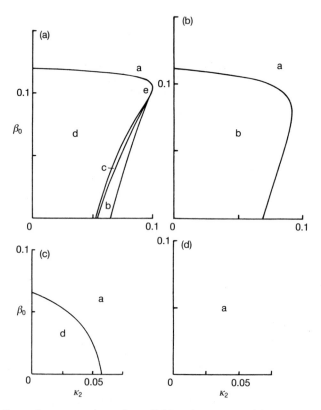

FIG. 6.21. Exact β_0–κ_2 parameter plane division for system with uncatalysed reaction: (a) $\kappa_u = 2 \times 10^{-3}$, all five stationary-state patterns from Fig. 6.19 are found; (b) $\kappa_u = 5.12 \times 10^{-3}$, unique or single hysteresis loop; (c) $\kappa_u = 2 \times 10^{-2}$, unique or single hysteresis loop; (d) $\kappa_u \geqslant \frac{1}{27}$, no multiplicity in stationary-state loci.

shows a series of slices, each corresponding to a constant non-zero value of the dimensionless uncatalysed reaction rate constant κ_u.

These figures bear similarities to Fig. 6.17 but have a significant difference. The line $\beta_0 = \frac{1}{8}$, corresponding to the change from mushroom to unique loci as the two hysteresis loops open up simultaneously, is now replaced by a continuous curve (Fig. 6.20(a)) with upper and lower boundaries. One hysteresis loop disappears as each of these is crossed, so within the region bounded by this curve the stationary-state locus has just one loop, i.e. a breaking wave. The isola cusp has a similar basic structure to that in Fig. 6.18, as shown in Fig. 6.20(b) except that its upper boundary does not now go to the origin. Also the cusp point does not lie exactly on either of the hysteresis boundaries: the two regions overlap (Figs 6.20(c) and, in more detail, 6.20(d)). In this way the β_0–κ_2 parameter plane is divided into five regions, one corresponding to each of the five possible stationary-state patterns.

As shown in the sequence of cross-sections in Fig. 6.21, the regions corresponding to the more 'exotic' patterns associated with multiple stationary states decrease as the dimensionless uncatalysed rate κ_u increases. When κ_u has increased to just over $1/200$, isolas and mushrooms have vanished, but breaking wave responses can still be found for small-enough autocatalyst inflow and decay rates (β_0 and κ_2). Further increases in the uncatalysed reaction rate constant see a contraction of the region corresponding to breaking waves. This disappears when $\kappa_u = \frac{1}{27}$, exactly, leaving only the possibility of unique dependences of the stationary-state extent of conversion on residence time.

6.6. Reversibility with catalyst decay

A second way of ensuring that the reaction does not become artificially switched off by the complete disappearance of the autocatalyst is to recognize the reversibility of the two reactions. Thus the extent of reaction can never proceed beyond the equilibrium state, which itself has a non-zero concentration for B. We will again neglect the uncatalysed step but include in our discussion those systems which may be fed not just by an inflow of A and B, but also by an inflow of C. The latter allows the reaction to proceed 'backwards', if the inflow composition has a product concentration exceeding its equilibrium value.

The model is thus

$$A + 2B \rightleftharpoons 3B \qquad k_1, k_{-1}$$
$$B \rightleftharpoons C \qquad k_2, k_{-2}.$$

The appropriate dimensionless equations for the mass balance on A and

B can be written in the form

$$\frac{d\alpha}{d\tau} = \frac{1 - \alpha}{\tau_{res}} - (\alpha - K_1^{-1}\beta)\beta^2 \tag{6.77}$$

$$\frac{d\beta}{d\tau} = \frac{\beta_0 - \beta}{\tau_{res}} + (\alpha - K_1^{-1}\beta)\beta^2 - \kappa_2[\beta - K_2^{-1}(1 + \beta_0 + \chi_0 - \alpha - \beta)] \tag{6.78}$$

where the concentration of C has been replaced by $(a_0 + b_0 + c_0) - (a + b)$ wherever it occurs. Here χ_0 is the dimensionless inflow concentration of the

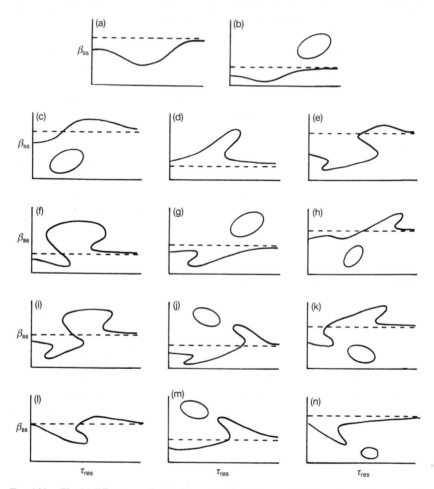

FIG. 6.22. The 14 different qualitative forms for the stationary-state locus for cubic autocatalysis with reversible reactions and inflow of all species, with $c_0 < \frac{1}{3}a_0$: the broken line represents the equilibrium composition which is approached at long residence times. (Reprinted with permission from Balakotaiah, V. (1987). *Proc. R. Soc.*, **A411**, 193.)

product c/a_0 and K_1 and K_2 are the equilibrium constants for the two steps: $K_1 = k_1/k_{-1}$; $K_2 = k_2/k_{-2}$.

If the reactor is fed by a stream relatively free from the product, so $a_0 > 3c_0$, a total of 14 different responses are possible for the dependence of the stationary-state extent of reaction on the residence time, as shown in Fig. 6.22. For higher inflow concentrations of C, so that a_0 is less than three times larger than c_0, only seven patterns are possible (Fig. 6.23). The remarkable detective work required to count and locate all of these bifurcation diagrams has been provided by Balakotaiah (1987). Exemplary values for the various parameters corresponding to each possible case are given in Tables 6.1 and 6.2. From the number of decimal places required for some of the patterns, it is clear that the ranges over which they exist can be very small. It is one of the most powerful features of singularity theory that it guides us directly to relevant locations in parameter space and enables us to be sure that all possible forms have been found. Again we can present this very useful mathematical technique in terms of an easy-to-apply recipe, with the minimum of background detail. We do this in the next chapter, where we also consider the stationary-state patterns of a non-isothermal reaction in a CSTR.

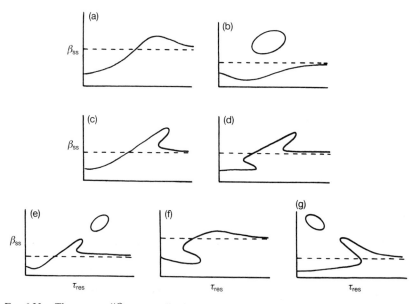

FIG. 6.23. The seven different qualitative forms for the stationary-state locus for cubic autocatalysis with reversible reactions and inflow of all species, with $c_0 > \frac{1}{3}a_0$: the broken line represents the equilibrium composition which is approached at long residence times. These patterns are the same as those found for the irreversible system with an uncatalysed step—see Fig. 6.19. (Reprinted with permission from Balakotaiah, V. (1987). *Proc. R. Soc.*, **A411**, 193.)

Table 6.1

Parameter values exemplifying the 14 stationary-state responses for the reversible autocatalator model with $\chi_0 = 0.05$ corresponding to Fig. 6.22

	β_0	κ_2	K_1^{-1}	K_2
a	0.1	0.1	0.4	11.7647
b	0.1	0.102	0.1	12
c	0.08	0.036	0.8	36
d	0.1	0.05	0.2	5.8823
e	0.8	0.04	1.0	40
f	0.1	0.098	0.3	11.529
g	0.08	0.08	0.4	16
h	0.08	0.073	0.4	14.6
i	0.08	0.75	0.4	150
j	0.08	0.073592485	1.105	14.718497
k	0.1	0.10119106435	0.35216	10.221309409
l	0.08	0.05	1.5	50
m	0.1	0.0976578384919	0.3930671	11.4892561079
n	0.08	0.0011674947051685	103.29361	1167.4947051685

Table 6.2

Parameter values exemplifying the 7 stationary-state responses for the reversible autoatalator model with $\beta_0 = 0.08$ and $\chi_0 = 1.0$ corresponding to Fig. 6.23

	$10^2 \kappa_2$	K_1^{-1}	K_2
a	2.0	1.0	200
b	8.0	0.1	800
c	2.0	0.2	200
d	5.0	0.2	500
e	7.967196216	0.58982	79.67196216
f	5.0	1.0	500
g	6.5831624	1.0456	658.31624

References

Aris, R., Gray, P., and Scott, S. K. (1988). Modelling cubic autocatalysis by successive bimolecular steps. *Chem. Eng. Sci.*, **43**, 207–11.

Balakotaiah, V. (1987). On the steady-state behaviour of the autocatalator model $A + 2B \rightleftharpoons 3B$, $B \rightleftharpoons C$ in a continuous-flow stirred-tank reactor *Proc. R. Soc.*, **A411**, 193–206.

Escher, C. and Ross, J. (1983). Multiple ranges of flow rate with bistability and limit cycles for Schlogl's mechanism in a CSTR. *J. Chem. Phys.*, **79**, 3773–7.

Gray, P. and Scott, S. K. (1983). Autocatalytic reactions in the isothermal, continuous stirred tank reactor: isolas and other forms of multistability. *Chem. Eng. Sci.*, **38**, 29–43.

Gray, P. and Scott, S. K. (1984). Autocatalytic reactions in the isothermal, continuous stirred tank reactor: oscillations and instabilities in the system $A + 2B \rightarrow 3B, B \rightarrow C$. *Chem. Eng. Sci.*, **39**, 1087–97.

Gray, P. and Scott, S. K. (1985). Sustained oscillations and other exotic patterns of behavior in isothermal reactions. *J. Phys. Chem.*, **89**, 22–32.

Lin, K. F. (1981). Multiplicity, stability and dynamics for isothermal autocatalytic reactions in CSTR. *Chem. Eng. Sci.*, **36**, 1447–52.

Scott, S. K. (1983). Reversible autocatalytic reactions in an isothermal CSTR. *Chem. Eng. Sci.*, **38**, 1701–8.

Scott, S. K. (1987). Oscillations in simple models of chemical systems. *Acc. Chem. Res.*, **20**, 186–91.

REACTION IN A NON-ISOTHERMAL CSTR: STATIONARY STATES AND SINGULARITY THEORY

Chapter 6 considered isothermal autocatalysis in an open system; here we study a classic case of thermal feedback. A rich variety of stationary-state patterns (bifurcation diagrams) are generated and considered here alongside those of the previous isothermal example. Flow diagrams are again illuminating and singularity theory provides a systematic approach. After study a reader should be able to:

(1) choose dimensionless forms with a sympathetic notation;

(2) recognize the value of the 'exponential approximation' to the Arrhenius temperature dependence and the ease of extending it;

(3) recognize the parallels between solutions for adiabatic operation here and the stable catalyst case of chapter 6, and between the broader circumstances in both cases;

(4) apply singularity theory to the enumeration of stationary-state patterns in this and other systems.

In chapters 2–5 two models of oscillatory reaction in closed vessels were considered: one based on chemical feedback (autocatalysis), the other on thermal coupling under non-isothermal reaction conditions. To begin this chapter, we again return to non-isothermal systems, now in a well-stirred flow reactor (CSTR) such as that considered in chapter 6.

We take just a single step

$$A \rightarrow B + \text{heat} \qquad \text{rate} = k(T)a \qquad (7.1)$$

where the reaction is exothermic and the rate constant depends on temperature, by means of the Arrhenius form

$$k = A \exp(-E/RT). \qquad (7.2)$$

An example of such a system is the first-order decomposition of the organic compound di-tertiarybutyl peroxide which has been studied experimentally in a non-adiabatic CSTR.

Comparing this with the model of chapter 4, we see that we now have no precursor reactant P: the continuous supply of A is provided here by the constant inflow to the reactor. We will consider a CSTR fed by a stream in which the concentration of A is a_0 and whose (inflow) temperature is T_0. The reactor is situated in a furnace or heat bath and maintained at an ambient temperature T_a (which may be different from the inflow temperature), as shown in Fig. 7.1.

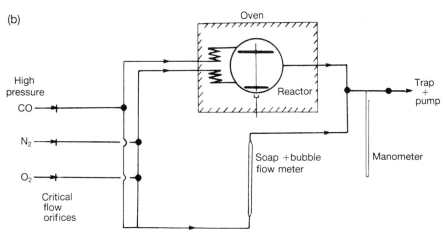

FIG. 7.1.   A CSTR appropriate to gas-phase reactions (a) and (b) a schematic representation of a typical experimental apparatus.

Again we can identify a mean residence time t_{res} given by the ratio of the reactor volume and the volumetric flow rate. The temperature of the reacting mixture within the CSTR is given by T, which may exceed T_a because of self-heating from the exothermic reaction. Heat transfer occurs by two processes: (i) by the outflow of material at a different temperature from the inflow and (ii) by Newtonian cooling across the walls. One point worth noting is that both of these processes can make either positive or negative contributions to the rate of heat flow from the reactor, depending on the relative magnitudes of T, T_0, and T_a.

7.1. Equations for mass and energy balance

The differential rate equations for the concentration of A and for the temperature within the reactor (the mass- and heat-balance equations) can be written in the form

$$\frac{da}{dt} = \frac{a_0 - a}{t_{res}} - k(T)a \tag{7.3}$$

$$c_p \sigma \frac{dT}{dt} = c_p \sigma \frac{(T_0 - T)}{t_{res}} + Qk(T)a - \frac{\chi S}{V}(T - T_a). \tag{7.4}$$

| rate of change | net inflow | reaction | Newtonian cooling |

Here c_p is the heat capacity ($JK^{-1}kg^{-1}$) and σ the total density ($kg\,m^{-3}$). If we divide the heat-balance equation throughout by $c_p\sigma$ the Newtonian cooling time $t_N = c_p\sigma V/\chi S$ emerges naturally in the denominator of the last term, as does the group $Q/c_p\sigma$ which is related to the adiabatic temperature rise appropriate to the system, $\Delta T_{ad} = Qa_0/c_p\sigma$, in the second term.

7.1.1. Dimensionless forms

The most obvious choice for dimensionless concentration is that used in the previous chapter, i.e.

$$\alpha = a/a_0 \tag{7.5}$$

so α varies between 0 (complete consumption) and 1 (no consumption). Again we will sometimes wish to present results in terms of the extent of reaction, which is given by $1 - \alpha$.

For temperature, we can again recognize that we are more interested in the temperature rise rather than in T itself. We can use the same dimensionless temperature excess θ as that introduced in chapters 4 and 5, with a slight modification. Instead of basing θ on the ambient temperature T_a we will base it on the inflow temperature T_0, so

$$\theta = \frac{E}{RT_0^2}(T - T_0). \tag{7.6}$$

In a similar way, the maximum temperature rise possible in the system (the adiabatic temperature rise ΔT_{ad}) has a dimensionless measure θ_{ad} given by

$$\theta_{ad} = \frac{E}{RT_0^2}\Delta T_{ad} = \frac{Qa_0 E}{c_p \sigma RT_0^2}. \tag{7.7}$$

The difference between the inflow and heat-bath temperatures can also be expressed in dimensionless terms as θ_c:

$$\theta_c = \frac{E}{RT_0^2}(T_a - T_0). \tag{7.8}$$

We can identify a chemical timescale t_{ch} from the value of the first-order rate constant k evaluated at our reference temperature T_0:

$$t_{ch} = 1/k(T_0) \tag{7.9}$$

and then introduce the groups

$$\tau = t/t_{ch} \quad = k(T_0)t \tag{7.10}$$

$$\tau_{res} = t_{res}/t_{ch} = k(T_0)t_{res} \tag{7.11}$$

$$\tau_N = t_N/t_{ch} \quad = c_p \sigma V k(T_0)/\chi S. \tag{7.12}$$

Thus τ_{res} is the dimensionless residence time, as in the previous chapter, and τ_N is the dimensionless Newtonian cooling time. High values of τ_N correspond to slow heat transfer across the reactor walls, indicating well-insulated vessels which approach adiabatc operation as τ_N tends to infinity. Small values of τ_N correspond to systems which have fast heat transfer and hence which would be expected not to have great departures from isothermal operation.

Finally, we need a measure of the activation energy

$$\gamma = RT_a/E \tag{7.13}$$

which, as in chapters 4 and 5, we would expect to be a relatively small quantity.

7.1.2. Dimensionless rate equations

The full dimensionless forms of the mass- and energy-balance equations are thus

$$\frac{d\alpha}{d\tau} = \frac{1 - \alpha}{\tau_{res}} - \alpha \exp\left(\frac{\theta}{1 + \gamma\theta}\right) \tag{7.14}$$

$$\frac{d\theta}{d\tau} = \theta_{ad}\, \alpha \exp\left(\frac{\theta}{1+\gamma\theta}\right) - \left(\frac{1}{\tau_{res}} + \frac{1}{\tau_N}\right)\theta + \frac{\theta_c}{\tau_N}. \qquad (7.15)$$

The exponential terms in these two equations correspond to the Arrhenius temperature dependence of the rate constant.

These equations describe the time dependence of the two variables—the concentration α and the temperature rise θ—in terms of five parameters. Of the latter, the residence time is again the one most easily varied during a given experiment, so we look to plot the stationary-state values of α and θ against τ_{res}. The remaining four parameters are θ_{ad}, τ_N, θ_c, and γ, and we might look to vary these independently between experiments.

Rather than tackle the full problem immediately, it will help to start with some simplification of the dimensionless equations and then build up to the full complexity in stages. At the first level of approximation we will make use of the typically small value of γ and replace the full Arrhenius dependence by the exponential form e^θ. We also begin with those systems for which the inflow and ambient temperatures are the same, so $\theta_c = 0$.

The mass- and energy-balance equations then become

$$\frac{d\alpha}{d\tau} = \frac{1-\alpha}{\tau_{res}} - \alpha e^\theta \qquad (7.16a)$$

$$\frac{d\theta}{d\tau} = \theta_{ad}\, \alpha e^\theta - \left(\frac{1}{\tau_{res}} + \frac{1}{\tau_N}\right)\theta. \qquad (7.16b)$$

7.2. Stationary states and flow diagrams with exponential approximation

Equations (7.16a) and (7.16b) correspond to our single first-order exothermic reaction occurring in a CSTR fed by reactants at the oven temperature, with the exponential approximation made to the Arrhenius temperature dependence of the reaction rate constant. Stationary-state solutions correpond to values of the dimensionless concentration α and temperature rise θ for which $d\alpha/d\tau$ and $d\theta/d\tau$ are simultaneously equal to zero, i.e.

$$\frac{1-\alpha_{ss}}{\tau_{res}} - \alpha_{ss} e^{\theta_{ss}} = 0 \qquad (7.17)$$

$$\theta_{ad}\, \alpha_{ss} e^{\theta_{ss}} - \left(\frac{1}{\tau_{res}} + \frac{1}{\tau_N}\right)\theta_{ss} = 0. \qquad (7.18)$$

Dividing eqn (7.18) throughout by the dimensionless adiabatic temperature rise θ_{ad} and then adding these two conditions gives the stationary-state relationship between the concentration and the temperature rise:

$$1 - \alpha_{ss} = \left(1 + \frac{\tau_{res}}{\tau_N}\right) \frac{\theta_{ss}}{\theta_{ad}}. \tag{7.19}$$

The left-hand side of this equation is the extent of reaction: the quotient θ_{ss}/θ_{ad} on the right-hand side is the fractional temperature rise compared with the maximum possible (adiabatic temperature rise).

At the extreme of *perfect adiabatic operation* ($\chi \to 0$ so τ_N becomes infinite and no heat is lost through the reactor walls), then eqn (7.19) reduces to

$$1 - \alpha_{ss} = \theta_{ss}/\theta_{ad}. \tag{7.20}$$

This says simply that at complete reaction ($\alpha_{ss} \to 0$), the temperature excess attains its adiabatic value: for 50 per cent reaction, $\theta_{ss} = \frac{1}{2}\theta_{ad}$ and so on. In fact, in this special adiabatic case the relationship between the extent of conversion and the temperature rise implied by eqn (7.20) holds for all conditions and not just at the stationary state. We will examine the adiabatic behaviour in some detail later, but first return to the more general case.

The extra factor $1 + (\tau_{res}/\tau_N)$ in eqn (7.19) for non-adiabatic operation allows for the transfer of heat through the vessel walls as well as by outflow. As Newtonian heat transfer becomes more important (τ_N decreases) or at long residence times (large τ_{res}), the value of this factor increases. Consequently, the fraction of the adiabatic temperature rise achieved for a given extent of reaction decreases.

Substituting for θ_{ss} in eqn (7.17) from (7.19) gives the stationary-state condition in terms of the concentration:

$$\frac{1 - \alpha_{ss}}{\tau_{res}} - \alpha_{ss} \exp\left(\frac{(1 - \alpha_{ss})\theta_{ad}}{1 + (\tau_{res}/\tau_N)}\right) = 0. \tag{7.21}$$

$$\qquad L \qquad\qquad\qquad\qquad R$$

As in the previous chapter we can identify a flow term L and a reaction term R. Figure 7.2 shows an example of the corresponding flow diagram in which R and L are plotted as a function of the stationary-state extent of reaction $1 - \alpha_{ss}$ for particular values of τ_{res}, τ_N, and θ_{ad}.

The flow term L is again a linear function of $1 - \alpha_{ss}$, with a gradient given by $1/\tau_{res}$. The reaction curve R has some similarities in shape with the cubic curve of the autocatalytic system in the previous chapter, although in Fig. 7.2 the maximum is sharper and lies closer to complete conversion. Intersections between R and L again give stationary-state solutions, and the tangencies between these two curves also locate ignition and extinction points. In general, now, both R and L depend on the residence time. Any flow diagram interpretation is complicated by the variations of both curves relative to each

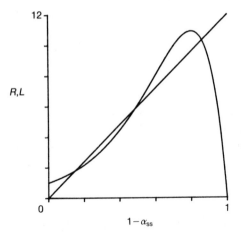

FIG. 7.2. Thermal or flow diagram for the first-order non-isothermal reaction (FONI) in a non-adiabatic CSTR: the rate curve R and the flow line L both depend on the dimensionless residence time, but their intersections still correspond to stationary-state solutions—and tangencies to points of ignition or extinction. Note that R has a non-zero value at zero conversion. Exact numerical values correspond to $\theta_{ad} = 10$, $\tau_{res} = \tau_N = \frac{1}{12}$.

other as τ_{res} is altered. Again, however, we find a simplification in the limit of adiabatic operation, for which R has the simpler form

$$R_{ad} = \alpha_{ss} \exp[\theta_{ad}(1 - \alpha_{ss})] \qquad (7.22)$$

and hence does not depend on the residence time.

7.2.1. Adiabatic operation: tangency approach

For a perfectly insulated reactor, with no heat loss through the walls, the Newtonian cooling time τ_N becomes infinite (because $\chi \to 0$). The mass- and heat-balance equations become

$$\frac{d\alpha}{d\tau} = \frac{1 - \alpha}{\tau_{res}} - \alpha e^{\theta} \qquad (7.23)$$

$$\frac{d\theta}{d\tau} = \theta_{ad}\alpha e^{\theta} - \frac{\theta}{\tau_{res}}. \qquad (7.24)$$

By dividing eqn (7.24) by θ_{ad}, adding the equations, and then integrating, we find that with suitable initial conditions eqn (7.20) holds for all times and not just at stationary states, i.e. for adiabatic systems

$$1 - \alpha = \theta/\theta_{ad}. \qquad (7.25)$$

This relationship between the extent of conversion and the temperature rise means that there is really only one independent variable in the system, i.e. we can reduce it to a single rate equation such as

$$\frac{d\alpha}{d\tau} = \frac{1 - \alpha}{\tau_{res}} - \alpha \exp[\theta_{ad}(1 - \alpha)]. \tag{7.26}$$

This reduction to a single variable is similar to that possible in the case of cubic autocatalysis when the species B is infinitely stable (i.e. $k_2 \to 0$) in chapter 6. In fact there are many qualitative parallels between the adiabatic non-isothermal reaction and autocatalysis without decay, as we shall see later.

The stationary-state condition becomes

$$\underbrace{\frac{1 - \alpha_{ss}}{\tau_{res}}}_{L} - \underbrace{\alpha_{ss} \exp[\theta_{ad}(1 - \alpha_{ss})]}_{R} = 0. \tag{7.27}$$

The flow diagram technique works very well for this adiabatic case. This is because, as noted above, the reaction curve R does not vary with the residence time, whereas the gradient of the flow line L does. The condition for an ignition or an extinction point is that R and L should become tangential, as shown in Fig. 7.3. We should simultaneously satisfy

$$R = L \quad \text{and} \quad dR/d(1 - \alpha_{ss}) = dL/d(1 - \alpha_{ss})$$

where R is given by eqn (7.22). Operating on these conditions yields the following expression for the extent of reaction at ignition or extinction, in terms of the dimensionless temperature excess:

$$(1 - \alpha_{ss})_{\pm} = \tfrac{1}{2}[1 \pm (1 - 4\theta_{ad}^{-1})^{1/2}] \tag{7.28}$$

with the upper root corresponding to extinction and the lower to ignition. The corresponding values of the residence time can then be obtained by rearranging eqn (7.27) and substituting from (7.28):

$$\tau_{res} = \frac{1 - \alpha_{ss}}{\alpha_{ss}} \exp[-\theta_{ad}(1 - \alpha_{ss})]. \tag{7.29}$$

For example, if θ_{ad} has the value 7.2, then the upper and lower roots of eqn (7.28) are $(1 - \alpha_{ss})_{+} = \tfrac{5}{6}$ and $(1 - \alpha_{ss})_{-} = \tfrac{1}{6}$. These then give: for extinction, $\alpha_{ss} = \tfrac{1}{6}$, $\tau_{res} = 5e^{-6} \approx 0.012$; and for ignition, $\alpha_{ss} = \tfrac{5}{6}$, $\tau_{res} = \tfrac{1}{5}e^{-6/5} \approx 0.06$.

In between these tangencies, the curves R and L have three intersections, so the system has multiple stationary states (Fig. 7.3(b)). We see the characteristic S-shaped curve, with a hysteresis loop, similar to that observed with cubic autocatalysis in the absence of catalyst decay (§ 4.2).

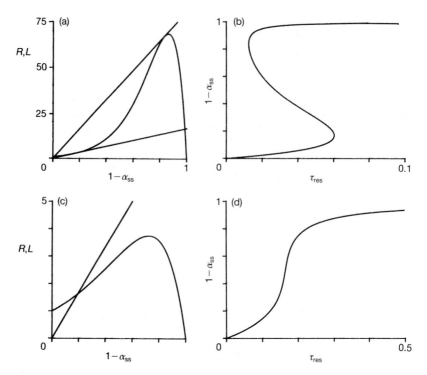

FIG. 7.3.   The existence and loss of multiplicity as the dimensionless adiabatic temperature θ_{ad} decreases through 4: (a) flow diagram for $\theta_{ad} = 7.2$; (b) corresponding stationary-state locus $(1 - \alpha_{ss})$–τ_{res}; (c) flow diagram for $\theta_{ad} = 3.6$; (d) corresponding stationary-state locus.

Equation (7.28) only has real roots, and hence ignition and extinction can only occur if the discriminant under the square root sign is positive. This then makes a requirement on the size of the dimensionless adiabatic temperature excess θ_{ad}. In particular, tangency only occurs if the reaction is sufficiently exothermic such that

$$\theta_{ad} \geqslant 4. \tag{7.30}$$

If the reaction is not sufficiently exothermic, so this inequality is not satisfied, then the reaction curve R is much flatter, as shown in Fig. 7.3(c). There are now no points of tangency as the residence time, and hence the gradient of L, varies. There is only ever one intersection and hence only ever one stationary state for any given τ_{res} (Fig. 7.3(d)).

7.2.2.   Further comparisons between non-isothermal and autocatalytic systems

The condition on the size of the dimensionless adiabatic temperature rise is similar to the condition for multiple stationary states in terms of the dimensionless autocatalyst inflow concentration $\beta_0 \leqslant \frac{1}{8}$ in §6.2.4. Other qualitative similarities between the present exothermic reaction in an adiabatic system and the cubic autocatalysis with the decay reaction having zero rate have been mentioned. In each case the reaction curve R has the same basic form, and there is a parameter θ_{ad} or β_0 whose value affects the curvature of R. Table 7.1 lists some of the properties of the two reaction curves for the two systems.

The first set of entries gives the values of the reaction rate at various significant points: R_{in} is the value at zero extent of reaction; R_{max} is the maximum value attained along the curve and R_{infl} is the value at the

Table 7.1

Comparisons between properties of the reaction rate curve R for cubic autocatalysis without decay and the first-order non-isothermal system in an adiabatic reactor

	Cubic autocatalysis	Non-isothermal
General features:		
$R_{in} = R_{(1-\alpha)=0}$	β_0^2	1
R_{max}	$\frac{4}{27}(1 + \beta_0)^3$	$\theta_{ad}^{-1}\exp(\theta_{ad} - 1)$
R_{infl}	$\frac{2}{27}(1 + \beta_0)^3$	$2\theta_{ad}^{-1}\exp(\theta_{ad} - 2)$
R_{infl}/R_{max}	$\frac{1}{2}$	$2e^{-1}$
R_{ign}	$\frac{1}{16}[1 + 2\beta_0 + 28\beta_0^2$	$\frac{1}{2}[1 + (1 - 4\theta_{ad}^{-1})^{1/2}]$
	$- (1 + 6\beta_0 - 4\beta_0^2)(1 - 8\beta_0)^{1/2}]$	$\times \exp\{\frac{1}{2}\theta_{ad}[1 - (1 - 4\theta_{ad}^{-1})^{1/2}]\}$
R_{ext}	$\frac{1}{16}[1 + 2\beta_0 + 28\beta_0^2$	$\frac{1}{2}[1 - (1 - 4\theta_{ad}^{-1})^{1/2}]$
	$+ (1 + 6\beta_0 - 4\beta_0^2)(1 - 8\beta_0)^{1/2}]$	$\times \exp\{\frac{1}{2}\theta_{ad}[1 + (1 - 4\theta_{ad}^{-1})^{1/2}]\}$
Asymptotic forms for small β_0 or large θ_{ad}:		
R_{ign}	$4\beta_0^2(1 + \beta_0 + \cdots)$	$\exp(1 + \frac{1}{2}\theta_{ad}^{-2} + \cdots)$
R_{ext}	$\frac{1}{8}(1 + 2\beta_0 - \cdots)$	$\theta_{ad}^{-1}\exp(\theta_{ad}^{-1})(1 - \theta_{ad}^{-2} - \cdots)$
R_{ext}/R_{max}	$\frac{27}{32}(1 - \beta_0 - \cdots)$	$(1 - \theta_{ad}^{-2} - \cdots)$
R_{ign}/R_{in}	$4(1 + \beta_0 + \cdots)$	$\exp(1 + \frac{1}{2}\theta_{ad}^{-2} + \cdots)$
R_{ign}/R_{max}	$27\beta_0^2(1 - 2\beta_0 - \cdots)$	$\theta_{ad}\exp(2 - \theta_{ad})(1 + \frac{1}{2}\theta_{ad}^{-2} + \cdots)$
Transition or loss of criticality:		
$(\beta_0, \theta_{ad})_{tr}$	$\beta_{0,tr} = \frac{1}{8}$	$\theta_{ad,tr} = 4$
R_{in}	$\frac{1}{64}$	1
$R_{ign} = R_{ext} = R_{infl}$	$3^3/4^4$	$\frac{1}{2}e^2$
R_{ign}/R_{in}	$\frac{27}{4}$	$\frac{1}{2}e^2$
R_{max}	$2(3^3/4^4)$	$\frac{1}{4}e^3$

inflection point; R_{ign} and R_{ext} are the values at the points of ignition and extinction respectively. In each case the corresponding forms are not dissimilar, and the equivalent roles of β_0 and $1/\theta_{ad}$ begin to emerge. We have seen that β_0 must be small for multiple stationary states and that θ_{ad} must be large (so $1/\theta_{ad}$ will be small). The second set of entries in Table 7.1 gives the leading-order terms for the various measures of R in the limit $\beta_0 \ll 1$, $\theta_{ad} \gg 1$. The third set of entries correspond to the conditions at which tangency and, hence, multiple stationary states just cease to exist. There the inflection point and the ignition and extinction tangents all merge together.

7.2.3. Full Arrhenius rate law

The multiplicity analysis described above can be extended to cope with the exact form of the Arrhenius temperature dependence (non-zero γ). The stationary-state condition has a slightly more cumbersome form:

$$\frac{1 - \alpha_{ss}}{\tau_{res}} - \alpha_{ss} \exp\left(\frac{\theta_{ad}(1 - \alpha_{ss})}{1 + \gamma\theta_{ad}(1 - \alpha_{ss})}\right) = 0 \qquad (7.31)$$

$$L \qquad\qquad\qquad R$$

but again the procedure of locating the tangencies of R and L leads to a quadratic for the extent of reaction at the points of ignition and extinction:

$$(1 - \alpha_{ss})_{\pm} = \frac{1 - 2\gamma \pm [1 - 4(\gamma + \theta_{ad}^{-1})]^{1/2}}{2(1 + \gamma^2\theta_{ad})}. \qquad (7.32)$$

The corresponding residence times can be obtained by substituting this result into eqn (7.31). More importantly, the requirements on the values of the two parameters γ and θ_{ad} can be obtained by the condition that the term under the square root sign must be positive for real solutions. This gives

$$\theta_{ad}^{-1} + \gamma \leqslant \tfrac{1}{4}. \qquad (7.33)$$

This requires at least that θ_{ad} must be larger than 4 and that the group γ, which is inversely proportional to the activation energy, must be less than $\tfrac{1}{4}$. Multistability is thus favoured by high exothermicity and by a high temperature coefficient for the rate constant.

7.2.4. Non-adiabatic operation

For finite, rather than infinite, values of the dimensionless Newtonian cooling time, the stationary-state condition is given by eqn (7.21). Thus, even with the exponential approximation, both R and L involve the residence time. The correspondence between tangency and ignition or extinction still holds, but the resulting expression for $(1 - \alpha_{ss})_\pm$ etc. are not particularly amenable to algebraic manipulation. In the next section we will introduce the technique of singularity theory for stationary-state curves which will allow us to find the limits of multiple solutions etc. in a systematic way. For now we present the results appropriate to the present model.

With the exponential approximation ($\gamma \approx 0$) and the assumption that the inflow and ambient temperatures are equal, we have a stationary-state equation which links α_{ss} to τ_{res} and which involves two other 'unfolding' parameters, θ_{ad} and τ_N. Depending on the particular values of the last two parameters the $(1 - \alpha_{ss})$ versus τ_{res} locus has one of five possible qualitative forms. These different patterns are shown in Fig. 7.4 as unique, single hysteresis loop, isola, mushroom, and hysteresis loop plus isola. The five corresponding regions in the θ_{ad}–τ_N parameter plane are shown in Fig. 7.5. This parameter plane is divided into these regions by a straight line and a cusp, which cut each other at two points.

The straight line is given by the equation

$$\theta_{ad} = 4\left(1 + \frac{e^{-2}}{\tau_N}\right). \tag{7.34}$$

It cuts the axis at $\theta_{ad} = 4$ as $1/\tau_N$ tends to zero (adiabatic limit). We have already seen that this is the condition for transition from multiple stationary states (hysteresis loop) to unique solutions for adiabatic reactors, so the line is the continuation of this condition to non-adiabatic systems. Above this line the stationary-state locus has a hysteresis loop: this loop opens out as the line is crossed and does not exist below it. Thus, as heat loss becomes more significant ($1/\tau_N$ increases), the requirement on the exothermicity of the reaction for the hysteresis loop to exist increases.

It is, however, possible to have multiple stationary states even for parameter values lying below the hysteresis line. These multiplicities are associated with the cusp in the parameter plane. The equation describing the full cusp is most easily presented in the form

$$\theta_{ad} = \frac{1}{x^2(1 - x)} \tag{7.35}$$

$$\frac{1}{\tau_N} = \frac{(1 - x)^2}{x^2} \exp\left(\frac{1}{1 - x}\right). \tag{7.36}$$

The cusp is drawn out as x is varied between 0 and 1.

The cusp curve begins at the cusp point which has coordinates

$$\theta_{ad} = 6\tfrac{3}{4} \qquad 1/\tau_N = \tfrac{1}{4}e^3 \approx 5.02 \tag{7.37}$$

and thus starts above the hysteresis line. Both the upper and the lower sections of the cusp eventually cross the hysteresis line at points given by

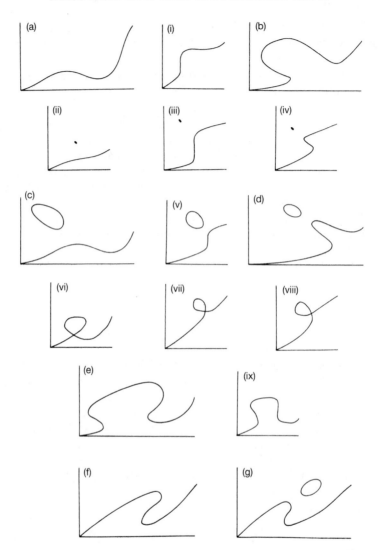

FIG. 7.4. The different qualitative forms for the stationary-state locus $(1 - \alpha_{ss})$–τ_{res} for the FONI model: (a) unique; (b) single hysteresis loop or breaking wave; (c) isola; (d) isola + hysteresis loop; (e) mushroom. With the full Arrhenius rate law and the provision pre-heating or cooling, two additional patterns are found; (f) reversed hysteresis loop; and (g) reversed hysteresis loop + isola. Also shown are various degenerate loci corresponding to parameter values on the boundaries or special points in the parameter plane (see Fig. 7.5).

$\theta_{ad} = 6.841$, $1/\tau_N = 5.249$ (lower branch), and by $\theta_{ad} = 8$, $1/\tau_N = e^2 \approx 7.39$ (upper branch).

At large values of $1/\tau_N$, which correspond to systems well away from the adiabatic limit, the upper branch tends to infinity and approaches the straight

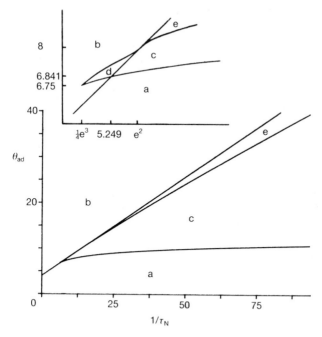

FIG. 7.5.   The θ_{ad}–τ_N^{-1} parameter plane showing the hysteresis line and the isola cusp described by eqns (7.34)–(7.36). The plane is divided into five regions (see inset for details in vicinity of cusp point) corresponding to the qualitative forms in Fig. 7.4. The numerical values are appropriate to the exponential approximation, $\gamma = 0$, but the qualitative form of the diagram holds for all $\gamma < \frac{1}{4}$.

line $\theta_{ad} \approx e^{-1}/(4\tau_N)$. Thus the limiting gradient of this branch in Fig. 7.5 is $\frac{1}{4}e^{-1} \approx 0.092$, which is less steep than the hysteresis line and which from eqn (7.34) has a slope of $4e^{-2} \approx 0.54$. The lower branch of the cusp also tends to infinity with $1/\tau_N$, but more slowly than the upper branch, tending to $1/\tau_N \approx \theta_{ad}^{-2}e^{1/\theta_{ad}}$.

If the parameters θ_{ad} and τ_N are varied so that the system crosses the lower boundary of the cusp, then an isola appears in the stationary-state locus. The isola has zero size exactly on the cusp branch but grows from this isolated point as we move further into the region. The isola can lie above either a unique curve (region c) or a hysteresis loop (region d). If we now continue and cross the upper boundary of the cusp, the isola grows and touches the branch of stationary states below it, and then opens out. Crossing from region c to region e thus gives rise to a mushroom pattern: going from region d to region b, the isola joins on to the end of the hysteresis loop, forming a larger loop. Some of the loci corresponding to parameter values on the various boundaries or at special points are also shown in Fig. 7.4.

7.2.5 Full Arrhenius form in non-adiabatic reactor

Figure 7.5 is quantitatively correct only for the special case of the exponential approximation to the Arrhenius rate law. However, the figure is also qualitatively correct for the exact Arrhenius form with non-zero γ, provided $\gamma < \frac{1}{4}$. No new stationary-state patterns are introduced.

The hysteresis line is given by

$$\theta_{ad} = \frac{4e^{-2}}{\tau_N} + \frac{4}{1 - 4\gamma}. \tag{7.38}$$

The line retains the same slope as that given above, but its intercept moves up the θ_{ad} axis as γ increases, tending to infinity as γ approaches $\frac{1}{4}$.

The equation for the cusp is slightly more complex and is again most easily expressed parametrically. The appropriate values for the adiabatic temperature excess must be obtained from a quadratic equation before it can be used to determine τ_N. Thus, for any given $\gamma < \frac{1}{4}$, the cusp is described by

$$\gamma^2 x^4 \theta_{ad}^2 - x^2(1 - x - 2\gamma)\theta_{ad} + 1 = 0 \tag{7.39}$$

$$\frac{1}{\tau_N} = \frac{(1 - x)^2}{x^2} \exp\left(\frac{\theta_{ad} x^2}{1 + \gamma \theta_{ad} x^2}\right) \tag{7.40}$$

with x taking values in the range $0 \leqslant x \leqslant 1 - 4\gamma$, so this range tends to zero as γ tends to $\frac{1}{4}$.

7.2.6. Inflow and ambient temperatures

If the temperature difference θ_c between the heat bath and the inflow is greater than zero, we can have the opposite effect to Newtonian cooling, with a net flow of heat into the reactor through the walls. With his possibility, two more stationary-state patterns can be observed, giving a total of seven different forms—the same seven seen before in cubic autocatalysis with the additional uncatalysed step (the two new patterns then required negative values for the rate constant) or with reverse reactions included and $c_0 > \frac{1}{2}a_0$ (§6.6).

By now we may reasonably begin to wonder whether the continued appearance of these seven patterns is typical in a wider context and stems from some cause common to all these models. We will see that this is indeed so.

7.3. Singularity theory approach to stationary-state loci

In this chapter, and in the previous one, we have frequently been faced with a stationary-state condition in the form of a polynomial or some other non-linear algebraic equation. Typical examples are eqns (6.30), (6.51), (6.55), (6.74), and (7.21). In all these cases, the condition can be written in the general form

$$F(x, \tau_{res}; p, q, r, s, \ldots) = 0. \tag{7.41}$$

Here F represents the functional form of the left-hand sides of the various stationary-state equations, x is the stationary-state solution such as the extent of reaction, the temperature excess, etc., and τ_{res} is the parameter we have singled out as the one which can be varied during a given experiment (the 'distinguished' or 'bifurcation' parameter). All the remaining parameters are represented by $p, q, r, s, \ldots$. For example, in eqn (7.21) the role of x could be played by the extent of reaction $1 - \alpha_{ss}$, with $p = \theta_{ad}$ and $q = \tau_N$: for isothermal autocatalysis, x can again be the extent of reaction, with $p = \beta_0$, $q = \kappa_2$, and $r = \kappa_u$.

Our general interest has been to find the conditions, in terms of the extra 'unfolding' parameters p, q, r, etc., at which the qualitative nature of the stationary-state locus changes (e.g. the appearance or disappearance of a hysteresis loop or an isola). In some cases we have been able to make use of special techniques such as factorization or the tangency condition. Now we seek a more widely applicable approach. This will involve the stationary-state condition $F = 0$ and also a series of equations obtained by differentiation of this expression with respect to the variable x and the parameter τ_{res}.

For instance, if we consider the simple case of the adiabatic non-isothermal CSTR, the stationary-state condition is given by eqn (7.27). Writing x for the extent of reaction, we have

$$F = \frac{x}{\tau_{res}} - (1 - x)\exp(\theta_{ad}x). \tag{7.42}$$

The partial derivative expressions we require will be of the form $\partial F/\partial x$, $\partial F/\partial \tau_{res}$, $\partial^2 F/\partial x \partial \tau_{res}$, etc. For the above equation these are simply evaluated:

$$\frac{\partial F}{\partial x} = F_x = \frac{1}{\tau_{res}} - [\theta_{ad}(1 - x) - 1]\exp(\theta_{ad}x) \tag{7.43}$$

$$\frac{\partial F}{\partial \tau_{res}} = F_\tau = -\frac{x}{\tau_{res}^2} \tag{7.44}$$

$$\frac{\partial^2 F}{\partial x \partial \tau_{res}} = F_{x\tau} = -\frac{1}{\tau_{res}^2}. \tag{7.45}$$

(Note that we use the shorthand notation F_x etc. Also for convenience, we denote $\partial F/\partial \tau_{res}$ by F_τ to save repeated subscripts.)

Of the different possible qualitative changes which can occur in the stationary-state locus, we will concentrate on three examples—those which we have seen before: (i) appearance/disappearance of a hysteresis loop, (ii) appearance of an isola, and (iii) growth of an isola into a mushroom. It will turn out that (ii) and (iii) emerge together.

7.3.1. Condition for appearance/disappearance of a hysteresis loop

We have seen examples of a hysteresis loop being unfolded, opening out to leave a monotonic stationary-state locus. The conditions for this change to occur can be written formally as

$$F = F_x = F_{xx} = 0 \tag{7.46}$$

so the stationary-state condition must be satisfied ($F = 0$) and the first two partial derivatives of F with respect to the stationary-state solution must also vanish. In addition, there is the constraint that a number of the higher derivatives must remain non-zero, so F_τ, $F_{x\tau}$, and F_{xxx} must all be non-zero.

As the simplest example, let us consider the isothermal cubic autocatalysis of §6.2, where there is catalyst inflow but no decay. In terms of the extent of conversion $x = 1 - \alpha_{ss}$, the appropriate stationary-state condition is

$$F(x, \tau_{res}; \beta_0) = \frac{x}{\tau_{res}} - (1 - x)(x + \beta_0)^2. \tag{7.47}$$

Then

$$F_x = \frac{1}{\tau_{res}} - (x + \beta_0)(2 - 3x - \beta_0) \tag{7.48}$$

and

$$F_{xx} = 2(3x + 2\beta_0 - 1). \tag{7.49}$$

We require all three right-hand sides to equal zero simultaneously. These three equalities fix the values of the three quantities x, τ_{res}, and β_0, at which the hysteresis loop just unfolds, uniquely. We find $\beta_0 = \frac{1}{8}$, which is the result found by more traditional analysis in §6.2.4, with the vertical inflection point in the stationary-state locus occurring at $x = \frac{1}{4}$ and $\tau_{res} = 64/27$.

We can also check the higher derivatives $F_\tau = -x/\tau_{res}^2 = -3^6/4^9$ and $F_{x\tau} = -1/\tau_{res}^2 = -3^6/4^8$, and $F_{xxx} = 3$, all of which are non-zero.

For all β_0 less than $\frac{1}{8}$, the ignition and extinction points in the x–τ_{res} locus are determined by the solutions of

$$F = F_x = 0 \quad \text{with} \quad F_{xx} \neq 0. \tag{7.50}$$

To see why this is a general recipe for locating turning points in the locus we can argue as follows. At *any* point along the locus, the function F has the same constant value (zero). Thus the derivative of F along the stationary-state curve, like the derivative of any constant, must be zero: $dF = 0$. We can write the total derivative dF in terms of the increments in x and τ_{res}:

$$dF = \left(\frac{\partial F}{\partial x}\right)dx + \left(\frac{\partial F}{\partial \tau_{res}}\right)d\tau_{res}$$

$$= F_x dx + F_\tau d\tau_{res} = 0. \tag{7.51}$$

The second equation can be rearranged to give the gradient of the stationary-state locus at any point along the curve, $dx/d\tau_{res}$:

$$dx/d\tau_{res} = -F_\tau/F_x. \tag{7.52}$$

As an example we can again take our cubic autocatalysis with inflow of B, so F_x and F_τ are as given above, leading to

$$\frac{dx}{d\tau_{res}} = \frac{x}{\tau_{res} - \tau_{res}^2(x + \beta_0)(2 - 3x - \beta_0)}. \tag{7.53}$$

This expression only gives the gradient of the locus provided the x and τ_{res} values actually correspond to stationary-state solutions, i.e. provided they satisfy $F = 0$. Then the gradient varies as x and τ_{res} vary along the locus.

If F_x becomes small with F_τ remaining finite, then the gradient of the stationary-state locus will become large in magnitude. In the limit $F_x \to 0$ the gradient becomes infinite and there is a vertical section of the locus.

If we turn the locus on its side (Fig. 7.6), and consider it as a plot of τ_{res} as a function of the extent of reaction x, then the gradient would be

$$d\tau_{res}/dx = -F_x/F_\tau. \tag{7.54}$$

The condition $F_x = 0$ (with $F_{xx} \neq 0$) then corresponds to a maximum or

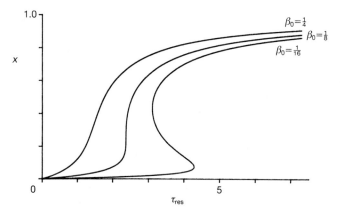

FIG. 7.6.   The unfolding of a hysteresis loop as condition (7.46) is satisfied. For the FONI model this occurs as θ_{ad} decreases through 4: for cubic autocatalysis the unfolding corresponds to β_0 increasing through $\frac{1}{8}$.

a minimum. Turning the locus back round again these extrema become vertical turning points, corresponding to ignition and extinctions. If F_{xx} is also equal to zero when F_x vanishes, then the two extrema become a point of inflection. This corresponds to the disappearance of a hysteresis loop and hence is the origin of the recipe in eqn (7.46).

7.3.2. Conditions for appearance of isola and growth to mushroom

The requirements for the two changes in the stationary-state locus associated with the appearance of an isola as an isolated point and the birth of a mushroom as an isola grows to touch a lower branch of solutions both have the form

$$F = F_x = F_\tau = 0 \qquad (7.55)$$

with the additional requirements

$$F_{x\tau} \neq 0 \qquad F_{xx} \neq 0 \qquad F_{\tau\tau} \neq 0. \qquad (7.56)$$

In terms of the gradient of the stationary-state locus, the condition expressed by eqn (7.55) locates a singular point as $dx/d\tau_{res} \rightarrow 0/0$. Loosely, we can think of the locus simultaneously having turning points both horizontally and vertically, as indicated in Fig. 7.7. This corresponds to situations such as (a) an isolated point or (b) the 'transcritical' touching of two branches.

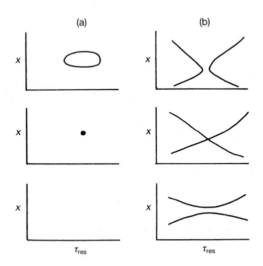

FIG. 7.7.   Representations of the two forms for the isola bifurcation which occurs as condition (7.55) becomes satisfied: (a) birth or death of an isola from an isolated point; (b) transition between mushroom and isola patterns.

For the simple autocatalytic model without decay, used as an example above, F_τ does not vanish, so isola and mushroom patterns are not possible (confirming our previous results). If we take the model of §6.4, where the decay rate constant κ_2 is non-zero, then from eqn (6.61) we obtain

$$F = \frac{(1 + \kappa_2 \tau_{res})^2}{\tau_{res}} x - (1 - x)(\beta_0 + x)^2 \tag{7.57}$$

so that

$$F_x = \frac{(1 + \kappa_2 \tau_{res})^2}{\tau_{res}} - (\beta_0 + x)(2 - 3x - \beta_0) \tag{7.58}$$

$$F_\tau = \frac{(1 + \kappa_2 \tau_{res})(\kappa_2 \tau_{res} - 1)}{\tau_{res}^2} x. \tag{7.59}$$

The higher derivatives are

$$F_{xx} = 2(3x + 2\beta_0 - 1) \tag{7.60}$$

$$F_{x\tau} = \frac{(1 + \kappa_2 \tau_{res})(\kappa_2 \tau_{res} - 1)}{\tau_{res}^2} \tag{7.61}$$

$$F_{\tau\tau} = \frac{(1 + \kappa_2 \tau_{res})(\kappa_2 \tau_{res} - 1)}{\tau_{res}^4} x. \tag{7.62}$$

Applying condition (7.55), we have three equalities. Two of these may be used to eliminate x and τ_{res}, the third then gives a relationship between the unfolding parameters κ_2 and β_0 (i.e. it gives an equation for the boundaries in the β_0–κ_2 plane at which the isola and mushroom transitions occur). In this way we find

$$x = \tfrac{1}{4}[1 \pm (1 - 8\beta_0)^{1/2}] \qquad \tau_{res} = \kappa_2^{-1}$$

and then

$$\kappa_2 = \tfrac{1}{32}[1 + 20\beta_0 - 8\beta_0^2 \pm (1 - 8\beta_0)^{3/2}]. \tag{7.63}$$

This gives the same result as the analysis of §6.45 (eqn (6.69)). As a slight word of warning, however, we should also note that whenever $F_\tau = 0$ then for the present scheme $F_{x\tau}$ will also be identically zero. This coincidence is rather unusual and is the cause of some subtlety in the behaviour of this particular model.

7.3.3. Highest-order singularities and unfoldings

The cusp described by eqn (7.60) meets the hysteresis line $\beta_0 = \tfrac{1}{8}$ at the point $\beta_0 = \tfrac{1}{8}$, $\kappa_2 = 27/256$, when simultaneously

$$F = F_x = F_{xx} = F_\tau = F_{x\tau} = 0. \tag{7.64}$$

Because all these derivatives (and the stationary-state condition) vanish at the same time, this point is highly singular. In fact, as no more of the higher derivatives can vanish, this is the highest-order singularity which occurs in the present model. An additional facet of singularity theory is that identifying the highest-order singularity also tells us how many, and which, patterns we may expect for the stationary-state locus.

The five equalities in eqn (7.64) above are characteristic of a 'winged cusp'. For the exact values of β_0 and κ_2 corresponding to the cusp point, the stationary-state locus has the form shown in Fig. 7.8(a). As the parameters are varied away from this cusp condition the locus is deformed and can take any of the seven qualitative forms shown in Figs 7.8(b–h). All of these patterns should, therefore, exist over some non-zero range of the parameters in the vicinity of the cusp point. In theory this 'unfolding' process requires that the model should have at least three unfolding parameters: p, q, and r in eqn (7.41). The cubic autocatalysis model, even with the decay step, only has two: this deficiency, and the coincidence that F_τ and $F_{x\tau}$ always vanish together, means that a full unfolding is not possible, and only three of the patterns arise.

For the model in the absence of decay, the highest-order singularity in the system is that for which

$$F = F_x = F_{xx} = 0. \tag{7.65}$$

This corresponds to an ordinary cusp, with an unfolding to a unique locus or a single hysteresis loop.

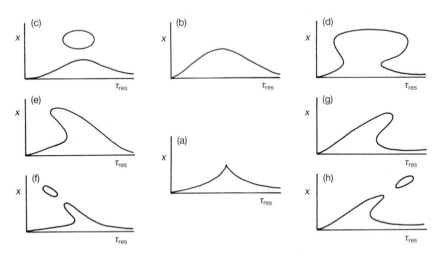

FIG. 7.8.   The full unfolding of the stationary-state locus (a) corresponding to a winged cusp singularity: (b) unique; (c) isola; (d) mushroom; (e) single hysteresis loop; (f) hysteresis loop + isola; (g) reverse hysteresis loop; (h) reverse hysteresis loop + isola.

Now that the recipes for locating the various changes in the qualitative form of the stationary-state locus have been presented, we can go on to examine the origin of the behaviour in the cubic autocatalytic system with the additional uncatalysed step, and for the non-adiabatic non-isothermal CSTR which has been asserted in previous sections.

7.4. Singularity theory for cubic autocatalysis with uncatalysed reaction

This version of the model, from §6.5, has three unfolding parameters: κ_2, β_0, and κ_u. If x again represents the stationary-state extent of conversion, $1 - \alpha_{ss}$, then the stationary-state condition (eqn (6.74)) can be written as

$$F(x, \tau_{res}; \kappa_2, \beta_0, \kappa_u) = \frac{(1 + \kappa_2 \tau_{res})^2}{\tau_{res}} x - (1 - x)[(\beta_0 + x)^2 + \kappa_u(1 + \kappa_2 \tau_{res})^2]$$

(7.66)

$$F_x = (1 + \kappa_2 \tau_{res})^2 \left(\frac{1}{\tau_{res}} + \kappa_u \right) - (\beta_0 + x)(2 - 3x - \beta_0)$$

(7.67)

$$F_{xx} = 2(3x + 2\beta_0 - 1)$$

(7.68)

$$F_\tau = (1 + \kappa_2 \tau_{res}) \left(\frac{\kappa_2 \tau_{res} - 1}{\tau_{res}^2} x - 2\kappa_u \kappa_2 (1 - x) \right)$$

(7.69)

$$F_{x\tau} = (1 + \kappa_2 \tau_{res}) \left(\frac{\kappa_2}{\tau_{res}} + 2\kappa_u \kappa_2 - \frac{1}{\tau_{res}^2} \right).$$

(7.70)

These five equations are all equal to zero simultaneously for one set of parameters only, i.e. for

$$x = \tfrac{1}{4} \qquad \tau_{res} = 4^4/3^3 \qquad \beta_0 = \tfrac{1}{8} \qquad \kappa_2 = 3^3/4^4 \qquad \kappa_u = 0. \quad (7.71)$$

The next highest partial derivatives, F_{xxx} etc., are all non-zero for these values, so the highest-order singularity is a winged cusp. With the three unfolding parameters, we expect to obtain all seven of the possible stationary-state patterns of Fig. 7.8. As we have seen previously, this is in fact the case, although those which appear only when κ_u is varied from zero to negative values will not have relevance for chemical systems.

We can find the hysteresis and isola boundaries from the above equations by applying conditions (7.46) and (7.55) respectively. In each case we have three equalities. Two of these can, in principle, be used to determine the values for x and τ_{res}. Eliminating these, the third equation then defines a surface in the $\kappa_2-\beta_0-\kappa_u$ space. Alternatively, we can also specify a particular value for, say, the uncatalysed rate constant κ_u and then use the third equation to give the boundaries in the $\kappa_2-\beta_0$ plane. This latter procedure has already been used in the special case $\kappa_u = 0$ in the previous section for the isola line (giving eqn (7.60)).

For the set of equations above, with non-zero κ_u, the algebra is not quite so rewarding, and explicit formulae such as eqn (7.60) cannot be obtained. Nevertheless, parametric expressions which can be readily used even with a non-programmable calculator (given patience) emerge.

(a) Hysteresis

For the onset or disappearance of a hysteresis loop, the condition $F = F_x = F_{xx} = 0$ can be manipulated to give the following set of equations:

$$\beta_0 = \tfrac{1}{2}(1 - 3x) \tag{7.72}$$

$$\tau_{res} = \frac{4x - 1}{4\kappa_u(1 - x)} \tag{7.73}$$

$$\kappa_2 = \frac{1 - x}{1 - 4x} \{4\kappa_u - [4\kappa_u(4x - 1)(1 - x)^2]^{1/2}\}. \tag{7.74}$$

Here, the extent of reaction plays a parametric role. For a given choice of κ_u, the hysteresis boundary is drawn out by eqns (7.72) and (7.74) as x varies. What range of values should x be varied over? Clearly, we wish the various quantities β_0, τ_{res}, and κ_2 to be positive. Equations (7.72) and (7.73) then limit x to the range $\tfrac{1}{4} \leqslant x \leqslant \tfrac{1}{3}$. In fact the lower limit can be further refined. The decay rate constant κ_2 becomes zero when

$$4\kappa_u = (4x - 1)(1 - x)^2. \tag{7.75}$$

This equation has a root lying between $\tfrac{1}{4}$ and $\tfrac{1}{3}$ provided κ_u is less than $\tfrac{1}{27}$. This root then gives the lower limit on the range of x. As κ_u tends to $\tfrac{1}{27}$, so the lower limit on x tends to $\tfrac{1}{3}$, which is the value at which β_0 becomes zero. Thus for $\kappa_u = \tfrac{1}{27}$, the hysteresis curve disappears from the physically acceptable region (positive quadrant) of the β_0–κ_2 parameter plane.

(b) Isola and mushroom

The equations for the isola boundaries, obtained from the condition $F = F_x = F_\tau = 0$, must also be expressed parametrically in terms of the extent of reaction and the uncatalysed reaction rate constant. These give

$$\tau_{res} = x \pm \left(x^2 - \frac{4\kappa_u[x - \kappa_u(1 - x)]^2}{(1 - x)^2[x - 2\kappa_u(1 - x)]^2}\right)^{1/2} \bigg/ 2\kappa_u(1 - x) \tag{7.76}$$

$$\kappa_2 = \frac{x}{\tau_{res}[x - 2\kappa_u(1 - x)]} \tag{7.77}$$

$$\beta_0 = x(1 - 2x) - 2(1 - x)^2 \kappa_u \tau_{res}. \tag{7.78}$$

These equations are easier to use than they are to analyse, but (7.77) shows that positive values for κ_2 require $x \geqslant 2\kappa_u/(1 + 2\kappa_u)$. Numerically we find that the boundaries disappear from the positive quadrant of the β_0–κ_2 parameter plane when κ_u is increased beyond about 0.005 12.

The loci described by eqns (7.72)–(7.74) and (7.76)–(7.78) have already been given for a variety of values of κ_u in the previous chapter (Figs 6.20 and 6.21).

7.5. Singularity theory for non-isothermal CSTR

We now turn to the non-isothermal reaction system in a non-adiabatic CSTR, as studied in §§ 7.2.4–6. We begin with the simplified model with exponential approximation to the Arrhenius law, and to systems for which the inflow and ambient temperatures are the same ($\gamma = 0$; and $\theta_c = 0$). This system has two unfolding parameters θ_{ad} and τ_N. The stationary-state equation and its various derivatives are

$$F(x, \tau_{res}; \theta_{ad}, \tau_N) = \frac{x}{\tau_{res}} - (1 - x)\exp\left(\frac{\theta_{ad}x}{1 + \tau_{res}/\tau_N}\right) \tag{7.79}$$

$$F_x = \frac{1}{\tau_{res}} + \left(1 - \frac{\theta_{ad}(1 - x)}{1 + \tau_{res}/\tau_N}\right)\exp\left(\frac{\theta_{ad}x}{1 + \tau_{res}/\tau_N}\right) \tag{7.80}$$

$$F_{xx} = \frac{\theta_{ad}}{1 + \tau_{res}/\tau_N}\left(2 - \frac{\theta_{ad}(1 - x)}{1 + \tau_{res}/\tau_N}\right)\exp\left(\frac{\theta_{ad}x}{1 + \tau_{res}/\tau_N}\right) \tag{7.81}$$

$$F_\tau = -\frac{x}{\tau_{res}^2} + \frac{(1 - x)}{\tau_N(1 + \tau_{res}/\tau_N)^2}\exp\left(\frac{\theta_{ad}x}{1 + \tau_{res}/\tau_N}\right) \tag{7.82}$$

$$F_{x\tau} = -\frac{1}{\tau_{res}^2} + \frac{1}{\tau_N(1 + \tau_{res}/\tau_N)^2}\left(\frac{\theta_{ad}(1 - x)}{1 + \tau_{res}/\tau_N} - 1\right)$$
$$\times \exp\left(\frac{\theta_{ad}x}{1 + \tau_{res}/\tau_N}\right). \tag{7.83}$$

Although these seem a formidable array of equations, they yield remarkably to elimination and simplification when various combinations are required to be zero simultaneously.

For the hysteresis limit we require, as usual, that $F = F_x = F_{xx} = 0$. Two equalities give x and τ_{res}. The third then leads to eqn (7.34) relating θ_{ad} to τ_N. The isola condition $F = F_x = F_\tau = 0$ is best handled parametrically, as x cannot be eliminated so readily (it is given by the solution of a cubic equation in terms of θ_{ad}). The parametric forms have been given as eqns (7.35) and (7.36).

The highest-order singularity in this system cannot satisfy the winged cusp condition $F = F_x = F_{xx} = F_\tau = F_{x\tau} = 0$. For one thing the system of equations only has four quantities x, τ_{res}, θ_{ad}, and τ_N, whereas the winged cusp singularity requires five equalities to be satisfied, making the problem over-determined.

With the full Arrhenius rate law, an extra unfolding parameter γ is introduced. Even then, however, the appropriate stationary-state condition and its derivatives for the winged cusp cannot be satisfied simultaneously (at least not for positive values of the various parameters). Thus we do not expect to find all seven patterns.

If we allow for the possibility of separate inflow and heat-bath temperatures, the system has four unfolding parameters, with θ_c augmenting the list. The full stationary-state equation is

$$F(x, \tau_{res}; \theta_{ad}, \tau_N, \gamma, \theta_c) = \frac{x}{\tau_{res}} - (1 - x)\exp\left(\frac{\theta_{ad}x + \theta_c\tau_{res}/\tau_N}{1 + \tau_{res}/\tau_N + \gamma(\theta_c\tau_{res}/\tau_N + x\theta_{ad})}\right)$$

(7.84)

with similarly more complex forms for the partial derivatives F_x etc.

The conditions for the hysteresis and isola boundaries on the $\theta_{ad}-\tau_N$ diagram can be obtained parametrically, although in the case of the isola curve we must solve a cubic equation for θ_{ad}. The expressions are: for *hysteresis*

$$\theta_{ad} = \frac{[1 - 4x(1 - x)]\theta_c}{4\gamma(1 + \gamma\theta_c)x(1 - x) - (1 - 2x)x}$$

(7.85)

$$\frac{1}{\tau_N} = \frac{(1 - x)}{x}\left(\frac{4\gamma^2\theta_{ad}x(1 - x)}{(1 - 2x)^2} - 1\right)\exp\left(\frac{1}{\gamma} - \frac{2}{1 - 2x}\right)$$

(7.86)

and for the *isola* curve we have to solve the cubic

$$\gamma^2 x^5 \theta_{ad}^3 - x^3(1 - x - \gamma^2 x\theta_c - 2\gamma)\theta_{ad}^2$$
$$+ x[1 - (1 - x)\theta_c - 2\gamma x\theta_c]\theta_{ad} - \theta_c = 0$$

(7.87)

$$\frac{1}{\tau_N} = \frac{\theta_{ad}(1 - x)^2}{(\theta_{ad}x^2 - \theta_c)}\exp\left(\frac{\theta_{ad}x^2}{1 + \gamma\theta_{ad}x^2}\right).$$

(7.88)

It will help if we deal separately with four separate cases, each corresponding to a different range of the activation energy parameter γ. These are: (a) $0 < \gamma < \frac{3}{8}$; (b) $\frac{3}{8} < \gamma < \frac{1}{2}$; (c) $\frac{1}{2} < \gamma < 1$; and (d) $\gamma > 1$.

(a) $0 < \gamma < \frac{3}{8}$

For small non-zero values of γ the stationary-state condition (7.84) has a winged cusp singular point. The usual condition for this singularity gives five equalities which determine the values of x, τ_{res}, θ_{ad}, τ_N, and θ_c in terms of

γ. The relevant expressions are

$$x = 1 - \gamma - (\tfrac{1}{4} + \gamma^2)^{1/2} \tag{7.89}$$

$$\theta_{ad} = \frac{4t^3}{\gamma[2t - \gamma(1 + t)^2]^2} \tag{7.90}$$

$$\theta_c = \frac{t^2}{\gamma[t - \gamma(1 + t)^2]} \tag{7.91}$$

$$\tau_{res} = \frac{[2t - \gamma(1 + t)^2]}{\gamma(1 + t)^2} \exp\left(\frac{-t}{\gamma(1 + t)}\right) \tag{7.92}$$

$$\frac{1}{\tau_N} = \frac{\gamma(1 + t)^2[\gamma(1 + t)^2 - t]}{[2t - \gamma(1 + t)^2]^2} \exp\left(\frac{t}{\gamma(1 + t)}\right) \tag{7.93}$$

where the group t is given by

$$t = \gamma^{-1}[(\tfrac{1}{4} + \gamma^2)^{1/2} - \tfrac{1}{2}]. \tag{7.94}$$

The important feature of these equations is that the winged cusp point exists for physically acceptable values of the various quantities (x, τ_{res}, θ_{ad}, and $\tau_N > 0$, $\theta_c > -\gamma^{-1}$) provided γ stays within the above range. The system can be unfolded from the singular point by varying θ_{ad}, τ_N, and θ_c, and in this way all seven of the stationary-state patterns shown in Fig. 7.8.

As γ approaches $\tfrac{3}{8}$, the coordinates of the singular point go to $x = \tau_{res} = \tau_N = 0$, $\theta_{ad} \to \infty$. Thus the winged cusp leaves the physically acceptable region if the activation energy for the reaction becomes too small.

(b) $\tfrac{3}{8} < \gamma < \tfrac{1}{2}$

With γ greater than $\tfrac{3}{8}$, the highest-order singularity in the region which is physically acceptable is that satisfying $F = F_x = F_{xx} = F_\tau$. This is known as a 'pitchfork' singularity. Such singular points are only found if the cooling temperature difference θ_c is negative and lies in the range

$$-\gamma^{-1} < \theta_c < (1 - 4\gamma)/4\gamma^2. \tag{7.95}$$

The pitchfork singularity can be unfolded to give five stationary-state patterns: unique, hysteresis loop, isola, mushroom, and hysteresis loop plus isola, as shown in Fig. 7.9. Note that, for this system, the hysteresis loop is reversed from the typical S-shape seen previously.

(c) $\tfrac{1}{2} < \gamma < 1$

As γ increases beyond $\tfrac{1}{2}$, corresponding to a very low value for the activation energy $E < 2RT_a$, even the pitchfork singularity leaves the

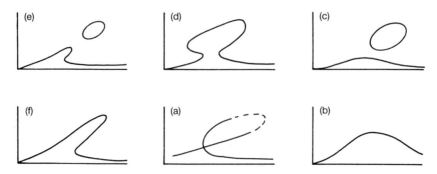

FIG. 7.9.   The full unfolding of the stationary-state locus (a) corresponding to a pitchfork singularity: (b) unique; (c) isola; (d) mushroom; (e) hysteresis loop + isola; (f) hysteresis loop.

physically acceptable region. Some variety remains in the multiple stationary-state behaviour, however, provided the bath temperature is sufficiently below that of the inflow. With $-\gamma^{-1} < \theta_c < -1$, the stationary-state locus can have four different forms, with only the mushroom pattern having been lost. With $-1 < \theta_c < (1-4\gamma)/4\gamma^2$, the system can still display the inverse hysteresis loop as well as the unique response.

(d) $\gamma > 1$

Even with the largest values of γ, corresponding to systems with very little temperature sensitivity, some limited multiplicity is possible. Condition (7.95) again gives the range within which the cooling temperature must lie and within which the system can display the inverse hysteresis loop for some combinations of θ_{ad} and τ_N.

7.6. Summary

Singularity theory clearly offers a powerful and systematic route for determining the range of stationary-state complexity possible with a given reaction system as the experimental parameters are varied. It has been successfully applied to an increasingly more complex set of simultaneous reactions and should eventually become a standard weapon in the chemical reaction engineers' armoury. However, even a full knowledge of the stationary-state patterns still leaves the considerable question of the dynamic bifurcations associated with the onset of oscillations etc. In the next chapter we turn our attentions to uncovering these aspects of the cubic autocatalysis model and of relating dynamic and stationary bifurcations so as to understand this system in great detail.

References

The first-order non-isothermal reaction in a CSTR has an extensive literature, especially in chemical reaction engineering, and the following list makes no attempt to be comprehensive:

Aris, R. (1969). *Elementary chemical reactor analysis*. Prentice-Hall, Englewood Cliffs, NJ.

Aris, R. and Amundson, N. R. (1958). An analysis of chemical reactor stability and control, parts I–III. *Chem. Eng. Sci.*, 7, 121–31, 132–47, 148–55.

Balakotaiah, V. and Luss, D. (1981). Analysis of the multiplicity patterns of a CSTR. *Chem. Eng. Commun.*, 13, 111–32.

Balakotaiah, V. and Luss, D. (1982). Structure of the steady-state solutions of lumped parameter chemically reacting systems. *Chem. Eng. Sci.*, 37, 1611–23.

Balakotaiah, V. and Luss, D. (1984). Multiplicity features of reacting systems. Dependence of the stationary-states of a CSTR on residence time. *Chem. Eng. Sci.*, 39, 1709–22.

Bilous, O. and Amundson, N. R. (1955). Chemical reactor stability and sensitivity. *Am. Inst. Chem. Eng. J.*, 1, 513–21.

D'Anna, A., Lignola, P.-G., and Scott, S. K. (1986). The application of singularity theory to isothermal autocatalytic open systems: the elementary scheme $A + mB = (m + 1)B$. *Proc. R. Soc.* A403, 341–63.

Denbigh, K. G. and Turner, J. C. R. (1984). *Chemical reactor theory*. Cambridge University Press.

Golubitsky, M. and Keyfitz, B. L. (1980). A qualitative study of the steady-state solutions for a continuous flow stirred tank chemical reactor. *SIAM J. Math. Anal.*, 11, 216–39.

Golubitsky, M. and Schaeffer, D. (1979). A theory for imperfect bifurcation via singularity theory. *Comm. Pure Appl. Math.*, 32, 21–98.

Golubitsky, M. and Schaeffer, D. (1985). *Singularities and groups in bifurcation theory*. Springer, New York.

Gray, P. and Mullins, J. (1987). Spontaneous ignition and stability in closed and open systems, part 1. *J. Chem. Soc. Faraday Trans. II*, 83, 301–18.

Gray, P. and Mullins, J. (1987). Spontaneous ignition and stability in closed and open systems, part 2. *J. Chem. Soc. Faraday Trans. II*, 83, 539–52.

Guckenheimer, J. (1986). Multiple bifurcation problem for chemical reactors. *Physica*, D20, 1–20.

Kay, S. R., Scott, S. K., and Lignola, P.-G. (1987). The application of singularity theory to isothermal autocatalytic reactions: the influence of uncatalysed reactions. *Proc. R. Soc.*, A409, 433–48.

Uppal, A., Ray, W. H., and Poore, A. B. (1974). On the dynamic behavior of continuous stirred tank reactors. *Chem. Eng. Sci.*, 29, 976–85.

Uppal, A., Ray, W. H., and Poore, A. B. (1976). The classification of the dynamic behavior of continuous stirred tank reactors—the influence of residence time. *Chem. Eng. Sci.*, **31**, 205–14.

Van Heerden, C. (1953). Autothermic processes. *Ind. Eng. Chem.*, **45**, 1242–47.

Zel'dovich, Ya. B. (1941). Towards the theory of combustion intensity. The evolution of an exothermic reaction in a flow, I. *Zh. Tekh. Fiz.*, **XI**, 493–500.

Zel'dovich, Ya. B and Zysin, Yu. A. (1941). Towards the theory of combustion intensity. The evolution of an exothermic reaction in a flow, II. *Zh. Tekh. Fiz.*, **XI**, 501–58.

OSCILLATORY BEHAVIOUR IN THE ISOTHERMAL CSTR: AUTOCATALYTIC SYSTEMS

The previous chapters focused on the existence and locus of stationary states. Here their stability to small perturbations is outlined, together with the systematic study of the onset of oscillations in isothermal, auto-catalytic open systems. After careful study the reader should be able to:

(1) understand the normal exponential response to transient perturbations—decay to stable states (relaxation) and exponential growth from unstable states;

(2) recognize the prolongation of transients near 'saddle node' bifurcations (points at which the stationary-state locus has a turning point);

(3) recognize the onset of non-exponential decay at such bifurcations;

(4) determine local stability in potentially oscillatory cases with two variables and interpret the elements of the Jacobian matrix;

(5) locate the points where stability changes, and establish whether limit cycles are born, and of what stability;

(6) understand the extinction of oscillations at points where homoclinic orbits are formed.

The previous two chapters have considered the stationary-state behaviour of reactions in continuous-flow well-stirred reactions. It was seen in chapters 2–5 that stationary states are not always stable. We now address the question of the local stability in a CSTR. For this we return to the isothermal model with cubic autocatalysis. Again we can take the model in two stages: (i) systems with no catalyst decay, $k_2 = 0$ and (ii) systems in which the catalyst is not indefinitely stable, so the concentrations of A and B are decoupled. In the former case, it was found from a qualitative analysis of the flow diagram in §6.2.5 that unique states are stable and that when there are multiple solutions they alternate between stable and unstable. In this chapter we become more quantitative and reveal conditions where the simplest exponential decay of perturbations is replaced by more complex time dependences.

When the catalyst decays we have a two-variable system and hence there is the potential for Hopf bifurcations and sustained oscillations. In our flow reactor, we have the possibility of oscillation about one stationary state interacting with other solutions on other branches of isolas or mushrooms, giving rise to a wide range of dynamic responses from which the system can choose.

8.1. Cubic autocatalysis without decay

We take the single cubic autocatalytic step

$$A + 2B \rightarrow 3B \qquad \text{rate} = k_1 ab^2 \tag{8.1}$$

in a CSTR with inflow of both A and B. For this example we will work in terms of the original rate constants etc. rather than going to dimensionless forms. This allows some effects to be interpreted more easily (at least at first sight), but means that we have more symbols to keep track of during the algebra. The mass-balance (rate) equation can be written as

$$\frac{da}{dt} = \frac{a_0 - a}{t_{res}} - k_1 ab^2 = \frac{a_0 - a}{t_{res}} - k_1 a(a_0 + b_0 - a)^2. \tag{8.2}$$

8.1.1. Stationary-state behaviour

The stationary states are given by the solutions of a cubic equation:

$$\frac{a_0 - a_{ss}}{t_{res}} - k_1 a_{ss}(a_0 + b_0 - a_{ss})^2 = 0. \tag{8.3}$$

As seen in chapter 6, this may have multiple solutions over some range of residence times if $b_0 < \frac{1}{8} a_0$. The range of residence times is given by

$$t_{res}^- \leqslant t_{res} \leqslant t_{res}^+ \tag{8.4}$$

where

$$\frac{8}{t_{res}^\pm} = k_1 a_0^2 \left[1 + 20 \frac{b_0}{a_0} - 8 \left(\frac{b_0}{a_0} \right)^2 \pm \left(1 - 4 \frac{b_0}{a_0} \right) \left(1 - 8 \frac{b_0}{a_0} \right)^{1/2} \right]. \tag{8.5}$$

If $b_0 > \frac{1}{8} a_0$ or if t_{res} lies outside this range, the system has just one stationary state. (Note that if $b_0 = \frac{1}{8} a_0$ exactly, then $t_{res}^- = t_{res}^+ = \frac{64}{27}(k_1 a_0^2)^{-1}$ and the value of a_{ss} at this residence time is $a_{ss} = \frac{3}{4} a_0$.)

8.1.2 Perturbation of the stationary state

Let us consider an experiment where a_0, b_0, k_1, and t_{res} are all specified and the concentration of a is then allowed to take up one of the corresponding

stationary-state values (the only one if $t_{res} < t_{res}^-$, $t_{res} > t_{res}^+$, or $b_0 > \frac{1}{8}a_0$).
What happens if there is then a small perturbation in the concentration, so
$a = a_{ss} + \Delta a$? How does Δa vary in time? The rate equation for Δa can be
obtained from eqn (8.2), which can be rewritten as

$$\frac{d(a_{ss} + \Delta a)}{dt} = \frac{a_0 - (a_{ss} + \Delta a)}{t_{res}} - k_1(a_{ss} + \Delta a)(a_0 + b_0 - a_{ss} - \Delta a)^2$$

$$= \frac{a_0 - a_{ss}}{t_{res}} - k_1 a_{ss}(a_0 + b_0 - a_{ss})^2$$

$$- \left(\frac{1}{t_{res}} + k_1(a_0 + b_0 - a_{ss})(a_0 + b_0 - 3a_{ss}) \right) \Delta a$$

$$+ k_1[2(a_0 + b_0) - 3a_{ss}](\Delta a)^2 - k_1(\Delta a)^3. \qquad (8.6)$$

From the definition of the stationary state, eqn (8.3), the first two terms on the
right-hand side of this expression (those not involving Δa) cancel identically.
Thus we can write the rate of change of the perturbation as

$$\frac{d\Delta a}{dt} = - \left(\frac{1}{t_{res}} + k_1(a_0 + b_0 - a_{ss})(a_0 + b_0 - 3a_{ss}) \right) \Delta a + O(\Delta a^2) \quad (8.7)$$

$$= - \frac{1}{t_{relax}} \Delta a + O(\Delta a^2) \qquad (8.8)$$

where we are assuming that the perturbation is small, so higher-order terms
will not determine the behaviour of the system. The relaxation time t_{relax} is
given by the term in large parentheses in eqn (8.7). This is also related to the
inverse of the eigenvalue λ of the (linearized) rate equation, $\lambda = -1/t_{relax}$.
Ignoring the $O(\Delta a^2)$ terms, the perturbation will grow or decay exponentially as

$$\Delta a(t) = \Delta a_0 e^{\lambda t} \qquad (8.9)$$

where Δa_0 is the initial size of the perturbation, depending on the sign of
λ (and hence t_{relax}). If $\lambda < 0$ (which occurs if $t_{relax} > 0$), then the perturbation
decays and the stationary state is locally stable to small perturbations. If
$\lambda > 0$ (with $t_{relax} < 0$) the stationary state is unstable.

8.1.3. Relaxation and chemical times

The relaxation time may be further interpreted as containing contributions
from flow (t_{res}) and chemistry ($t_{ch,ss}$). Then

$$\frac{1}{t_{relax}} = \frac{1}{t_{res}} + \frac{1}{t_{ch,ss}} \qquad (8.10)$$

where the chemical time $t_{ch,ss}$ is not $(k_1 a_0^2)^{-1}$ as used previously, but is given by

$$1/t_{ch,ss} = k_1(a_0 + b_0 - a_{ss})(a_0 + b_0 - 3a_{ss}). \tag{8.11}$$

The sign of $t_{ch,ss}$ is determined by the sign of the factor $(a_0 + b_0 - 3a_{ss})$, which may be either negative or positive. The sign and value of the relaxation time and the eigenvalue also then depend on the value of the particular stationary-state concentration of A.

8.1.4. No inflow of autocatalyst

In the especially simple case of a reactor fed only with the reactant A, so $b_0 = 0$, the stationary states can be obtained explicitly. One solution of eqn (8.3) is $a_{ss} = a_0$, corresponding to no reaction. The other solutions are given by the roots of a quadratic

$$a_{ss} = \frac{1}{2} a_0 \left[1 \pm \left(1 - \frac{4}{k_1 a_0^2 t_{res}} \right)^{1/2} \right] \tag{8.12}$$

and exist for $t_{res} > 4(k_1 a_0^2)^{-1}$.

For the no reaction state $a_{ss} = a_0$, the relaxation time given by eqn (8.10) is simply equal to the residence time. In terms of the eigenvalue, we have $\lambda = -1/t_{res}$, which is negative. The stationary state is always stable, irrespective of a_0 and k_1. Chemistry makes no contribution (formally we have $1/t_{ch,ss} = 0$, so the 'chemical time' goes to infinity); the perturbation of a does not introduce any B to the system, so no reaction is initiated. The recovery of the stationary state is achieved only by the inflow and outflow.

For the non-zero states, substitution of a_{ss} from eqn (8.12) into (8.10) leads to the expression

$$\frac{1}{t_{relax}} = \frac{1}{2} k_1 a_0^2 \left(1 - \frac{4}{k_1 a_0^2 t_{res}} \right)^{1/2} \left[\left(1 - \frac{4}{k_1 a_0^2 t_{res}} \right)^{1/2} \pm 1 \right] \tag{8.13}$$

with the upper root here corresponding to the lower root in (8.12). These non-zero solutions, and hence the relaxation time, are only defined for $4/k_1 a_0^2 t_{res} < 1$, so the final term in the above expression is positive when the plus sign is taken and negative with the minus sign.

Thus the upper root of eqn (8.10), which gives the middle branch of stationary-state solutions and requires the minus sign above, has a negative value for t_{relax}. It then follows that the eigenvalue λ for this branch is positive, so perturbations grow. This is an unstable state.

For the other non-zero solution, with the lowest value for a_{ss} (highest extent of conversion), the relaxation time has the plus sign in eqn (8.13) and so is positive. The corresponding eigenvalue λ is negative, so the solutions along this branch are stable.

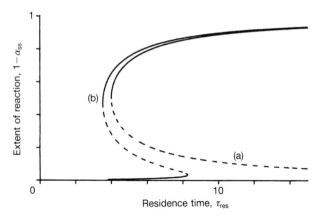

FIG. 8.1. Indication of local stability or instability for the simple cubic autocatalytic step without decay: solid curves indicate branches of stable stationary-state solutions, broken curves correspond to unstable states. (a) Stationary-state locus with no autocatalyst inflow, $\beta_0 = 0$, with one stable solution, $1 - \alpha_{ss} = 0$, corresponding to zero reaction; (b) stationary-state locus with $\beta_0 = \frac{1}{32}$, again showing alternation in stability between the three branches.

These arguments thus confirm the alternation of stability with the three branches of stationary-state solutions, as shown in Fig. 8.1. We can also make quantitative comments. As the residence time becomes very long, so the relaxation time for the unstable branch tends to $-\infty$: for the stable non-zero state t_{relax} decreases as t_{res} increases, tending to the value $1/k_1 a_0^2$ as t_{res} tends to ∞.

8.1.5. Inflow of autocatalyst

For systems with inflow of both reactant and autocatalyst, the station-ary-state solutions cannot be obtained explicitly. The alternation in stability when there are multiple solutions described above is, however, quite general and is strictly followed even when $b_0 \neq 0$.

8.2. Relationship between relaxation time and flow diagram: non-exponential decay (slowing down)

We are dealing at present with systems which have only one independent concentration, and hence which are governed by a single rate equation. Thus eqn (8.2) can be written in the general from

$$da/dt = f(a, t_{res}, \text{etc.}). \tag{8.14}$$

In this form we can think in similar terms to those used in chapter 3 where we examined the local stability of general two-variable systems. In this way the eigenvalue λ defined above is given in terms of the function f by

$$\lambda = -1/t_{\text{relax}} = (\partial f/\partial a)_{a_{ss}} \tag{8.15}$$

where the partial derivative is evaluated at the stationary state whose stability is being tested.

We have also found it helpful to think of the rate equation for CSTR systems in terms of a flow rate L and a reaction rate R, and eqn (8.2) has the form

$$da/dt = L - R. \tag{8.16}$$

The eigenvalue λ can then be expressed as

$$\lambda = -1/t_{\text{relax}} = (\partial L/\partial a - \partial R/\partial a)_{a_{ss}}. \tag{8.17}$$

Thus λ can be interpreted in terms of the difference in the slopes of the flow and reaction curves on the flow diagram, as illustrated in Fig. 8.2.

8.2.1. Relaxation times near ignition and extinction points

When the residence time is varied so that we approach an ignition or extinction point in the stationary-state locus, then the flow and reaction curves L and R become tangential. The condition for tangency is $R = L$ and $\partial R/\partial a = \partial L/\partial a$. Thus the difference between the slopes of R and L decreases to zero. From eqn (8.17) we see that the tangency condition also causes the value of the eigenvalue λ to tend to zero. An alternative interpretation, in

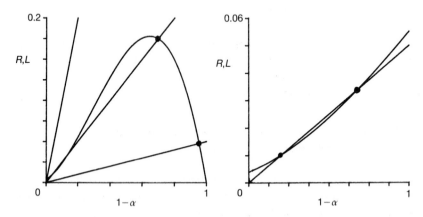

FIG. 8.2. The different relative slopes of the reaction rate curve, R, and the flow line L at stable and unstable stationary states: (a) for stable states the gradient of L is steeper (more positive) than that of R; (b) enlargement of intersections near origin showing that R is steeper than L for the middle unstable intersection.

terms of the relaxation time, has t_{relax} tending to infinity as tangency is approached. Long relaxation times correspond to perturbations which decay (or grow) only slowly. Relaxation times can be measured experimentally, and the lengthening predicted in the vicinity of ignition or extinction points has been observed and called 'slowing down'.

The equivalence between a vanishing eigenvalue and a turning point in the stationary-state locus can be made firmer using the ideas of the singularity theory introduced in the previous chapter. The stationary-state condition, in general terms, for eqn (8.14) is

$$f(a_{ss}, t_{res}, \text{etc.}) = L - R = 0. \qquad (8.18)$$

Comparing this with eqn (7.41), we can see that here f plays the role of the function F, and that a_{ss} is equivalent to x. The condition for a turning point, eqn (7.50), then becomes

$$f = f_a = 0 \qquad (8.19)$$

where $f_a = (\partial f/\partial a)_{a_{ss}}$, which is also the definition of λ. This is an important result which can be generalized to systems with many independent concentrations: the condition for a turning point in the stationary-state locus also corresponds to the condition for an eigenvalue λ becoming zero.

Equation (8.13) gives the appropriate form for $1/t_{relax}(= -\lambda)$ for the cubic autocatalysis model with no inflow of autocatalyst. The condition for the turning point in the stationary-state locus (there is only one) is $k_1 a_0^2 t_{res} = 4$. When this is satisfied, the right-hand side becomes zero, as predicted from above.

If λ becomes zero, the decay of the perturbation expressed in eqn (8.9) would appear to cease, with $\Delta a(t) = \Delta a_0$ for all times. In fact, even before λ becomes exactly zero, the system adjusts from its simple exponential behaviour. This change arises because higher-order terms in eqn (8.8) can become important if $1/t_{relax}$ is a small quantity.

8.2.2. Non-exponential growth or decay

To determine the response of the cubic autocatalytic system to perturbations in the vicinity of a turning point in the locus, we must return to eqn (8.6). The first two terms (not involving Δa) again cancel exactly, because of the stationary-state condition. If we are also at an ignition or extinction point, the tangency condition in any of its forms discussed above ensures that the coefficient of the Δa term is also zero. Thus the first non-zero term is that involving $(\Delta a)^2$:

$$d\Delta a/dt = k_1[2(a_0 + b_0) - 3a_{ss}](\Delta a)^2 + O(\Delta a^3). \qquad (8.20)$$

Provided $b_0 < \frac{1}{8}$, the coefficient of $(\Delta a)^2$ will be non-zero, so by specifying infinitesimal perturbations the time dependence is determined by terms of order Δa^2.

Integrating the leading-order equation for Δa gives an inverse dependence of the perturbation with time:

$$\Delta a(t) = \frac{\Delta a_0}{1 - k_1[2(a_0 + b_0) - 3a_{ss}]\Delta a_0 t}. \qquad (8.21)$$

The exact behaviour now depends on the sign of the term $k_1[2(a_0 + b_0) - 3a_{ss}]$, i.e. the coefficient of $(\Delta a)^2$ in the rate equation (8.20). If this term is positive (as it is for an extinction point) and if the initial perturbation is such that Δa_0 is negative (i.e. if a is momentarily reduced below its stationary-state value), then the perturbation will decay as the time increases. If, however, the initial perturbation at the extinction point is positive (a *increases* above a_{ss}), there will be a growth away from the stationary state. (Formally, Δa would become infinite after a finite time as the denominator goes to zero, but before then the higher-order terms in the rate law (8.20) would become important.) The turning point thus has a 'one-sided' stability, as indicated in Fig. 8.3.

A similar scenario holds at the ignition point, only there the coefficient is negative, so positive initial perturbations decay back; perturbations which decrease the concentration of A grow in time.

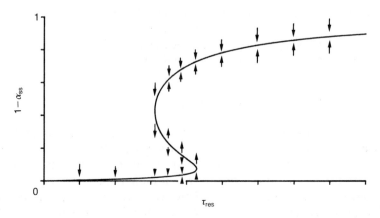

FIG. 8.3.   The approach to, or departure from, stationary-state solutions following small perturbations for simple cubic autocatalysis again showing the instability of the middle branch. The turning points (ignition and extinction) have 'one-sided' stability as perturbations in one direction decay back to the saddle–node point, but those of the opposite sign depart for the other branch of solutions.

8.2.3. Relaxation with disappearance of ignition and extinction

There is another type of time dependence possible in this system. If the inflow concentration of the autocatalyst is adjusted so that $b_0 = \frac{1}{8}a_0$, then the ignition and extinction points merge at $t_{res} = \frac{64}{27}(k_1 a_0^2)^{-1}$, with $a_{ss} = \frac{3}{4}a_0$. Under these special conditions, the coefficient of the term in $(\Delta a)^2$ in the rate equation, and hence in the denominator of eqn (8.21), becomes zero as well as those of the lower powers in Δa. Thus the inverse time dependence disappears, and the only non-zero term governing the decay of perturbation is that in $(\Delta a)^3$:

$$d\Delta a/dt = -k_1(\Delta a)^3. \qquad (8.22)$$

Integrating this equation gives a decay (always) which follows an inverse dependence on the square root of the time:

$$\Delta a(t) = \Delta a_0/(1 + 2k_1\Delta a_0^2 t)^{1/2}. \qquad (8.23)$$

This time dependence is typical of all one-variable systems at a point of disappearance of a hysteresis loop (i.e. for which $F = F_x = F_{xx} = 0$ in the terms of §7.3.1).

8.2.4. Exothermic reaction in adiabatic CSTR

The many qualitative similarities between the cubic autocatalytic system without decay and the adiabatic non-isothermal CSTR have been noted previously. Thus all of the behaviour described above also applies to the latter. In general, small perturbations decay or begin to grow exponentially, as $e^{\lambda t}$. When there is a single stationary state it is stable; when there are three, those corresponding to the highest and lowest extents of reaction are stable $(\lambda < 0)$ and the middle solution is unstable $(\lambda > 0)$. In the vicinity of ignition and extinction points, there is a lengthening of the relaxation time (slowing down) as λ tends to zero, with a departure from exponential to inverse time dependence. If the adiabatic temperature rise decreases (in dimensionless terms $\theta_{ad} = 4$), the ignition and extinction points merge and there is an inverse square root decay.

8.3. Cubic autocatalysis with decay

We have seen in earlier chapters that kinetic systems with two independent concentrations can show additional complexities of dynamic behaviour beyond those of one-variable systems. Of particular interest are undamped oscillations. The cubic autocatalysis with the additional decay step

$$B \rightarrow C \qquad \text{rate} = k_2 b \qquad (8.24)$$

is an example of such a system, as is the non-isothermal non-adiabatic reactor. Because we have a constant inflow of fresh material into our CSTR, oscillations can truly become indefinitely sustained—without the need for the artificial neglect of reactant consumption, or the corresponding slow variation and ultimate decay when a slow consumption is admitted, which we used for closed systems in chapters 2–5. We can analyse for the onset of oscillatory behaviour through Hopf bifurcations etc., using the same recipes introduced in chapter 5. We begin with the simplest case: the isothermal system with no catalysed reaction and no inflow of autocatalyst ($\beta_0 = 0$). It will be most convenient now to use dimensionless variables, as introduced in chapter 6.

8.3.1. Cubic autocatalysis with no inflow of B

We should first recall the stationary-state behaviour for this case. If the reaction rate constant for the catalyst decay step is large compared with that for the autocatalytic step, so that $\kappa_2 > \frac{1}{16}$, the system can only ever have one stationary state. This state corresponds to no net conversion of A to B, so $\alpha_{ss} = 1$. For slower decay rates, $\kappa_2 < \frac{1}{16}$, non-zero stationary states exist over a range of residence times $\tau_{res}^- \leqslant \tau_{res} \leqslant \tau_{res}^+$ in the form of an isola. The extents of conversion along the branches of the isola are given by

$$(1 - \alpha_{ss})_\pm = \frac{1}{2}\left[1 \pm \left(1 - \frac{4(1 + \kappa_2 \tau_{res})^2}{\tau_{res}}\right)^{1/2}\right] \tag{8.25}$$

$$\tau_{res}^- \leqslant \tau_{res} \leqslant \tau_{res}^+ \tag{8.26}$$

$$\tau_{res}^\pm = \frac{1 - 8\kappa_2 \pm (1 - 16\kappa_2)^{1/2}}{8\kappa_2^2}. \tag{8.27}$$

The higher extent of conversion lies on the top of the isola as shown in Fig. 8.4, with $(1 - \alpha_{ss})_+ > \frac{1}{2}$; the lower shore of the isola has $(1 - \alpha_{ss})_- < \frac{1}{2}$. The corresponding stationary-state values for the concentration of the autocatalyst are $\beta_{ss} = 0$ (for $\alpha_{ss} = 1$) and

$$(\beta_{ss})_\pm = \frac{1}{2}\left\{\left[1 \pm \left(1 - \frac{4(1 + \kappa_2 \tau_{res})^2}{\tau_{res}}\right)^{1/2}\right]\middle/(1 + \kappa_2 \tau_{res})\right\} \tag{8.28}$$

with the upper root in eqn (8.28) corresponding to the upper root in (8.25), so the highest concentrations of B occur with the highest extents of conversion.

It will be convenient to identify the three possible stationary states with subscripts 1, 2, and 3 in order of increasing extent of conversion, as shown in Table 8.1 where the group $T = (1 + \kappa_2 \tau_{res})^2/\tau_{res}$.

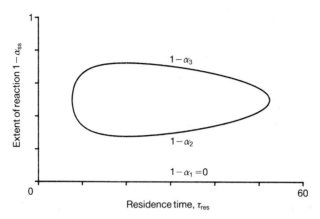

FIG. 8.4.   A typical isola for cubic autocatalysis with decay, but for no catalyst inflow $\beta_0 = 0$, identifying the notation $(1 - \alpha_1)$, $(1 - \alpha_2)$, and $(1 - \alpha_3)$ in order of increasing extent of reaction. (Exact numerical values correspond to $\kappa_2 = \frac{1}{20}$.)

Table 8.1

Stationary-state concentrations and extents of reaction for cubic autocataly-sis with no inflow of B

Stationary state	α_{ss}	$1 - \alpha_{ss}$	β_{ss}
$1\ (\alpha_1, \beta_1)$	1	0	0
$2\ (\alpha_2, \beta_2)$	$\frac{1}{2}[1 + (1 - 4T)^{1/2}]$	$\frac{1}{2}[1 - (1 - 4T)^{1/2}]$	$\frac{1}{2}\dfrac{1 - (1 - 4T)^{1/2}}{(1 + \kappa_2 \tau_{res})}$
$3\ (\alpha_3, \beta_3)$	$\frac{1}{2}[1 - (1 - 4T)^{1/2}]$	$\frac{1}{2}[1 + (1 - 4T)^{1/2}]$	$\frac{1}{2}\dfrac{1 - (1 - 4T)^{1/2}}{(1 + \kappa_2 \tau_{res})}$

8.3.2. Local stability with no inflow of B

The route to determining the local stability of a given stationary state begins with the Jacobian matrix for this system:

$$\mathbf{J} = \left. \begin{pmatrix} \dfrac{\partial}{\partial \alpha}\left(\dfrac{d\alpha}{d\tau}\right) & \dfrac{\partial}{\partial \beta}\left(\dfrac{d\alpha}{d\tau}\right) \\[2mm] \dfrac{\partial}{\partial \alpha}\left(\dfrac{d\beta}{d\tau}\right) & \dfrac{\partial}{\partial \beta}\left(\dfrac{d\beta}{d\tau}\right) \end{pmatrix} \right|_{ss} \tag{8.29}$$

where the two rate equations are

$$\frac{d\alpha}{d\tau} = \frac{1 - \alpha}{\tau_{res}} - \alpha\beta^2 \tag{8.30}$$

$$\frac{d\beta}{d\tau} = \frac{\beta_0 - \beta}{\tau_{res}} + \alpha\beta^2 - \kappa_2\beta. \tag{8.31}$$

given previously as eqns (6.46) and (6.47) (we have the special case $\beta_0 = 0$, for the moment). From these, we then obtain

$$\mathbf{J} = \begin{pmatrix} -(\beta_{ss} + 1/\tau_{res}) & -2\alpha_{ss}\beta_{ss} \\ \beta_{ss} & 2\alpha_{ss}\beta_{ss} - (\kappa_2 + 1/\tau_{res}) \end{pmatrix}. \tag{8.32}$$

The four elements of this matrix define values of the eigenvalues λ_1 and λ_2 by

$$\lambda^2 - \text{tr}(\mathbf{J})\lambda + \det(\mathbf{J}) = 0 \tag{8.33}$$

in terms of the trace (sum of the terms on the leading diagonal) and the determinant. These depend on the values of the stationary-state concentrations, which are different for each of the three solutions and which also vary along the various loci. Thus, we may expect different stabilities and variations in stability along each of the different branches.

8.3.3. Local stability for state of no conversion

The zero conversion state, $\alpha_1 = 1$, $\beta_1 = 0$, exists for all residence times τ_{res} and catalyst decay rates κ_2. The elements of the Jacobian matrix are especially simple for this solution:

$$\mathbf{J} = \begin{pmatrix} -1/\tau_{res} & 0 \\ 0 & -(\kappa_2 + 1/\tau_{res}) \end{pmatrix}. \tag{8.34}$$

Thus, the trace is given by $\text{tr}(\mathbf{J}) = -(\kappa_2 + 2/\tau_{res})$ which is always negative, and the determinant by $\det(\mathbf{J}) = (1/\tau_{res})(1/\tau_{res} + \kappa_2)$ which is always positive. The eigenvalues themselves are

$$\lambda_1 = -1/\tau_{res} \qquad \lambda_2 = -(\kappa_2 + 1/\tau_{res}). \tag{8.35}$$

Both are always real and both are always negative. This stationary state is thus always stable to small perturbations which decay monotonically to zero in an exponential manner. The relaxation times (there are now two and the decay occurs as the sum of two exponentials) are related to the residence time and to the inverse of the rate constant for catalyst decay. For instance, the decay of the perturbation in A can be written in the form

$$\Delta\alpha(\tau) = \exp(-\tau/\tau_{res})[c_1 + c_2\exp(-\kappa_2\tau)] \tag{8.36}$$

where the constants c_1 and c_2 will depend on the initial perturbations in α and β.

The zero conversion state is always a stable node.

8.3.4. Local stability for states of non-zero conversion

The solutions α_2 and α_3 are real when τ_{res} lies in the range given by eqn (8.26). The general forms for the trace and determinant of the Jacobian are

$$\text{tr}(\mathbf{J}) = 2\alpha_{ss}\beta_{ss} - (\beta_{ss}^2 + \kappa_2 + 2/\tau_{res}) \tag{8.37}$$

$$\det(\mathbf{J}) = 1/\tau_{res}[(1 + \kappa_2\tau_{res})(\beta_{ss}^2 + 1/\tau_{res}) - 2\alpha_{ss}\beta_{ss}]. \tag{8.38}$$

Substituting for the stationary-state concentrations from Table 8.1, the determinant becomes

$$\det(\mathbf{J}) = \frac{(1 - 4T)^{1/2}[(1 - 4T)^{1/2} \pm 1]}{2\tau_{res}(1 + \kappa_2\tau_{res})} \tag{8.39}$$

with the upper root (plus sign) for α_3 and the lower root (minus sign) for α_2. The determinant, like the two stationary states, is only real provided $4T < 1$.

Considering the middle solution α_2 first, then with the minus sign, the determinant is negative. We thus have real eigenvalues λ_1 and λ_2 with opposite signs, irrespective of $\text{tr}(\mathbf{J})$.

This stationary state is always a saddle point and hence basically unstable to small perturbations. (One exponential term, corresponding to the negative eigenvalue, decays; the other, with the positive eigenvalue, grows. If we arrange a very special perturbation such that the pre-exponential coefficient associated with the positive eigenvalue is exactly zero, then we would see a return to the saddle point. General perturbations, however, always show an ultimate divergence from this solution.)

For the stationary state with the highest extent of conversion α_3, β_3 across the top of the isola, with the upper root in eqn (8.39), the determinant is always positive. The local stability is thus determined by the sign of the trace. We can write $\text{tr}(\mathbf{J})$ in terms of τ_{res}, κ_2, and T:

$$\text{tr}(\mathbf{J}) = \frac{1}{\tau_{res}}\left((1 + \kappa_2\tau_{res}) - \frac{1 + (1 - 4T)^{1/2}}{2T}\right). \tag{8.40}$$

If $\text{tr}(\mathbf{J})$ is negative, which will be favoured by short residence times, this upper stationary state is stable. Under such conditions the system as a whole has three stationary states, two of which are stable and one unstable. The concentrations of A and B will adjust to one of the stable states and will remain at the chosen one (which will depend on the previous history of our experiment) provided it receives only small perturbations. The middle state, or saddle point, again acts as a watershed or potential barrier between the stable solutions. If the system is given a large perturbation, so that it is forced across the saddle, it moves to the other stable state.

8.3.5. Hopf bifurcation condition

The condition for a change in stability along the uppermost branch of solutions, by means of a Hopf bifurcation, is $\text{tr}(\mathbf{J}) = 0$. This generally occurs as the residence time is increased, when

$$2T(1 + \kappa_2 \tau_{\text{res}}) = 1 + (1 - 4T)^{1/2}. \tag{8.41}$$

This condition is satisfied when $T = \kappa_2^{1/2}$, so that

$$\tau_{\text{res}} = \tau_{\text{res}}^* = \frac{1 - 2\kappa_2^{1/2} \pm (1 - 4\kappa_2^{1/2})^{1/2}}{2\kappa_2^{3/2}}. \tag{8.42}$$

In fact only the upper root corresponds to a Hopf bifurcation point (the lower solution to the condition $\text{tr}(\mathbf{J}) = 0$ being satisfied along the saddle point branch of the isola where the system does not have complex eigenvalues).

Thus we have an explicit formula in this case for the Hopf bifurcation points as a function of the decay rate constant: for $\kappa_2 = \frac{1}{20}$, $\tau_{\text{res}} = 39.25$; for $\kappa_2 = \frac{1}{40}$, $\tau_{\text{res}} = 163.2$. Figure 8.5 shows how the bifurcation point moves to longer residence times as κ_2 decreases, along with the locations of the extinction points $\tau_{\text{res}}^{\pm}$ from eqn (8.27).

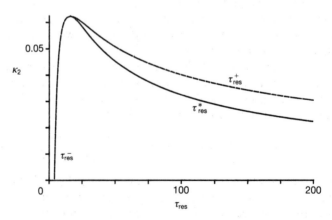

FIG. 8.5. The locus of Hopf bifurcation points $\tau_{\text{res}}^* - \kappa_2$ described by eqn (8.42) for cubic autocatalysis with decay but no catalyst inflow. Also shown are the loci of the extinction points $\tau_{\text{res}}^{\pm}$, marking the ends of the isola, given by eqn (8.27). The three curves meet at the common point $\tau_{\text{res}} = 16$, $\kappa_2 = \frac{1}{16}$.

We can see from eqn (8.42) that κ_2 must be small if there are to be real solutions for the Hopf point. In fact the condition $\kappa_2^{1/2} < \frac{1}{4}$ is exactly the same as that for the existence of isolas ($\kappa_2 < \frac{1}{16}$). Thus all isolas have a point of Hopf bifurcation along their upper branch.

8.3.6. Stability of emerging limit cycle

The formulae of §5.2 seem to be less algebraically tractable for the equations appropriate to flow reactors than for those from closed systems. (This is basically because we can eliminate all the terms involving one of the variables by adding the two rate equations together. For flow systems this cancellation does not occur because of the outflow terms.) Nevertheless, the various coefficients can be evaluated implicitly without much problem and the stability parameter β_2 calculated. The following situation is found for the particular case of interest here, $\beta_0 = 0$.

If κ_2 is greater than $\frac{1}{16}$, we know there will be no isola and no Hopf bifurcation point. For $\kappa_2 < \frac{1}{16}$, but greater than $9/256$, β_2 is positive. This means that the emerging limit cycle will be *unstable*. The limit cycle grows as the residence time is reduced below the bifurcation point τ_{res}^* surrounding the upper stationary state which is stable.

We now know that if a system on the upper branch of the isola, just below the Hopf bifurcation point, is given a small perturbation which remains within the unstable limit cycle, it will decay back to the upper solution. If, however, the perturbation is larger, so we move to a point outside the cycle, we will not be able to get back to the upper solution: the system must move to the other stable state, with no reactant consumption.

Figure 8.6 shows how the limit cycle grows as the residence time is reduced. At some point τ_{res}^h, it hits the lower shore of the isola. Here the limit cycle manages to join up with the saddle point, forming a saddle loop or 'homoclinic orbit'. Below τ_{res}^h the limit cycle does not exist, so we see here a new way in which oscillations may be extinguished from the system. As indicated by Fig. 8.6, the amplitude does not shrink to zero at the homoclinic point. If, however, we follow the period of the motion around the cycle, we find that this grows to infinity logarithmically as τ_{res} tends to τ_{res}^h. (Strictly speaking, the period approaches $-\infty$: the limit cycle is unstable so we have to run time backwards to get on to it.)

The point at which the homoclinic orbit is formed must be calculated numerically, but once it has been located we can show that the limit cycle is still unstable as it approaches the loop formation. We do this by evaluating the trace of the Jacobian matrix for the saddle point solution α_2, β_2 corresponding to τ_{res}^h: if tr($\mathbf{J}$) is positive, the limit cycle is unstable (as we always find for this special case of the present model); if tr($\mathbf{J}$) is negative for the saddle

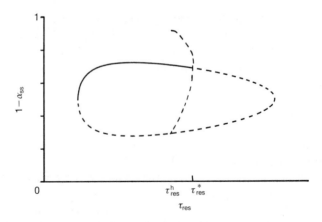

FIG. 8.6. Typical arrangement of local stabilities and development of unstable limit cycle, from a subcritical Hopf bifurcation, appropriate to cubic autocatalysis with decay and no autocatalyst inflow and with $9/256 < \kappa_2 < 1/16$. The unstable limit cycle grows as τ_{res} decreases below τ^*_{res}, and terminates by means of the formation of a homoclinic orbit at τ^h_{res}. Stable stationary states, including the zero conversion branch $1 - \alpha_1 = 0$, are indicated by solid curves, unstable states and limit cycles by broken curves.

at τ^h_{res}, the limit cycle forming the orbit would be stable (so we must have had a change in stability in between the birth at the Hopf point and the death at the homoclinic orbit).

We should also consider the behaviour along the top of the isola, on the part of the branch lying at longer residence times than the Hopf point. For $\tau_{res} > \tau^*_{res}$, and with κ_2 still in the above range, the uppermost stationary state is unstable and is not surrounded by a stable limit cycle. The system cannot sit on this part of the branch, so it must eventually move to the only stable state, that of no conversion. Thus we fall off the top of the isola not at the long residence time turning point, but earlier as we pass the Hopf bifurcation point.

If the dimensionless rate constant for catalyst decay is smaller, so $\kappa_2 < 9/256$, then the stability parameter β_2 changes sign and becomes negative. The emerging limit cycle is now stable and grows as the residence time increases above τ^*_{res}. There is still only one Hopf bifurcation point along the isola, so the limit cycle must again terminate in a homoclinic orbit. Calculation of τ^h_{res} and then of $\mathrm{tr}(\mathbf{J})$ for the corresponding saddle point solution shows that the limit cycle is still unstable as it forms the orbit. The variation of the limit cycle amplitude with the residence time must thus be qualitatively as shown in Fig. 8.7. There is a turning point in the locus where the limit cycle changes stability. The unstable cycle grows as the residence time decreases. Over a range of τ_{res}, the upper stationary state is surrounded by both stable and unstable cycles. For residence times longer than that corresponding to

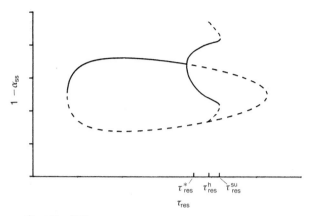

FIG. 8.7. Supercritical Hopf bifurcation for cubic autocalysis with decay and $\beta_0 = 0$, appropriate for small dimensionless decay rate constant $\kappa_2 < 9/256$. A stable limit cycle emerges and grows as the residence time is increased above τ^*_{res}. At higher residence times, this disappears at τ^{su}_{res} by merging with an unstable limit cycle born from a homoclinic orbit at τ^h_{res}. (With non-zero autocatalyst inflow, $\beta_0 > 0$, the stable limit cycle itself may form a homoclinic orbit at long τ_{res}.)

the turning point in the amplitude curve, there are no limit cycles: the upper stationary state is unstable and so the system must eventually move to the stable state of no conversion. Thus we again fall off the isola before the extinction point in stationary-state locus and, now, from an oscillatory state.

8.3.7. Phase portraits

Even for the present simple case, for which the inflow does not contain the autocatalyst, we have seen a variety of combinations of stable and unstable stationary states with or without stable and unstable limit cycles. Stable limit cycles offer the possibility of sustained oscillatory behaviour (and because we are in an open system, these can be sustained indefinitely). A useful way of cataloguing the different possible combinations is to represent the different possible qualitative forms for the 'phase plane'. The phase plane for this model is a two-dimensional surface of α plotted against β. As these concentrations vary in time, they also vary with respect to each other. The projection of this motion onto the α–β plane then draws out a 'trajectory'. Stationary states are represented as points, to which or from which the trajectories tend. If the system has only one stationary state for a given combination of κ_2 and τ_{res}, there is only one such stationary point. (For the present model the only unique state is the no conversion solution: this would have the coordinates $\alpha_{ss} = 1$, $\beta_{ss} = 0$.) If the values of κ_2 and τ_{res} are such that the system is lying at some point along an isola, there will be three stationary states on the phase plane. These will all lie on the straight line given by eqn (6.50):

$$\beta_{ss} = \frac{1 - \alpha_{ss}}{1 + \kappa_2 \tau_{res}}. \tag{8.43}$$

Figure 8.8(a) shows a typical example of the phase plane for a system with three stationary solutions, chosen such that there are two stable states and the middle saddle point. The trajectories drawn on to the diagram indicate the direction in which the concentrations will vary from a given starting point. In some cases this movement is towards the state of no conversion ($\alpha_{ss} = 1$, $\beta_{ss} = 0$), in others towards the stable non-zero solution. Only two trajectories approach the saddle point: these divide the plane into two and separate those initial conditions which move to one stable state from those which move to the other. These two special trajectories are known as the 'separatrices' of the saddle point.

In some special cases, the two separatrices join up to give a loop which corresponds, in fact, to the formation of a homoclinic orbit. A limit cycle may

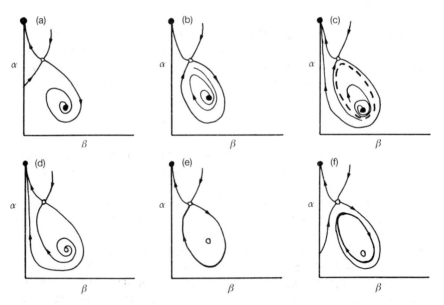

FIG. 8.8.   Phase plane representations of the birth (or death) of limit cycles through homoclinic orbit formation. In the sequence (a)–(b)–(c) the system has two stable stationary states (solid circles) and a saddle point. As some parameter is varied, the separatrices of the saddle join together to form a closed loop or homoclinic orbit (b); this loop develops as the parameter is varied further to shed an unstable limit cycle surrounding one of the stationary states. The sequence (d)–(e)–(f) shows the corresponding formation of a stable limit cycle which surrounds an unstable stationary state. (In each sequence, the limit cycle may ultimately shrink on to the stationary state it surrounds—at a Hopf bifurcation point.)

develop from this loop as we vary, say, the residence time. Two examples are given in Fig. 8.8: through the sequence (a)–(b)–(c) an unstable limit cycle emerges as the separatrices join up; and in (d)–(e)–(f) the lowest stationary state is unstable and a stable limit cycle emerges. Limit cycles are represented in the phase plane by closed loops, and oscillations in the concentrations correspond to motions around these loops. If the sequences in Fig. 8.8 are traversed in the opposite order, they allow for the death of a limit cycle and hence of oscillations.

8.3.8. Degenerate bifurcations

We have seen one type of degenerate bifurcation in our travel through previous chapters—that where two Hopf points come together and merge. In the present circumstance ($\beta_0 = 0$) this does not occur, since we only have one bifurcation on a given isola. We do have, however, a different degeneracy at which the stability of the emerging limit cycle changes, occurring for $\kappa_2 = 9/256$.

We also have the hint of a new type of degeneracy associated with systems possessing multiple stationary states. It is possible for both the trace and the determinant of the Jacobian matrix to become zero simultaneously: this gives the system two eigenvalues which are both equal to zero. These 'double-zero eigenvalue' situations are important because they represent conditions at which a Hopf bifurcation point with an associated homoclinic orbit first appears. In the present case, $\mathrm{tr}(\mathbf{J}) = \det(\mathbf{J}) = 0$ only when $\kappa_2 = \frac{1}{16}$, but then the isola has shrunk to a point.

In the next section we will investigate the behaviour shown by systems which contain both reactant A and autocatalyst B in the inflow. Our aim is to map out the regions of the parameter plane (i.e. the combinations of κ_2 and β_0) for which qualitatively different responses are found. The boundaries between these regions are the loci of degenerate bifurcations, so we wish to follow these as β_0 varies.

8.4. Influence of autocatalyst inflow on stationary-state and dynamic behaviour

In the course of any given experiment we may vary the residence time. In between experiments there are now two parameters which we can alter: the decay rate constant κ_2 and the inflow concentration of autocatalyst β_0. We thus wish to divide up the parameter plane into different regions, within each of which our experiments will reveal qualitatively different responses. We have already achieved this for the stationary-state behaviour, yielding regions of unique, isola, and mushroom patterns (see Fig. 6.18). We will now add the loci of degenerate Hopf bifurcations to this diagram, which will further subdivide the plane into smaller regions.

8.4.1. Double-zero eigenvalues

Figure 8.9 shows the various degenerate stationary-state boundaries from Fig. 6.18 as broken curves, and also (solid curve) a locus corresponding to $\beta_0\text{–}\kappa_2$ values for which the system can have a stationary-state solution with two eigenvalues simultaneously and exactly equal to zero. Mathematically the conditions for this are

$$\mathrm{tr}(\mathbf{J}) = \det(\mathbf{J}) = 0 \qquad (8.44)$$

as well as the stationary-state conditions. This gives four simultaneous equations:

$$\frac{(1 - \kappa_2 \tau_{\mathrm{res}})^2}{\tau_{\mathrm{res}}}(1 - \alpha_{\mathrm{ss}}) - \alpha_{\mathrm{ss}}(1 + \beta_0 - \alpha_{\mathrm{ss}})^2 = 0 \qquad (8.45)$$

$$\beta_{\mathrm{ss}} = \frac{1 + \beta_0 - \alpha_{\mathrm{ss}}}{1 + \kappa_2 \tau_{\mathrm{res}}} \qquad (8.46)$$

$$2\alpha_{\mathrm{ss}}\beta_{\mathrm{ss}} - (\beta_{\mathrm{ss}}^2 + \kappa_2 + 2/\tau_{\mathrm{res}}) = 0 \qquad (8.47)$$

$$(1/\tau_{\mathrm{res}})\left[(1 + \kappa_2\tau_{\mathrm{res}})(\beta_{\mathrm{ss}}^2 + 1/\tau_{\mathrm{res}}) - 2\alpha_{\mathrm{ss}}\beta_{\mathrm{ss}}\right] = 0. \qquad (8.48)$$

Three of these can be used to eliminate the 'unwanted' quantities α_{ss}, β_{ss}, and

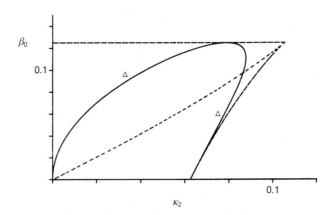

FIG. 8.9. The locus Δ of double-zero eigenvalue degeneracies of the Hopf bifurcation for cubic autocatalysis with decay. Also shown, as broken curves, are the loci of stationary-state degeneracies, corresponding to the boundaries for isola and mushroom patterns. The curve Δ lies completely within the parameter regions for multiple stationary states.

τ_{res}, the remaining equation giving the locus of the degenerate bifurcation path in the β_0–κ_2 parameter plane.

In fact these expressions are most conveniently written in the form

$$\beta_0 = x(1 - 2x) \tag{8.49}$$

$$\kappa_2 = 4x^2(1 - x)^4 \tag{8.50}$$

where x is the extent of conversion $(1 - \alpha_{ss})$ and hence varies between 0 and $\frac{1}{2}$. These equations describe the curve $\triangle$ in Fig. 8.9. This curve emerges from the origin and cuts through the region of mushroom stationary-state patterns. It touches the hysteresis line tangentially at the point $\beta_0 = \frac{1}{8}$, $\kappa_2 = 3^4/4^5 \approx 0.079$ and then turns down through the isola region: $\triangle$ cuts the $\beta_0 = 0$ axis at the point $\kappa_2 = \frac{1}{16}$, as mentioned in the previous section. In this way this locus of degeneracy has already divided the mushroom and isola regions into five subregions.

The condition $\mathrm{tr}(\mathbf{J}) = \det(\mathbf{J}) = 0$ corresponds to a Hopf bifurcation point moving exactly onto the saddle–node turning point (ignition or extinction point) on the stationary-state locus. Above the curve $\triangle$ the system may have two Hopf bifurcations, or it may have none as we will see in the next subsection. Below $\triangle$ there are two points at which $\mathrm{tr}(\mathbf{J}) = 0$, but only one of

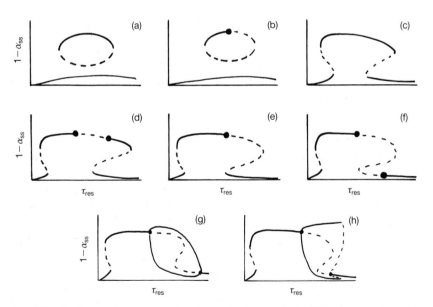

FIG. 8.10. Stationary-state patterns showing multiplicity and Hopf bifurcation points distinguished by the curve $\triangle$ in Fig. 8.9: stable states are indicated by solid curves, unstable states by broken curves and Hopf bifurcation points by solid circles.

these is a Hopf bifurcation (with $\det(\mathbf{J}) > 0$); the other lies on the branch of saddle points.

We will see that, in fact, the isola region above the curve $\triangle$ does not have any Hopf points: the isola region below $\triangle$ shows similar behaviour to that observed when $\beta_0 = 0$, with one Hopf point on the upper shore.

With mushroom patterns, below $\triangle$ there is one Hopf point, on the uppermost branch (Fig. 8.10(e)). Above $\triangle$, to the left of the tangency between this curve and the hysteresis line, the mushroom has two points of Hopf bifurcation: one on the upper branch, the other on the lowest branch at long residence times (Fig. 8.10f). Above $\triangle$, to the right of the tangency, both Hopf points are on the uppermost branch of the mushroom. If we move too far away from $\triangle$, these two Hopf points may merge and disappear. This latter behaviour is typical of the second form of degeneracy which we consider now.

8.4.2. Merging of Hopf bifurcation points (transversality)

When two Hopf bifurcation points lie on the same branch of a mushroom etc., as in Fig. 8.10(d), between them lies a region of instability for which the stationary state has eigenvalues with positive real parts. These real parts pass through zero exactly at each of the Hopf points, and so must attain a maximum value somewhere in the range of residence time between them. As the Hopf points move together, the range over which the real parts are positive decreases. Generally, the maximum positive value attained will also decrease during this process. When the Hopf points finally merge, the real part of the eigenvalues has a maximum value of zero as τ_{res} varies at the degenerate Hopf point. The stationary state never quite loses stability in this case.

We may use the above description to write down the condition for two Hopf points to merge, by requiring that

$$\mathrm{tr}(\mathbf{J}) = \mathrm{d}\,\mathrm{tr}(\mathbf{J})/\mathrm{d}\tau_{res} = 0 \qquad (8.51)$$

be satisfied together with the stationary-state conditions. We will also require that the determinant of the Jacobian should be positive at such a point.

Again these conditions specify a curve, H in Fig. 8.11, in the β_0–κ_2 parameter plane. This locus starts from the special point $\kappa_2 = \frac{1}{16}$ when $\beta_0 = 0$, but along the first part of the curve the corresponding value for $\det(\mathbf{J})$ is negative. When $\beta_0 = \frac{1}{2}(3^{3/2} - 5)$, at which point H would cross the line $\triangle$ in Figure 8.9 where $\det(\mathbf{J}) = 0$: this crossing occurs for $\kappa_2 = \frac{9}{2}(2 - \sqrt{3})^3$. Beyond this point (for higher β_0), curve H satisfies all the conditions for the merging of two Hopf points. The locus tends to $\beta_0 = 1$ as κ_2 tends to zero.

We can now draw the following conclusions. If we choose values of the parameters β_0 and κ_2 which lie to the right of or above the line H, Hopf bifurcation will not occur as we vary the residence time. Hopf bifurcations are favoured by small values for the decay rate and the autocatalyst inflow

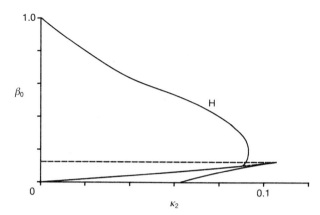

FIG. 8.11. The locus H of degenerate Hopf bifurcation points described by the transversality condition (merging of two Hopf points), eqn (8.51). Below this curve, the stationary-state locus exhibits Hopf bifurcation (dynamic instability) at some residence times; above it, the system does not.

concentration, i.e. conditions which lie below H. If we are below H but above $\triangle$, the stationary-state locus will show two Hopf points; if we are below both H and $\triangle$, there will be only one.

8.4.3. Degenerate stability of Hopf bifurcations

We have seen that the emerging limit cycle can be stable or unstable, depending on the value of κ_2, for the case $\beta_0 = 0$. The condition for the change in stability is that the exponent β_2 describing the stability of a limit cycle passes through zero at the Hopf point. We can follow this third type of degeneracy as a curve across the parameter plane by specifying that

$$\text{tr}(\mathbf{J}) = \beta_2 = 0 \qquad (8.52)$$

together with the stationary-state conditions.

These requirements specify two loci: one of them, labelled DH_1 in Fig. 8.12, emanates from the points $\beta = 0$, $\kappa_2 = 9/256$, as located in §8.3.6. This curve cuts through the parameter space for isola and mushroom patterns, but always lies below the curve $\triangle$. (In fact it intersects $\triangle$ at the common point $\beta_0 = \frac{1}{2}(3^{3/2} - 5)$, $\kappa_2 = \frac{9}{2}(2 - \sqrt{3})^3$ where the locus H also crosses.) In the vicinity of DH_1, the stationary-state curve has only one Hopf point. This changes from a subcritical bifurcation (unstable limit cycle emerging) for conditions to the right of the curve of supercritical (stable limit cycle emerging) to the left.

The second locus, DH_2, emerges from the point of tangency between the double-zero eigenvalue curve $\triangle$ and the hysteresis line at $\beta_0 = \frac{1}{8}$, $\kappa_2 = 3^4/4^5$.

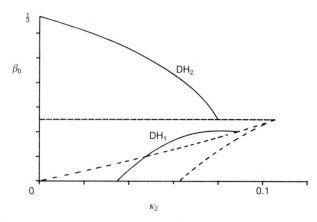

FIG. 8.12. The loci DH_1 and DH_2 corresponding to degenerate Hopf bifurcation points at which the stability of the emerging limit cycle is changing. Again, these are shown relative to the loci for stationary-state multiplicity (broken curves).

It passes through a region of the parameter plane in which the stationary-state locus does not have multistability and is always above $\triangle$. There are two Hopf points along the locus in this region: it is the bifurcation at longer residence times which changes as we cross DH_2. For parameter values to the right of DH_2, both Hopf bifurcations are supercritical, with a stable limit cycle growing and dying as shown in Fig. 8.10(g). To the left of DH_2, the Hopf bifurcation at longer residence times changes to subcritical. An unstable limit cycle now emerges here, growing as τ_{res} increases until it collides with the stable limit cycle which has grown out of the lower supercritical bifurcation (Fig. 8.10(h)).

The locus DH_2 tends to $\beta_0 = \frac{1}{3}$ as κ_2 tends to zero.

8.4.4. Bifurcation diagrams

Putting all the stationary-state and Hopf degeneracy loci together on one diagram, Fig. 8.13(a), we find the parameter plane divided into a total of 11 regions. In each of these the pattern drawn out by the stationary-state curve and limit cycles as the residence time varies is qualitatively different. Typical forms for these bifurcation diagrams ((i)–(xi)) are shown in Fig. 8.13(b).

As we move along any one of these stationary-state loci, varying τ_{res}, the number of solutions, limit cycles, and their relative orientations change, giving rise to corresponding changes in the phase portraits. It is difficult to be sure that we have ever completely counted the number of different phase portraits which occur even for a system as simple as this. Those which have been confirmed for this model (so far) are shown in Fig. 8.14.

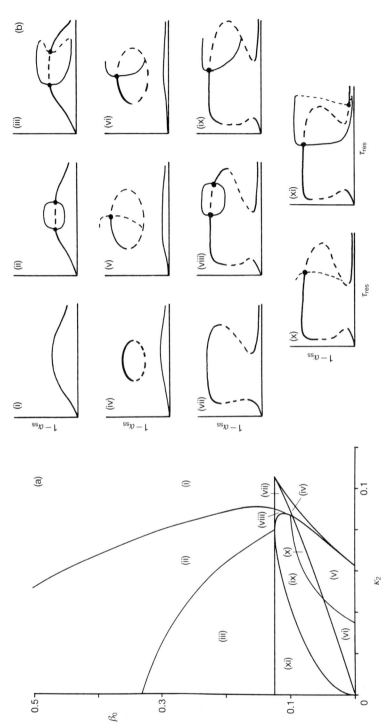

Fig. 8.13. (a) The division of the $\beta_0 - \kappa_2$ parameter region into 11 regions by the various loci of stationary-state and Hopf bifurcation degeneracies. The qualitative forms of the bifurcation diagrams for each region are given in (i)–(xi) in (b), where solid lines represent stable stationary states or limit cycles and broken curves correspond to unstable stationary states or limit cycles. (i) unique solution, no Hopf bifurcation; (ii) unique solution, two supercritical Hopf bifurcations; (iii) unique solution, one supercritical and one subcritical Hopf; (iv) isola, no Hopf points; (v) isola with one subcritical Hopf; (vi) isola with one supercritical Hopf; (vii) mushroom with no Hopf points; (viii) mushroom with two supercritical Hopf points; (ix) mushroom with one supercritical Hopf; (x) mushroom with one supercritical and subcritical Hopf bifurcations on separate branches.

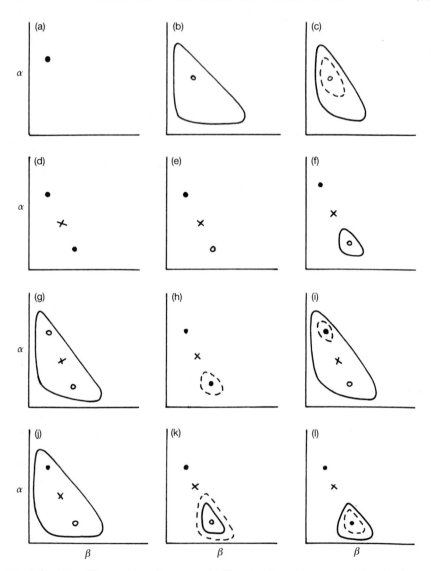

FIG. 8.14.   The different phase plane portraits identified for cubic autocatalysis with decay: (a) unique stable state; (b) unique unstable stationary state with stable limit cycle; (c) unique stable state with unstable and stable limit cycles; (d) two stable stationary states and saddle point; (e) stable and unstable states with saddle point; (f) stable state, saddle point, and unstable state surrounded by stable limit cycle; (g) two unstable states and a saddle point, all surrounded by stable limit cylcle; (h) two stable states, one surrounded by an unstable limit cycle, and a saddle point; (i) stable state surrounded by unstable limit cycle, unstable state, and saddle point, all surrounded by stable limit cycle; (j) stable state, unstable state, and saddle point, all surrounded by stable limit cycle; (k) stable state, saddle point, and unstable state, the latter surrounded by concentric stable and unstable limit cycles; (l) two stable states, one surrounded by concentric unstable and stable limit cycles, and a saddle point.

8.5. Conclusion

The analyses applied to the simplest two-variable autocatalytic system in the previous sections can obviously be brought to bear on other systems. Much effort has been expended on the first-order non-isothermal model of chapter 7, and very similar ranges of complexity are found. Up to 35 phase portraits have been predicted for the full system with the Arrhenius temperature dependence and forced cooling, with different combinations of one or three stationary states and up to three limit cycles of varying stability.

References

Dewel, G., Borckmans, P., and Walgraef, D. (1984). Relaxation kinetics in a bistable chemical system. *J. Phys. Chem.*, **88**, 5442–5.

Dewel, G., Borckmans, P., and Walgraef, D. (1985). Relaxation kinetics and critical point behaviour. *J. Phys. Chem.*, **89**, 4670–2.

Ganapathisubramanian, N. and Showalter, K. (1986). Relaxation behaviour in a bistable chemical system near the critical point and hysteresis limits. *J. Chem. Phys.*, **84**, 5427–36.

Golubitsky, M. and Langford, W. F. (1981). Classification and unfoldings of degenerate Hopf bifurcations. *J. Differ. Equations*, **41**, 373–415.

Gray, B. F. and Roberts, M. J. (1988). A method for the complete qualitative analysis of two coupled ordinary differential equations dependent on three parameters. *Proc. R. Soc.*, **416**, 361–89.

Gray, P. and Scott, S. K. (1983). Autocatalytic reactions in the isothermal, continuous stirred tank reactor: isolas and other forms of multistability. *Chem. Eng. Sci.*, **38**, 29–43.

Gray, P. and Scott, S. K. (1984). Autocatalytic reactions in the isothermal, continuous stirred tank reactor: oscillations and instabilities in the system A + 2B → 3B, B → C. *Chem. Eng. Sci.*, **39**, 1087–97.

Gray, P. and Scott, S. K. (1985). Sustained oscillations and other exotic patterns of behaviour in isothermal reactions. *J. Phys. Chem.*, **89**, 22–32.

Heinrichs, M. and Schneider, F. W. (1980). On the approach to steady states of reacting systems in the continuous stirred tank reactor. *Ber. Bunsenges. Phys. Chem.*, **84**, 857–65.

Heinrichs, M. and Schneider, F. W. (1981). Relaxation kinetics of steady-states in the continuous flow stirred tank reactor. Response to small and large perturbations: critical slowing down. *J. Phys. Chem.*, **85**, 2112–16.

Kaas-Petersen, C. and Scott, S. K. (1988). Homoclinic orbits in a simple model of an autocatalytic reaction. *Physica*, D **32**, 461–70.

Scott, S. K. and Farr, W. W. (1988). Dynamic fine structure in the cubic autocatalator. *Chem. Eng. Sci.*, **43**, 1708–10.

AUTOCATALYTIC REACTIONS IN PLUG-FLOW AND DIFFUSION REACTORS

Well-stirred systems are particularly convenient for the theoretician but are often less easy to realize in practice. Indeed, spatial inhomogeneities, with consequent molecular diffusion or thermal conduction processes, arise in many important situations—as varied as a single biological cell and a hay-stack. In the next three chapters we turn to unstirred systems, again seeking to determine 'bifurcation' phenomena driven by non-linear kinetics.

After careful study of this chapter, the reader should be able to:

(1) construct the governing rate equations appropriate to a plug-flow reactor and recognize their equivalence with the forms for a well-stirred closed system;

(2) derive the reaction–diffusion equations appropriate to a reaction–diffusion cell with their boundary conditions;

(3) recognize multiplicity as a feature of autocatalytic reaction–diffusion systems with one variable;

(4) understand the origins of isola and mushroom patterns for autocatalysis with decay;

(5) discover sustained spatial and temporal oscillations in two-variable reaction–diffusion systems;

(6) see how the corresponding well-stirred systems often give good qualitative guides to the behaviour of reaction–diffusion systems.

The previous chapters have discussed the behaviour of non-linear chemical systems in the two most familiar experimental contexts: the well-stirred closed vessel and the well-stirred continuous-flow reactor. Now we turn to a number of other situations. First we introduce the plug-flow reactor, which has strong analogies with the classic closed vessel and which will also lead on to our investigation of chemical wave propagation in chapter 11. Then we relax the stirring condition. This allows diffusive processes to become important and to interact with the chemistry. In this chapter, we examine one form of the reaction–diffusion cell, whose behaviour can be readily understood by comparison with the responses observed in the CSTR.

9.1. The plug-flow reactor (PFR)

We can imagine a PFR as a relatively long tube through which a reactive mixture or solution flows at a constant velocity c, e.g. from left to right in Fig. 9.1. The tube has a uniform cross-section along its whole length, and although the exact shape is not particularly important a circular cross-section is readily envisaged. There are two important directions: one axial, i.e. along the tube in the direction of the flow; the other radial, perpendicular to the flow direction. The basic principle of the PFR is that the flowing fluid is *well-mixed* in the radial direction, i.e. all concentrations are uniform across a given cross-section but not mixed at all axially. The latter means that there is no diffusion of chemical species along the tube: the flow can be considered as a series of separate plugs flowing along the tube with the same velocity but not interacting with each other.

Given a particular chemical reaction, which obeys a known set of reaction rate equations, we now seek to find out how the concentrations of the various species vary in space along the reactor, i.e. we wish to find the concentrations $c(x)$. To show how this problem is equivalent to that of determining the time-dependent behaviour in a well-stirred closed vessel, we can take a general example for which a reactant A is converted to a product B. Let the rate law appropriate to a well-stirred closed vessel be

$$- \mathrm{d}a/\mathrm{d}t = f(a, b) \tag{9.1}$$

where we have allowed the possibility of autocatalysis etc. in the form of the function f. Now f specifies how the rate at any given instant depends on the concentrations at that time. If we conduct the reaction in a PFR, the dependence of the reaction rate at any given point also depend only on the local concentrations at that point, and hence will have the same function $f(a, b)$. We can relate the time that a given plug has had to react to the position of the plug along the reactor through the velocity $c = \mathrm{d}x/\mathrm{d}t$. Thus

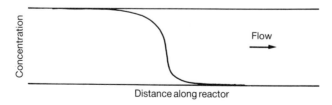

FIG. 9.1. A plug-flow reactor (PFR). Reactants flow from left to right at a constant velocity c. There is no diffusion of reactants or products along the axial direction, but radial mixing is assumed to be perfect.

$x = ct$, and we can write the rate law in various equivalent forms:

$$\frac{da}{dx} = \frac{da}{dt}\frac{dt}{dx} = \frac{1}{c}\frac{da}{dt} = \frac{f(a, b)}{c}. \tag{9.2}$$

Thus, apart from the constant factor c, the rate of change of concentration with position along the tube has the same form as the rate of change of concentration in time. With autocatalytic processes, for instance, this allows for a clock reaction in space rather than in time—if the reaction has an associated colour change, there can be a sharp band at a point x_{Cl} related to the clock time t_{Cl} by $x_{Cl} = ct_{Cl}$.

All of the temporal behaviour for the different autocatalytic models discussed in chapters 1–4 can be paralleled in a PFR. If the autocatalytic step A + 2B → 3B is coupled with a decay process B → C, the reaction may die out before all the reactant is consumed. Again, the inclusion of an uncatalysed step will be significant, ensuring complete conversion of A eventually (i.e. as $X \to \infty$) to C. With a precursor step P → A, oscillatory reaction will be possible. The oscillations in concentration will now be manifest as maxima and minima in a and b along the reactor. These spatial oscillations will begin and end at distances x_1^* and x_2^* given by $x_{1,2}^* = ct_{1,2}^*$, where t_1^* and t_2^* are as given in eqns (2.22) and (2.23). Thermokinetic oscillations in the non-isothermal model of chapter 4 require that there can be heat transfer (Newtonian cooling) radially, but not axially, along the reactor.

9.2. Reaction–Diffusion cells

The previous section introduced the first situation in this book where spatial distribution has been relevant. Even so, the only transport process was that imposed by the fluid flow: no molecular transport mechanisms were invoked, except perhaps through the hoped-for perfect radial mixing.

In this section we move on to examine the behaviour of an open system in which the transport of reactants and products relies on molecular diffusion processes. The situation we envisage is that of a 'reaction zone' in which various species diffuse and react. Outside this zone there exists an external reservoir in which the reactants have fixed concentrations. The reservoir provides a source of reactants which can diffuse across the boundary into the reaction zone, and either a source or a sink for the intermediates and products. The reaction–diffusion cell is sketched in Fig. 9.2.

9.2.1. Reaction–diffusion equations

We can again start by considering a single autocatalytic process of the form

$$A + 2B \to 3B \qquad \text{rate} = k_1 ab^2. \tag{9.3}$$

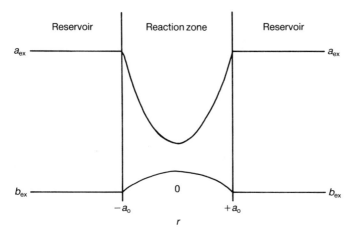

FIG. 9.2. A reaction–diffusion cell. The reactants and products, here denoted A and B, have constant concentrations a_{ex} and b_{ex} in the surrounding reservoir, but these vary across the reaction zone, $-a_0 \leqslant r \leqslant +a_0$.

The reaction–diffusion equation describes how the local concentration of the reactant A varies within some infinitesimal volume element at some point in the reaction zone. For the above reaction kinetics, this equation will have the following form:

$$\frac{\partial a}{\partial t} = D_A \nabla^2 a - k_1 ab^2 . \tag{9.4}$$

$$\underset{\substack{\text{rate of} \\ \text{change}}}{} \quad \underset{\text{diffusion}}{} \quad \underset{\text{chemistry}}{}$$

Here, the first term on the right-hand side gives the net diffusive inflow of species A into the volume element. We have assumed that the diffusive process follows Fick's law and that the diffusion coefficient does not vary with position. The spatial derivative term $\nabla^2 a$ is the Laplacian operator, defined for a general three-dimensional body in x, y, z coordinates by

$$\nabla^2 a = \frac{\partial^2 a}{\partial x^2} + \frac{\partial^2 a}{\partial y^2} + \frac{\partial^2 a}{\partial z^2}$$

i.e. it contains the second partial derivatives of the species concentration with respect to each of the three coordinates.

For three special geometries, the Laplacian operator can be written in a simpler form. These are the 'class A' geometries: the infinite slab, the infinite circular cylinder, and the sphere. For these we can write

$$\nabla^2 a = \frac{\partial^2 a}{\partial r^2} + \frac{j}{r} \frac{\partial a}{\partial r} . \tag{9.5}$$

The parameter j is a shape factor and has the value $j = 0, 1$, and 2 for the slab, cylinder, and sphere respectively. For the cylinder and sphere, then, the Laplacian involves both the first and second derivatives, but we now have only one spatial coordinate, r. (An infinite slab has a finite width in the x direction but extends to infinity in the y and z directions: with this geometry, then, no concentration gradients are established in the y and z directions and so $\partial^2 a/\partial y^2 = \partial^2 a/\partial z^2 = 0$ and $x = r$. Alternatively, we can specify that there is no transport of A across the boundary between the reaction zone and the reservoir in the y and z directions: this will again avoid concentration gradients in these directions. For a cylinder with circular cross-section in the x and y directions and extending to an infinite extent in z, or with impermeable boundaries in the z direction, and for a sphere we can again simplify the equation, i.e. we use circular or spherical coordinates.)

The behaviour of the simpler autocatalytic models in each of these three class A geometries seems to be qualitatively very similar, so we will concentrate mainly on the infinite slab, $j = 0$. For the single step process in eqn (9.3) the two reaction–diffusion equations, for the two species concentrations, have the form

$$\frac{\partial a}{\partial t} = D_A \frac{\partial^2 a}{\partial r^2} - k_1 ab^2 \tag{9.6}$$

$$\frac{\partial b}{\partial t} = D_B \frac{\partial^2 b}{\partial r^2} + k_1 ab^2 . \tag{9.7}$$

We must also specify the boundary conditions for the cell. If the boundaries of the reaction zone are at $r = \pm a_0$, and the concentrations in the external reservoir are a_{ex} and b_{ex}, we can take

$$a = a_{ex} \qquad b = b_{ex} \qquad \text{at} \quad r = \pm a_0 . \tag{9.8}$$

As in previous chapters, there is a certain economy of style to be achieved by introducing dimensionless groups and terms. Thus we take the following transformations:

concentration $\alpha = a/a_{ex} \quad \beta = b/a_{ex} \quad \beta_{ex} = b_{ex}/a_{ex}$

time $\tau = k_1 a_{ex}^2 t$

distance $\rho = r/a_0$

diffusion coefficient $D = D_A/(k_1 a_{ex}^2 a_0^2) = D_B/(k_1 a_{ex}^2 a_0^2).$

Note that the distance scale is chosen so that the reaction zone corresponds to $-1 \leqslant \rho \leqslant +1$. We have also made the reasonable assumption that both chemical species A and B have the same diffusion coefficient ($D_A = D_B$).

The dimensionless equations then become

$$\frac{\partial \alpha}{\partial \tau} = D \frac{\partial^2 \alpha}{\partial \rho^2} - \alpha \beta^2 \tag{9.9}$$

$$\frac{\partial \beta}{\partial \tau} = D \frac{\partial^2 \beta}{\partial \rho^2} + \alpha \beta^2 \tag{9.10}$$

in $-1 < \rho < +1$, with the boundary conditions

$$\alpha = 1 \qquad \beta = \beta_{\text{ex}} \qquad \text{at} \quad \rho = \pm 1. \tag{9.11}$$

One further simplification to these equations can be made if we choose suitable initial conditions. Equations (9.9) and (9.10) can be added to give

$$\frac{\partial (\alpha + \beta)}{\partial \tau} = D \frac{\partial^2 (\alpha + \beta)}{\partial \rho^2} \tag{9.12}$$

with $(\alpha + \beta) = 1 + \beta_{\text{ex}}$ on the boundary. This can be integrated to give what is effectively the stoichiometric relationship

$$\alpha + \beta = 1 + \beta_{\text{ex}} \tag{9.13}$$

holding at all times and at all points r within the reaction zone. Thus, for this one-step model, the concentrations of A and B are not independent, but linked by eqns (9.13). We can thus eliminate β, say, and work with the single reaction–diffusion equation

$$\frac{\partial \alpha}{\partial \tau} = D \frac{\partial^2 \alpha}{\partial \rho^2} - \alpha (1 + \beta_{\text{ex}} - \alpha)^2. \tag{9.14}$$

9.2.2. Solutions of the reaction–diffusion equations

A 'solution' of the reaction–diffusion equation (9.14) subject to the boundary condition on the reactant A will have the form $\alpha = \alpha(\rho, \tau)$, i.e. it will specify how the spatial dependence of the concentration (the concentration profile) will evolve in time. This differs in spirit from the solution of the same reaction behaviour in a CSTR only in the sense that we must consider position as well as time. In the analysis of the behaviour for a CSTR, the natural starting point was the identification of stationary states. For the reaction–diffusion cell, we can also examine the stationary-state behaviour by setting $\partial \alpha / \partial \tau$ equal to zero in (9.14). Thus we seek to find a concentration profile $\alpha_{\text{ss}} = \alpha_{\text{ss}}(\rho)$ which satisfies

$$D \frac{\mathrm{d}^2 \alpha_{\text{ss}}}{\mathrm{d}\rho^2} - \alpha_{\text{ss}} (1 + \beta_{\text{ex}} - \alpha_{\text{ss}})^2 = 0 \tag{9.15}$$

with $\alpha_{\text{ss}} = 1$ for $\rho = \pm 1$.

For a CSTR the stationary-state relationship is given by the solution of an algebraic equation; for the reaction–diffusion system we still have a (non-linear) differential equation, albeit ordinary rather than partial as in eqn (9.14). The stationary-state profile can be determined by standard numerical methods once the two parameters D and β_{ex} have been specified. Figure 9.3 shows two typical profiles for two different values of $D(0.1157$ and $0.0633)$ with $\beta_{ex} = 0.04$. In the upper profile, the stationary-state reactant concentration is close to unity across the whole reaction zone, reflecting only low extents of reaction. The profile has a minimum exactly at the centre of the reaction zone $\rho = 0$ and is symmetric about this central line. This symmetry with the central minimum is a feature of all the profiles computed for the class A geometries with these symmetric boundary conditions. With the lower diffusion coefficient, $D = 0.0633$, much greater extents of conversion—in excess of 50 per cent—are possible in the stationary state.

The dimensionless diffusion coefficient D can be regarded in some sense as the reaction–diffusion equivalent of the flow rate, or the inverse of the residence time, in a CSTR. In fact, we can interpret D as the quotient of the chemical and diffusional timescales

$$D = t_{ch}/t_{diff}$$

so systems with high diffusion rates relative to the characteristic chemical rate

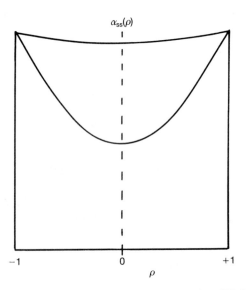

FIG. 9.3. Stationary-state concentration profiles $\alpha_{ss}(\rho)$ for a reaction–diffusion cell with a single cubic autocatalytic reaction: (a) $D = 0.1157$, only small extents of reactant consumption arise; (b) $D = 0.0633$, a higher extent of reactant consumption occurs, particularly towards the centre of the reaction zone.

will have large D, and those with fast chemistry relative to diffusion have low D. Just as we considered how the stationary-state solutions for a CSTR depended on the residence time in chapters 6–8, so we can treat D as a bifurcation parameter in the present situation. There is one additional problem here, though, in presenting this dependence. For the CSTR there is a unique concentration which represents the solution at every point within the reactor. In the reaction–diffusion cell, even for the stationary state, the concentration is different at every point. We need to choose a sensible representative of a given profile in order to construct the stationary-state

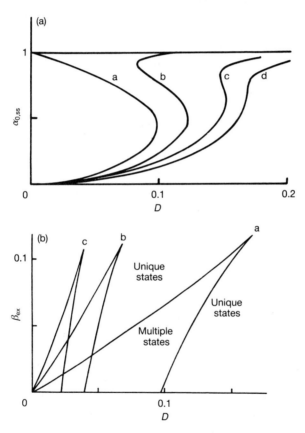

FIG. 9.4. (a) The dependence of the stationary-state concentration of reactant A at the centre of the reaction zone, $\alpha_{ss}(0)$, on the dimensionless diffusion coefficient D for systems with various reservoir concentrations of the autocatalyst B: curve a, $\beta_{ex} = 0$, so one solution is the no reaction states $\alpha_{0,ss} = 0$, whilst two other branches exist for low D; curves b and c show the effect of increasing β_{ex}, unfolding the hysteresis loop; curve d corresponds to $\beta_{ex} = 0.1185$ for which multiplicity has been lost. (b) The region of multiple stationary-state profiles forms a cusp in the β_{ex}–D parameter plane: the boundary a corresponds to the infinite slab geometry, with b and c appropriate to the infinite cylinder and sphere respectively.

locus. Clearly the natural choice is the concentration at the centre of the reaction zone, at the minimum in the profile, $\alpha_{ss}(0)$.

Figure 9.4(a) shows the stationary-state loci $\alpha_{ss}(0)$–D for different values of the 'unfolding parameter' $\beta_{ex} = 0$, 0.05, 0.10, and 0.1185. There are great qualitative similarities between these curves and the corresponding loci for the same cubic autocatalytic reaction without decay in a CSTR (Figs 1.13 and 4.8). In the absence of autocatalyst in the reservoir ($\beta_{ex} = 0$), one solution of the stationary-state equation is simply $\alpha_{ss} = 1$ for all positions ρ. This is the no reaction state discussed in previous chapters. However, even in this case there can be other stationary states, provided the dimensionless diffusion coefficient is small, i.e. $D < 0.0974$. When this inequality is satisfied, there are two other solutions. These merge at $D = 0.0974$ with the corresponding dimensionless reactant concentration at the centre of the reaction zone then being given by $\alpha_{ss}(0) = 0.4312$. For small non-zero reservoir concentrations of the autocatalyst, the stationary-state locus has a characteristic hysteresis loop. This multiplicity of solutions is unfolded as β_{ex} is increased to the value 0.1185. Recalling the behaviour in a CSTR, multiplicity is lost there when the dimensionless inflow concentration of the autocatalyst has the value $\frac{1}{8}$. Thus there are small quantitative changes between the two systems.

The unfolding of the hysteresis loop gives rise to a cusp in the D–β_{ex} parameter plane, as shown in Fig. 9.4(b). Also shown there are the cusps for infinite cylinder and spherical geometries. For the latter, multiple stationary states cease for $\beta_{ex} = 0.1129$ and 0.1078 respectively, values still smaller than the $\frac{1}{8}$ for the CSTR.

9.2.3. Stability of stationary states

The local stability of a given stationary-state profile can be determined by the same sort of test applied to the solutions for a CSTR. Of course now, when we substitute in $\alpha = \alpha_{ss} + \Delta\alpha$ etc., we have the added complexity that the profile is a function of position, as may be the perturbation. Stability and instability again are distinguished by the decay or growth of these small perturbations, and except for special circumstances the governing reaction–diffusion equation for $\partial\Delta\alpha/\partial\tau$ will be a linear second-order partial differential equation. Thus the time dependence of $\Delta\alpha$ will be governed by an infinite series of exponential terms:

$$\Delta\alpha(\rho, \tau) = \alpha_{ss}(\rho) + \sum A_i(\rho)\exp(\lambda_i\tau) \tag{9.16}$$

($i = 1, \infty$) where the coefficients A_i are dependent on the initial perturbation and are functions of position. The qualitative behaviour of the perturbation will, however, depend only on the *principal eigenvalue*, λ_{pr}. For this system, all the eigenvalues will be real: the principal eigenvalue is then the most positive one. If λ_{pr} (and hence all other λ_i) is negative, all the exponential terms will tend to zero, the $A_{pr}(\rho)\exp(\lambda_{pr}\tau)$ terms going to zero most slowly. The stationary-state profile is thus stable to infinitesimal perturbations. If, however, λ_{pr} passes through zero and becomes positive then the associated exponential term will grow in time and the system will diverge from the unstable stationary state. The special case $\lambda_{pr} = 0$, therefore, characterizes a change in stability and a turning point in the stationary-state locus.

For the present scheme, when there is a unique stationary state, we find $\lambda_{pr} < 0$ and local stability. Under circumstances with multiple solutions, the highest and lowest states always have $\lambda_{pr} < 0$ and hence are stable; the middle branch of solutions has $\lambda_{pr} > 0$ and hence is a branch of unstable saddle points.

This alternation between stable and unstable states follows that found for the CSTR. Other responses such as slowing down described in chapter 8 also occur for the reaction–diffusion model.

9.3. Reaction–diffusion systems with decaying catalyst

When the autocatalyst is not indefinitely stable, but instead undergoes further reaction such as the first-order decay process

$$\text{B} \to \text{C} \qquad \text{rate} = k_2 b \qquad (9.17)$$

the concentrations of A and B are decoupled. We now have two, coupled, reaction–diffusion equations governing the concentration profiles. In dimensionless form, these can be written as

$$\frac{\partial \alpha}{\partial \tau} = D\frac{\partial^2 \alpha}{\partial \rho^2} - \alpha\beta^2 \qquad (9.18)$$

$$\frac{\partial \beta}{\partial \tau} = D\frac{\partial^2 \beta}{\partial \rho^2} + \alpha\beta^2 - \kappa_2\beta \qquad (9.19)$$

where $\kappa_2 = k_2/k_1 a_{ex}^2$ is the dimensionless decay rate constant. Again these equations have similar structure to the corresponding CSTR forms, with the inflow and outflow replaced by the diffusion terms.

9.3.1. Stationary-state solutions

The stationary states of eqns (9.18) and (9.19) are concentration profiles of the form $\alpha_{ss}(\rho)$, $\beta_{ss}(\rho)$ satisfying the ordinary differential equations

$$D\frac{d^2\alpha_{ss}}{d\rho^2} - \alpha_{ss}\beta_{ss}^2 = 0 \qquad (9.20)$$

$$D\frac{d^2\beta_{ss}}{d\rho^2} + \alpha_{ss}\beta_{ss}^2 - \kappa_2\beta_{ss} = 0. \qquad (9.21)$$

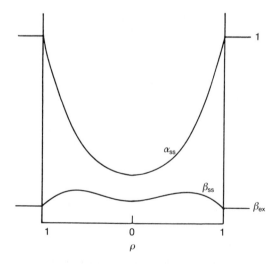

FIG. 9.5. A typical stationary-state solution for the dimensionless concentration profiles $\alpha_{ss}(\rho)$ and $\beta_{ss}(\rho)$ for cubic autocatalysis with decay. The reactant concentration shows simply a central minimum, but the autocatalyst profile has three extrema, including two non-central maxima.

A typical stationary-state solution is shown in Fig. 9.5 for the specific choice of the three parameters $D = 5.2 \times 10^{-3}$, $\beta_{ex} = 0.08$, and $\kappa_2 = 0.05$. There is considerable consumption of the reactant near to the centre of the reaction zone, but the α_{ss} has a similar form to that found in the absence of catalyst decay—a 'hanging chain' with a central minimum and no inflection point. The concentration profile for the autocatalyst has β_{ss} increasing beyond its reservoir concentration near to the edge, but falling again as it approaches the centre. Thus the autocatalyst profile can have non-central extrema, but is always symmetric about $\rho = 0$.

The concentration of the reactant A at the reaction zone centre can again be used to characterize a given stationary-state profile, and we can then investigate the dependence of $\alpha_{ss}(0)$ on the bifurcation parameter D. For the CSTR, the present model of cubic autocatalysis with the decay step allowed only three different patterns of stationary-state response: a unique solution, an isola and a mushroom. With the reaction–diffusion form of interest here, however, four different forms have been found: unique, isola, mushroom, and a single hysteresis loop (breaking wave). In fact, by analogy with the arguments from chapter 7 we can suppose that a fifth pattern is also possible—the isola plus breaking wave—but this has proven elusive so far. These five forms for the stationary-state locus are shown in Figs 9.6(a–e). Each form is found in a different region of the β_{ex}–κ_2 parameter plane, as shown in Fig. 9.6(f). The postulated, and probably vanishingly small, region corresponding to the isola plus breaking wave pattern is shown in the enlargement.

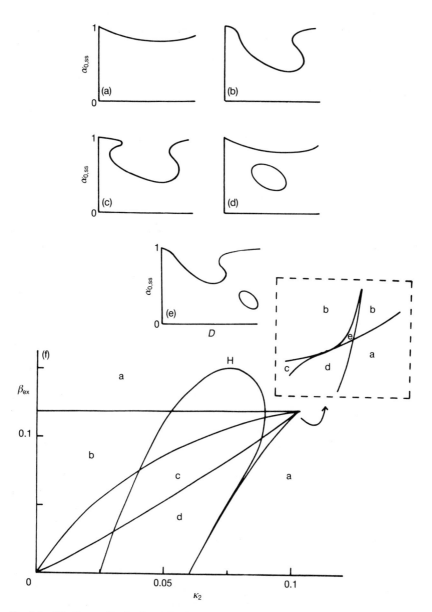

FIG. 9.6.　The five qualitative forms for the stationary-state locus $\alpha_{ss}(0)$–D for cubic autocatalysis with decay in a reaction–diffusion cell: (a) unique; (b) single hysteresis loop; (c) mushroom; (d) isola; (e) isola + hysteresis loop. (f) The division of the β_{ex}–κ_2 parameter plane giving the five regions corresponding to the stationary-state forms in (a)–(e): note that the region for response (e), shown inset, is particularly small and has not yet been successfully located.

Here, then, we have found some similarities and also some differences between the response of cubic autocatalysis with decay in the CSTR and reaction–diffusion situations. In the former, the two parameters β_0 and κ_2 are not sufficient to unfold fully the winged cusp singularity which occurs as a degenerate point in the system: a third unfolding parameter such as the rate constant for the uncatalysed reaction $A \rightarrow B$ will allow the extra response curves for the well-stirred system of equations. In the present case, for which the stationary states must satisfy differential rather than algebraic equations, the winged cusp again occurs but is now apparently not degenerate: an unfolding is achieved even with only the two parameters β_{ex} and κ_2.

9.3.2. Local stability and oscillations

Now that we have two governing reaction–diffusion equations and two independent concentrations we can hope for a more varied range of local stabilities and perhaps also for sustained oscillatory solutions. The latter may then be manifest as solutions which are distributed in both space and time.

Introducing the perturbations $\alpha = \alpha_{ss} + \Delta\alpha$ and $\beta = \beta_{ss} + \Delta\beta$ into eqns (9.18) and (9.19) we obtain, at leading order, a pair of linear equations for $\Delta\alpha$ and $\Delta\beta$. The solutions of these forms are infinite sums of exponential terms:

$$\Delta\alpha(\rho, \tau) = \alpha_{ss}(\rho) + \sum A_i(\rho)\exp(\lambda_i \tau) \tag{9.22}$$

$$\Delta\beta(\rho, \tau) = \beta_{ss}(\rho) + \sum B_i(\rho)\exp(\lambda_i \tau) \tag{9.23}$$

($i = 1, \infty$). Now, the decay or growth of these perturbations is determined by a real or complex pair of principal eigenvalues—the pair whose real parts are most positive. The local stability and character of a given stationary-state profile varies with the character of this pair of eigenvalues in exactly the same way as that described in § 3.2.1. If the pair are both real, with one positive and one negative, the solution is a saddle point (unstable); this situation holds for the middle branch in any stationary-state diagram in Fig. 9.6. The highest and lowest solutions, or a unique stationary state have a principal pair which are either real and of the same sign (negative gives stable nodal character, positive an unstable node) or a complex pair with negative real parts (stable focus) or positive real parts (unstable focus).

Of particular interest is the special case of a complex pair of principal eigenvalues whose real parts are passing through zero. This is the situation which we have seen corresponding to a Hopf bifurcation in the well-stirred systems examined previously. Hopf bifurcation points locate the conditions for the emergence of limit cycles. Using the CSTR behaviour as a guide it is relatively easy to find conditions for Hopf bifurcations, and then locally values of the diffusion coefficient for which a unique stationary state is unstable. Indeed the stationary-state profile shown in Fig. 9.5 is such a

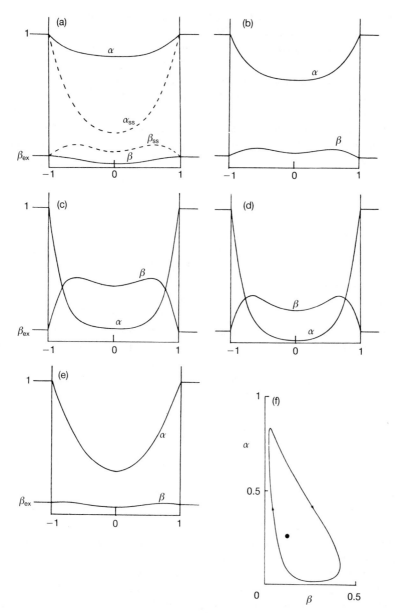

Fig. 9.7. Non-stationary behaviour in the diffusive autocatalysis model showing sustained temporal and spatial oscillations with $D = 5.2 \times 10^{-3}$, $\beta_{ex} = 0.08$, and $\kappa_2 = 0.05$: (a) $\tau = 0$ or 465 (the oscillatory period); (b) $\tau = 115$; (c) $\tau = 140$; (d) $\tau = 160$; (e) $\tau = 235$. The limit cycle obtained by plotting the concentrations at the centre of the reaction zone, $\alpha_{ss}(0)$ and $\beta_{ss}(0)$, versus each other is shown in (f). The broken curve in (a) is the unstable stationary-state profile about which the system is oscillating.

solution—an unstable focus. When the full time-dependent equations for this system are solved numerically, the concentrations do not approach the stationary profile. Instead there is a continuous and regular oscillation in α and β at every point within the reaction zone. Five 'snapshots' of this spatio-temporal oscillation are shown in Fig. 9.7, along with the corresponding limit cycle formed by plotting the concentrations of the two species at the centre of the zone against each other.

The range of parameter values in the β_{ex}–κ_2 plane for which Hopf bifurcation is possible in the present system is that lying below the line H in Fig. 9.6(f).

9.3.3. Asymmetric boundary conditions

So far almost all aspects of the stationary-state and even the time-dependent behaviour of this reaction–diffusion system differ only qualitatively from that found in the corresponding CSTR. In this section, however, we can consider a variation for which there can be no parallel in the well-stirred system—that of a reaction–diffusion cell set up with asymmetric boundary conditions. Thus we might consider our infinite slab with separate reservoirs on each side, with different concentrations of the autocatalyst in each reservoir. (For simplicity we will take the reactant concentration to be equal on each side.) Thus if we identify the reservoir concentration for $\rho \leqslant -1$ as β_L and on the other side ($\rho \geqslant +1$) as β_R, the simple boundary conditions in eqn (9.11) are replaced by

$$\begin{aligned} \alpha = 1 \quad & \beta = \beta_L \quad \text{at} \quad \rho = -1 \\ \alpha = 1 \quad & \beta = \beta_R \quad \text{at} \quad \rho = +1. \end{aligned} \tag{9.24}$$

By varying β_L and β_R, we can arrange for a situation where the whole system has to compromise between an oscillatory solution with one frequency corresponding to β_L and another corresponding to β_R. In this way we achieve a form of internal coupling between different oscillations. We will discuss the theoretical aspects of forced and coupled oscillators in chapter 13, but can display the results for the present system here. The idea of coupling two oscillators is particularly relevant if the dimensionless diffusion coefficient D is small. In such cases, diffusional effects are relatively unimportant in much of the central region of the reaction zone, over which the concentration profiles become almost uniform. There are then boundary layers near the edge of the reaction zone (whose dimensionless thickness will be of order $D^{1/2}$). The behaviour in the left-hand boundary layer (near $\rho = -1$) will be determined primarily by β_L and that in the right-hand boundary layer (near $\rho = +1$) by β_R. The uniform central region then provides a weak coupling between the two boundary layers.

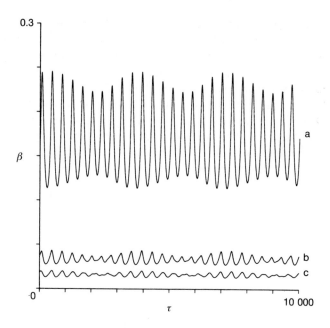

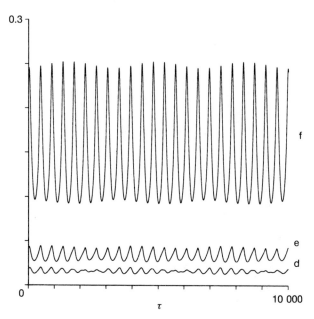

FIG. 9.8. The diffusive autocatalysis model with asymmetric boundary conditions showing a 'bi-periodic' response with $D = 10^{-3}$, $\kappa_2 = 0.04$, $\beta_L = 0.044$, and $\beta_R = 0.041$. The curves show the concentration of autocatalyst B as a function of time at various locations across the cell: (a) $\rho = -0.8$; (b) $\rho = -0.4$; (c) $\rho = 0$; (d) $\rho = 0$; (e) $\rho = +0.4$; (f) $\rho = +0.8$.

With this scenario it is possible to find parameter values such that degenerate Hopf bifurcations can arise where more than one pair of eigenvalues have real parts passing through zero at the same time. Close to such degeneracies, the system behaves as though it has more than two independent variables, and complex periodicities can be found. Figure 9.8 shows how the concentration of the autocatalyst varies in time for a series of positions across the reaction zone when $D = 0.001$, $\kappa_2 = 0.04$, $\beta_L = 0.044$ and $\beta_R = 0.041$. The large amplitude oscillation at $\rho = -0.8$ in particular shows a biperiodic form, with the amplitude of the 'normal' oscillation modulated by a second oscillation with a lower frequency. In the centre of the reaction zone, the

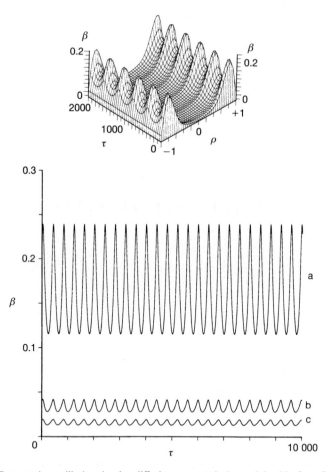

FIG. 9.9.   Symmetric oscillation in the diffusive autocatalysis model with $\beta_L = \beta_R$ showing three-dimensional representation $\beta(\rho, \tau)$ and the temporal evolution of the autocatalyst concentration at (a) $\rho = \pm 0.8$, (b) $\rho = \pm 0.4$, and (c) $\rho = 0$.

concentration shows only a very small amplitude, and rather erratic, variation. On the right-hand side of the reactor, there is again a fundamental oscillation which is also modulated.

Figure 9.9 shows the symmetric oscillations which are recovered if β_R is increased to 0.044, to be equal to β_L.

9.4. Pool chemical model with diffusion

We may also briefly consider the behaviour of the simple autocatalytic model of chapters 2 and 3 under reaction–diffusion conditions. In a thermodynamically closed system this model has no multiplicity of (pseudo-) stationary states. We now consider a reaction zone surrounded by a reservoir of pure precursor P. Inside the zone, the following reactions occur:

$$P \to A$$

$$A \to B$$

$$A + 2B \to 3B$$

$$B \to C.$$

Using the dimensionless concentrations and rate constants introduced in chapter 3, the governing rate equations for this scheme can be written in the form

$$\frac{\partial \mu}{\partial \tau} = D\frac{\partial^2 \mu}{\partial \rho^2} - \varepsilon\mu \tag{9.25}$$

$$\frac{\partial \alpha}{\partial \tau} = D\frac{\partial^2 \alpha}{\partial \rho^2} + \mu - \alpha\beta^2 - \kappa_u\alpha \tag{9.26}$$

$$\frac{\partial \beta}{\partial \tau} = D\frac{\partial^2 \beta}{\partial \rho^2} + \alpha\beta^2 + \kappa_u\alpha - \beta \tag{9.27}$$

where the distance scale $\rho = r/a_0$, with $-1 \leqslant \rho \leqslant +1$. The diffusion coefficients for the four species are assumed to be equal. The boundary conditions are

$$\mu(\rho = \pm 1) = \mu_0 \qquad \alpha(\rho = \pm 1) = \beta(\rho = \pm 1) = 0 \tag{9.28}$$

where μ_0 represents the dimensionless concentration of P in the reservoir.

The stationary-state profiles correspond to $\partial\mu/\partial\tau = \partial\alpha/\partial\tau = \partial\beta/\partial\tau = 0$. The resulting condition for the precursor

$$D\frac{\partial^2 \mu}{\partial \rho^2} = \varepsilon\mu \tag{9.29}$$

can be integrated analytically to give

$$\mu_{ss}(\rho) = \mu_0 \frac{\cosh[\rho(\varepsilon/D)^{1/2}]}{\cosh[(\varepsilon/D)^{1/2}]}. \tag{9.30}$$

This profile has the form of a caternary. The dimensionless reactant concentration at any point depends on three parameters: ρ, ε, and D. If, however, we follow the same arguments which underlie the pool chemical approximation in chapters 2 and 3, with $\varepsilon \to 0$, there is a significant simplification. Now $\mu(\rho)$ is simply a constant over the whole reaction zone, with $\mu = \mu_0$. Thus the remaining two reaction–diffusion equations become

$$\frac{\partial \alpha}{\partial \tau} = D\frac{\partial^2 \alpha}{\partial \rho^2} + \mu_0 - \alpha\beta^2 - \kappa_u \alpha \tag{9.31}$$

$$\frac{\partial \beta}{\partial \tau} = D\frac{\partial^2 \beta}{\partial \rho^2} + \alpha\beta^2 + \kappa_u \alpha - \beta. \tag{9.32}$$

9.4.1. Stationary states

Typical values for the parameters D and κ_u might be $D = 0.05$ and $\kappa_u = 0.01$. We now examine how the stationary-state concentration profiles $\alpha_{ss}(\rho)$ and $\beta_{ss}(\rho)$ depend on the dimensionless concentration of the precursor reactant, μ_0. Figure 9.10 shows the stationary-state concentrations at the centre of the reaction zone $\alpha_{ss}(0)$ and $\beta_{ss}(0)$ as functions of μ_0. These loci each draw out a hysteresis loop, with a range of corresponding multiplicity of solutions.

At low μ_0, the system has a high stationary-state concentration of A relative to that of the autocatalyst. Typically, both profiles have a maximum at the centre of the reaction zone, $\rho = 0$, as shown in Fig. 9.11(a). High reactant concentrations favour larger concentrations of the autocatalyst B and lower

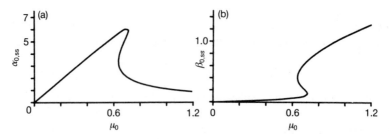

FIG. 9.10.   Stationary-state loci (a) $\alpha_{ss}(0)$–μ_0 and (b) $\beta_{ss}(0)$–μ_0 for diffusive model of autocatalysis with decaying precursor, showing hysteresis with $D = 0.05$ and $\kappa_u = 0.01$. The extinction and ignition points occur at $\mu_0 = 0.636$ and 0.72 and there is a Hopf bifurcation at $\mu_0 = 1.105$.

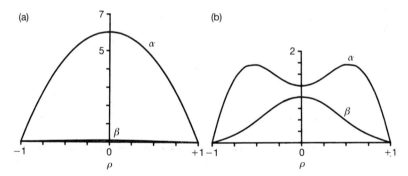

FIG. 9.11.   Two stationary-state profiles for the diffusive model with precursor, $D = 0.05$, and $\kappa_u = 0.01$: (a) $\mu_0 = 0.6885$, from the lowest branch on the stationary-state locus; (b) $\mu_0 = 0.9226$, from the uppermost branch.

values for A. The profile $\alpha_{ss}(\rho)$ now has two non-central maxima and a central minimum. The autocatalyst profile still has its maximum at the centre, but the profile may become quite flat at the middle of the reaction zone (Fig. 9.11(b)). Another feature of the central region is that the station-ary-state concentrations begin to take values close to the 'well-stirred' forms of chapter 3:

$$\alpha_{ss} = \mu/(\mu^2 + \kappa_u) \quad \text{and} \quad \beta_{ss} = \mu \qquad (9.33)$$

These features become even more accentuated if the dimensionless diffusion coefficient D is made smaller. In the limit $D \to 0$, as may be expected, the profiles tend to be virtually uniform across most of the region, with α_{ss} and β_{ss} given by eqn (9.33). At the edges of the reaction zone the system develops thin boundary layers, with thickness of order $D^{1/2}$.

9.4.2. Local stability and time-dependent solutions

The local stability of a given pair of stationary-state profiles for this model can be determined by the methods discussed in the previous section. Refer-ring to the $\beta_{ss}(0)-\mu_0$ locus, in all cases examined so far the lowest branch has stable nodal character and the middle branch is one of saddle points. There is, however, a variety in both stability and character along the upper branch. Just after the left-hand turning point, the stationary state is typically an unstable node, having two real and positive principal eigenvalues. These eigenvalues soon become a complex pair, but the real parts remain positive, so the solution is an unstable focus. As μ_0 is increased, the real part passes through zero and we have a point of Hopf bifurcation. For higher reactant concentrations the stationary state is first a stable focus and later a stable node.

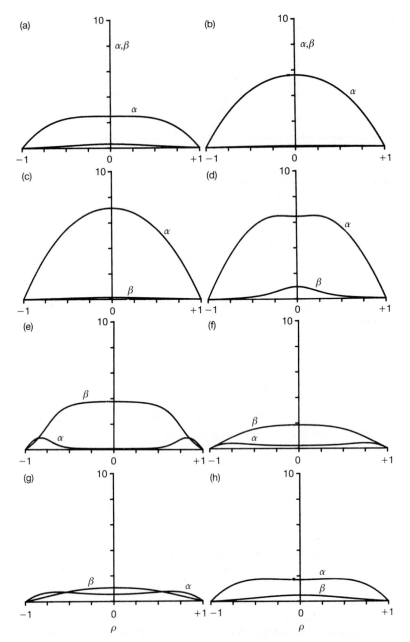

FIG. 9.12. Time-dependent concentration profiles showing sustained oscillations with $D = 0.05$, $\kappa_u = 0.01$, and $\mu_0 = 1.0$: (a) $\tau = 0$ and 118; (b) $\tau = 5$; (c) $\tau = 10$; (d) $\tau = 12$; (e) $\tau = 13$; (f) $\tau = 14$; (g) $\tau = 15$; (h) $\tau = 17$.

For the particular example in Fig. 9.10, the Hopf point occurs for $\mu_0 \approx 1.105$. The two turning points are located at $\mu_0 = 0.72$ and 0.636. This means that for reactant concentrations in the range $0.72 < \mu_0 < 1.105$, the system has a unique stationary-state profile which is unstable. Under such conditions, the reaction will exhibit time-dependent as well as spatially dependent solutions, i.e. there is a limit cycle. Some representative non-stationary profiles are shown in Fig. 9.12.

9.5. The non-isothermal catalyst pellet (smouldering combustion)

A classic chemical engineering problem of the form under consideration here is that of a non-isothermal reaction occurring in a catalytic particle or packed bed into which a single gaseous participant diffuses from a surrounding reservoir (Hatfield and Aris 1969; Luss and Lee 1970; Aris 1975; Burnell *et al.* 1983). This scenario is also appropriate to the technologically important problem of spontaneous combustion of stockpiled, often cellulosic, material in air (Bowes 1984). If we represent the concentration of the gaseous species as c, the mass- and heat-balance equations for reaction in an infinite slab are

$$\frac{\partial c}{\partial t} = D \frac{\partial^2 c}{\partial r^2} - Ac \exp\left(-\frac{E}{RT} \right) \tag{9.34}$$

$$C_{\mathrm{p}} \frac{\partial T}{\partial t} = \kappa \frac{\partial^2 T}{\partial r^2} + QAc \exp\left(-\frac{E}{RT} \right). \tag{9.35}$$

Here C_{p} is a heat capacity, κ the thermal conductivity, and Q the reaction exothermicity. The term $A \exp[-E/RT]$ is simply the Arrhenius form of the reaction rate constant.

Equations (9.34) and (9.35) can be re-expressed in dimensionless form as

$$(\mathrm{Le}) \frac{\partial \lambda}{\partial \tau} = \nabla^2 \lambda - \alpha \delta \lambda \exp\left(\frac{\theta}{1 + \varepsilon \theta} \right) \tag{9.36}$$

$$\frac{\partial \theta}{\partial \tau} = \nabla^2 \theta + \delta \lambda \exp\left(\frac{\theta}{1 + \gamma \theta} \right). \tag{9.37}$$

The dimensionless temperature excess and activation energy have the same forms as those used in chapter 4: $\theta = (T - T_{\mathrm{a}}) E / R T_{\mathrm{a}}^2$ and $\gamma = R T_{\mathrm{a}} / E$, where T_{a} is the ambient (reservoir) temperature. The dimensionless concentration λ is simply c/c_0, where c_0 is the reservoir concentration of the gaseous reactant. The group δ is given by

$$\delta = \frac{A E r_0^2 Q c_0 \exp(-E/RT_{\mathrm{a}})}{\kappa R T_{\mathrm{a}}^2}$$

and is large for large particles, high ambient temperatures, high reactant concentrations, and slow heat transfer. The Lewis number (Le) is given by (Le) $= \kappa/DC_p$ and will typically be of order unity, although departures from such values may lead to interesting non-stationary responses. Finally, α is the dimensionless heat content per unit diffusivity:

$$\alpha = \frac{\kappa R T_a^2}{QDEc_0}$$

and is small for strongly exothermic reactions.

9.5.1. Stationary-state solutions: Dirichlet boundary conditions

The simplest boundary conditions for the catalyst pellet are those for which the concentration and temperature at the edges of the slab are specified as being equal to the respective reservoir values. These Dirichlet boundary conditions then give

$$T = T_a \qquad c = c_0 \qquad \text{at} \quad r = \pm r_0$$

or

$$\theta = 0 \qquad \lambda = 1 \qquad \text{at} \quad \rho = \pm 1. \tag{9.38}$$

We now consider the dependence of the stationary-state solution on the parameter δ. To represent a given stationary-state solution we can take the dimensionless temperature excess at the middle of the slab, $\theta_{ss}(\rho = 0)$ or $\theta_{0,ss}$.

With the above boundary conditions, two different qualitative forms for the stationary-state locus $\theta_{0,ss} - \delta$ are possible. If γ and α are sufficiently small (generally both significantly less than $\frac{1}{4}$), multiplicity is a feature of the system, with ignition on increasing δ and extinction at low δ. For larger values of α or γ, corresponding to weakly exothermic processes or those with low temperature sensitivity, the hysteresis loop becomes unfolded to provide

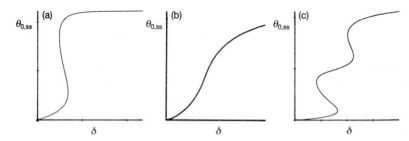

FIG. 9.13.   Stationary-state loci for the non-isothermal catalyst pellet: (a) three stationary-state branches; (b) unique response; (c) five stationary-state branches, appropriate to Robin boundary conditions with $\sigma < 1$.

a unique dependence of $\theta_{0,\mathrm{ss}}$ on δ. These fairly familiar forms are shown in Figs 9.13(a) and (b).

9.5.2. Stationary-state solutions: Robin boundary conditions

A different set of boundary conditions is that for which the concentration and temperature excess at the edge of the slab are determined by the two fluxes from the pellet. Thus, these Robin conditions have the form

$$\frac{\kappa \partial T}{\partial r} + \chi(T - T_\mathrm{a}) = 0 \quad \text{and} \quad \frac{D \partial c}{\partial r} + h(c - c_0) = 0 \quad \text{at} \quad r = \pm r_0$$

or

$$\frac{\partial \theta}{\partial \rho} + (\mathrm{Bi})_\mathrm{T} \theta = 0 \quad \text{and} \quad \frac{\partial \lambda}{\partial \rho} - (\mathrm{Bi})_\mathrm{M}(1 - \lambda) = 0 \quad \text{at} \quad \rho = \pm 1.$$

There are two new parameters—the Biot numbers for heat and mass transfer at the surface, $(\mathrm{Bi})_\mathrm{T}$ and $(\mathrm{Bi})_\mathrm{M}$. We can also distinguish their quotient $\sigma = (\mathrm{Bi})_\mathrm{T}/(\mathrm{Bi})_\mathrm{M}$. The two cases $\sigma > 1$ and $\sigma < 1$ have different stationary-state possibilities.

$\sigma > 1$. If the Biot number for heat transfer exceeds that for mass exchange, the stationary-state multiplicity is not qualitatively different from that found with Dirichlet boundary conditions. The system exhibits either a single hysteresis loop or a monotonic dependence of $\theta_{0,\mathrm{ss}}$ on δ.

$\sigma < 1$. When the magnitude of the surface transfer coefficients is reversed, a new stationary-state response arises for some parameter values. Now up to five branches of stationary states are possible (Kapila et al. 1980; Burnell et al. 1983), as shown in Figure 9.13(c). The second and fourth branches are saddle point solutions and hence unstable. The remaining three branches, however, are not saddles and thus may possess stability. The locus in Fig. 9.13(c) thus presents the possibility of three different types of accessible stationary state: one corresponding to virtually no reaction and two which would probably be referred to as diffusion limit. Of these latter two, one has a very high temperature excess and so might be identified as an 'ignited' state. The middle branch, however, has only a moderate temperature excess and a significantly lower overall reaction rate. It is tempting to assign the name 'smouldering' state to this solution, although smouldering in real chemical systems is certainly a very complex and probably non-stationary process.

9.5.3. Non-stationary states: effect of Lewis number

We will not pursue the question of local stability of the different stationary-state solutions in any great depth here. Qualitatively, the non-isothermal and cubic autocatalytic models have shown remarkable degrees of similarity in all the situations investigated so far, so it is no great surprise that many of the results from the previous section can be recreated here. There are, however, a great many parameters in this model. Luss and Lee (1970) investigated the particular effect of the Lewis number (Le). For (Le) = 1, they showed that unique stationary-state solutions are locally stable, which virtually rules out oscillatory behaviour under those circumstances. However, as (Le) is reduced, so stability can be lost. Luss and Lee found a Hopf bifurcation point and produced time-dependent temperature-position profiles similar to those shown previously in Fig. 9.12.

9.6. Summary

In this chapter we have seen that reaction–diffusion cells behave in many ways like typical open, non-linear systems. They exhibit multiplicity of stationary states, varying local stabilities and character, and allow for sustained oscillatory responses. We can even get spatial distribution of concentrations thrown in to the bargain. Hardly any of these features is new, however, almost all forms having been seen before in the corresponding well-stirred models. The latter is a significant point. Reaction–diffusion systems, with their governing partial differential equations, are less easy to handle than the ordinary differential equation for the CSTR, and fewer analytical techniques exist. It is often felt, incorrectly, that solely because of this mathematical complexity these systems can show a great deal more complex behaviour. In fact, much qualitative information can almost always be gained for a new reaction–diffusion situation by first studying the equivalent equations for the well-stirred analogue. In the next chapter we go on to examine responses, such as pattern formation, which are fundamentally non-uniform in space. Even then it will be vital to refer back constantly to the well-stirred schemes.

References

Aris, R. (1975). *The mathematical theory of diffusion and reaction in permeable catalysts*, Vols I and II. Clarendon Press, Oxford.

Bowes, P. C. (1984). *Self-heating: evaluating and controlling the hazards*. HMSO, London.

Brindley, J., Kaas-Petersen, C., Merkin, J. H., and Scott, S. K. (1988). Biperiodic behaviour in the diffusive autocatalator. *Phys. Lett.*, A **128**, 260–5.

Brown, K. J. and Eilbeck, J. C. (1982). Bifurcation, stability diagrams and varying diffusion coefficient in reaction-diffusion equations. *Bull. Math. Biol.*, **44**, 87–102.

Burnell, J. G., Lacey, A. A., and Wake, G. C. (1983). Steady-states of the reaction–diffusion equations, part 1: questions of existence and continuity of solution branches. *J. Aust. Math. Soc.*, **B24**, 374–91.

Hatfield, B. and Aris, R. (1969). Communications on the theory of diffusion and reaction, part (iv), combined effects of internal and external diffusion in the non-isothermal case. *Chem. Eng. Sci.*, **24**, 1213–22.

Kaas-Petersen, C. and Scott, S. K. (1988). Stationary-state and Hopf bifurcation patterns in isothermal, autocatalytic reaction-diffusion equations. *Chem. Eng. Sci.*, **43**, 391–2.

Kapila, A. K., Matkowsky, B. J., and Vega, J. (1980). Reactive-diffusive systems with Arrhenius kinetics: the Robin problem. *SIAM J. Appl. Math.*, **38**, 391–401.

Kay, S. R. and Scott, S. K. (1988). Multiple stationary states, sustained oscillations and transient behaviour in autocatalytic reaction diffusion equations. *Proc. R. Soc.*, **A418**, 345–64.

Luss, D. and Lee, J. C. M. (1970). The effect of Lewis number on the stability of a catalytic reaction. *Am. Inst. Chem. Eng. J.*, **16**, 620–5.

Scott, S. K. (1987). Isolas, mushrooms and oscillations in isothermal, autocatalytic reaction-diffusion equations. *Chem. Eng. Sci.*, **42**, 307–15.

CHEMICAL DIFFUSION PATTERN FORMATION
(TURING STRUCTURES)

Waves of chemical reaction may travel through a reaction medium, but the ideas of important stationary spatial patterns are due to Turing (1952). They were at first invoked to explain the slowly developing stripes that can be exhibited by reactions like the Belousov–Zhabotinskii reaction. This (rather mathematical) chapter sets out an analysis of the physically simplest circumstances but for a system (P → A → B + heat) with thermal feedback in which the internal transport of heat and matter are wholly controlled by molecular collision processes of thermal conductivity and diffusion. After a careful study the reader should be able to:

(1) choose dimensionless forms that preserve good features developed for simpler circumstances and incorporate a new length scale (based on chemical times and diffusivities) and an important new unfolding parameter β—like a Lewis number—comparing thermal and mass diffusivities (Turing structures require $\beta \neq 1$);

(2) appreciate both the simplifications offered by one-dimensional cases and their limitations;

(3) establish the local stability of stationary-state solutions under small perturbations recognizing analogies to the well-stirred systems;

(4) progress towards understanding which of a given set of possible patterns will be selected.

In this chapter we investigate some of the ways in which spatial patterns can arise in closed chemical systems. Coloured patterns can arise when a reaction is carried out in thin layers, in a petri dish for instance, corresponding to spatial variations in chemical concentrations. Thus in the Belousov–Zhabotinskii reaction, blue concentric circles indicating local regions of high Ce^{4+} concentration travel through a red background of Ce^{3+} to form target patterns. The latter are moving in time, i.e. they are a form of travelling wave, about which we will say more in the next chapter. Here we are interested in the possible spontaneous formation of stationary patterns.

The analysis is simplest algebraically for the non-isothermal scheme of chapters 4 and 5. We imagine the following experimental arrangement, illustrated in Fig. 10.1. The reaction is carried out in a long, thin rectangle. The ends of this region are perfectly insulated, so there is no heat or mass transfer (so-called 'zero-flux' boundary conditions). Along the top and bottom of the region there is again no mass transfer, but heat can be lost by Newtonian cooling. Thus we have no concentration or temperature gradients perpendicular to the long axis, but in the absence of stirring such gradients may develop along the region, parallel to the long axis, as indicated in Fig. 10.2. Although this is rather a contrived arrangement it will serve to illustrate more general principles.

From the results of chapters 4 and 5, we can predict the behaviour of the system if it is well stirred. For some experimental conditions, represented by particular values for the dimensionless reactant concentration μ and the rate constant κ, the system will have a uniform, stable stationary state (really only a pseudo-stationary state as μ is decreasing slowly because of the inevitable consumption of the reactant discussed previously). For other conditions, the stationary state loses its stability and stable uniform oscillations can be

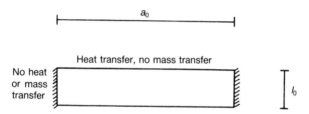

FIG. 10.1. Schematic representation of the reaction zone appropriate to our simple thermokinetic model. The end walls ($r = 0$ and $r = a_0$) are perfectly insulated against both mass and heat transfer. The side walls are impermeable to mass, but allow heat transfer such that there are no spatial gradients perpendicular to the long axis.

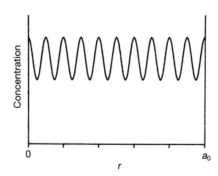

FIG. 10.2. A typical non-uniform stationary-state profile. Note the vanishing spatial derivative at the end walls ($r = 0$ and $r = a_0$) appropriate to 'zero-flux' boundary conditions.

observed. In this chapter we seek to answer two questions about the behaviour which might follow if we stop stirring.

First, we ask whether it is possible that the diffusion of the intermediate A and the conduction of heat along the box might destabilize a stable uniform state. An important condition for this is that the diffusion and conduction rates should proceed at different rates (i.e. be characterized by different timescales). Secondly, if the well-stirred system is unstable, can diffusion stabilize the system into a time-independent spatially non-uniform state? Here we find a qualified 'yes', although the resulting steady patterns may be particularly fragile to some disturbances.

10.1. Reaction–diffusion equation

The mass- and energy-balance equations for our new system, allowing for diffusion and conduction along one spatial dimension r, can be written as

$$\frac{\partial p}{\partial t} = D_P \frac{\partial^2 p}{\partial r^2} - k_0 p \tag{10.1}$$

$$\frac{\partial a}{\partial t} = D_A \frac{\partial^2 a}{\partial r^2} + k_0 p - k_1(T)a \tag{10.2}$$

$$c_v \sigma \frac{\partial T}{\partial t} = \kappa \frac{\partial^2 T}{\partial r^2} + Q k_1(T)a - \frac{\chi}{l_0}(T - T_a). \tag{10.3}$$

If one end of the box is situated at $r = 0$ and the other at $r = a_0$, then the zero-flux boundary conditions are

$$\frac{\partial p}{\partial r} = \frac{\partial a}{\partial r} = \frac{\partial T}{\partial r} = 0 \quad \text{at} \quad r = 0, a_0 \text{ for all } t. \tag{10.4}$$

In all cases we will assume that the initial distribution of the precursor P is uniform, $p(r, t = 0) = p_0$ for all $0 \leqslant r \leqslant a_0$. There is no mechanism by which a concentration gradient in P can then appear spontaneously, so the concentration of the precursor is then determined by the ordinary differential equation

$$dp/dt = -k_0 p \quad \text{for all } r. \tag{10.5}$$

This equation can be integrated to give the exponential decay

$$p = p_0 \exp(-k_0 t) \quad \text{for all } r. \tag{10.6}$$

This form can be substituted into eqn (10.2), so we have two coupled equations:

$$\frac{\partial a}{\partial t} = D_A \frac{\partial^2 a}{\partial r^2} + k_0 p_0 \exp(-k_0 t) - k_1(T)a \tag{10.7}$$

$$c_v \sigma \frac{\partial T}{\partial t} = \kappa \frac{\partial^2 T}{\partial r^2} + Q k_1(T) a - \frac{\chi}{l_0}(T - T_a). \tag{10.8}$$

The temperature and the concentration of the intermediate A are thus functions of both time and position, as will be the local value of the reaction rate constant k_1 given by the Arrhenius law

$$k_1(T) = A \exp(-E/RT). \tag{10.9}$$

10.2. Dimensionless forms

Equations (10.7) and (10.8) can be cast in dimensionless form using the groups α, θ, μ, and κ introduced in chapter 4. Thus α is the dimensionless concentration of A; θ, the dimensionless temperature rise; μ, a dimensionless measure of the concentration of the precursor reactant P; and κ is the dimensionless value of the reaction rate constant k_1 evaluated at the ambient temperature T_a. The equations then become

$$\frac{\partial \alpha}{\partial \tau} = \beta \frac{\partial^2 \alpha}{\partial x^2} + \gamma(\mu - \kappa \alpha e^\theta) \tag{10.10}$$

$$\frac{\partial \theta}{\partial \tau} = \frac{\partial^2 \theta}{\partial x^2} + \gamma(\alpha e^\theta - \theta) \tag{10.11}$$

where we have already adopted the exponential approximation $f(\theta) = e^\theta$ for the temperature dependence of the reaction rate constant k_1.

There are two new dimensionless groups, β and γ. The first of these

$$\beta = \frac{D_A c_v \sigma}{\kappa} = \frac{\text{thermal diffusion time}}{\text{mass diffusion time}} \tag{10.12}$$

represents the ratio of mass and thermal diffusivities. If heat and mass transfer occur on similar timescales, β will be close to unity. If the diffusion of A can be achieved quickly compared with conduction of heat, then β will be large. If diffusion is slow compared with conduction, β will be small.

The quantity γ is related to the length of the reaction zone a_0:

$$\gamma = \frac{\chi}{l_0 \kappa} a_0^2. \tag{10.13}$$

Thus, γ increases as the square of the length of the reaction zone.

The dimensionless length coordinate x, defined as

$$x = r/a_0 \tag{10.14}$$

is chosen so that $0 \leqslant x \leqslant 1$, irrespective of a_0. This particular formulation is especially useful if we imagine a series of experiments in systems of different size (each having a different γ).

The appropriate boundary conditions are that the fluxes of mass and heat should be zero at $x = 0$ and $x = 1$:

$$\frac{\partial \alpha}{\partial x} = \frac{\partial \theta}{\partial x} = 0 \qquad \text{at} \quad x = 0, 1 \text{ for all } \tau. \tag{10.15}$$

The dimensionless time τ in eqns (10.10) and (10.11) differs from that used in chapters 4 and 5 by the factor γ^{-1}:

$$\tau = t/\gamma t_N = \kappa t/c_v \sigma a_0^2 = t/t_F \tag{10.16}$$

where $t_F = c_v \sigma a_0^2/\kappa$ is the Fourier timescale, used here rather than the Newtonian timescale t_N. (An alternative formulation of the dimensionless equations scales the term γ into the length, taking $\rho = \gamma^{1/2} x$, so

$$\frac{\partial \alpha}{\partial \tau} = \beta \frac{\partial^2 \alpha}{\partial \rho^2} + \mu - \kappa \alpha e^\theta \tag{10.17}$$

$$\frac{\partial \theta}{\partial \tau} = \frac{\partial^2 \theta}{\partial \rho^2} + \alpha e^\theta - \theta \tag{10.18}$$

but now requiring $0 \leqslant \rho \leqslant \gamma^{1/2}$. Thus the parameter γ is not removed from the problem, just moved to the boundary condition. In this form, the dimensionless time τ is given by t/t_N.)

10.3. Diffusion-driven instabilities (Turing structures)

The idea that spatial patterns may arise spontaneously in a system such as that under consideration here, driven by the different diffusivities of the participants (here concentration and temperature), goes back to Turing. Many of the applications of this approach have been to isothermal systems of biological importance, and the basic principles apply to all two-variable schemes.

We start with a well-stirred system, so the diffusion terms $\partial^2 \alpha/\partial x^2$ and $\partial^2 \theta/\partial x^2$ make no contribution. The stationary states of this spatially uniform case satisfy

$$\theta_{ss} = \mu/\kappa \tag{10.19}$$

$$\alpha_{ss} = (\mu/\kappa)e^{-\mu/\kappa} = \theta_{ss} e^{-\theta_{ss}} \tag{10.20}$$

as in chapter 4. We know from our previous analyses that this uniform state will be stable if μ and κ lie outside the closed region of Hopf bifurcation points, whose boundary is determined by the equations

$$\kappa^* = (\theta_{ss}^* - 1)e^{-\theta_{ss}^*} \tag{10.21}$$

$$\mu^* = \kappa^* \theta_{ss}^*. \tag{10.22}$$

The division of the parameter plane into regions of stability and instability is reproduced in Fig. 10.3. If μ and κ lie within the closed region, the stationary state is unstable and spatially uniform oscillations exist.

Now let us return to the reaction–diffusion model and imagine that the stirring is somehow stopped. The question we address is whether the region of instability can be extended beyond the closed region of Fig. 10.3, particularly when the participants diffuse at different rates so $\beta \neq 1$.

Returning to eqns (10.10) and (10.11), we look for stationary-state solutions

$$\beta\frac{\partial^2\alpha}{\partial x^2} + \gamma(\mu - \kappa\alpha e^\theta) = 0 \tag{10.23}$$

$$\frac{\partial^2\theta}{\partial x^2} + \gamma(\alpha e^\theta - \theta) = 0 \tag{10.24}$$

subject to the boundary conditions (10.15).

In general we expect solutions where α and θ may vary with position as well as with the parameters μ, κ, and γ:

$$\alpha = \alpha_{ss}(x; \mu, \kappa, \gamma) \qquad \theta = \theta_{ss}(x; \mu, \kappa, \gamma). \tag{10.25}$$

However, these stationary-state equations are still satisfied by the uniform solutions (10.19) and (10.20) for which the spatial derivatives are zero and the reaction terms cancel. We denote this uniform state $\bar{\alpha}$, $\bar{\theta}$.

Thus, when the stirring stops, the uniform state remains a stationary solution of the system. Diffusion does not affect the existence of the uniform state, but it may influence its stability. In particular we are interested in determining whether this state can become unstable to spatially non-uniform perturbations.

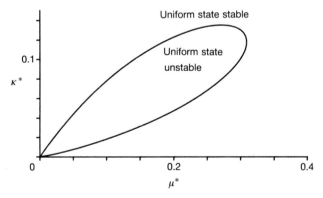

FIG. 10.3. The locus of Hopf bifurcation points indicating the conditions for loss of local stability for the spatially uniform stationary-state solution. Inside this region the system may show spatially uniform time-dependent oscillations.

As a simple example, we might impose a perturbation with a cosine distribution as illustrated in Fig. 10.4. If the uniform state is stable to such a perturbation, the amplitude will decay to zero; if the uniform state is unstable, the amplitude will grow. We could ask this question of stability with respect to any specific spatial pattern, but non-uniform solutions will also have to satisfy the boundary conditions. This latter requirement means that we should concentrate on perturbations composed of cosine terms, with different numbers of half-wavelengths between $x = 0$ and $x = 1$.

We will look at perturbations whose spatial form can be expressed in terms of the following sums (Fourier components):

$$\alpha(x) = \bar{\alpha} + \sum_{n=0}^{\infty} A_n(\tau) \cos(n\pi x) \qquad (10.26)$$

$$\theta(x) = \underset{\substack{\text{uniform} \\ \text{state}}}{\bar{\theta}} + \underset{\substack{\text{spatially and time} \\ \text{dependent terms}}}{\sum_{n=0}^{\infty} T_n(\tau) \cos(n\pi x)}. \qquad (10.27)$$

Each component of the perturbations has been separated into two terms: a time-dependent amplitude A_n and T_n, and a time-dependent spatial term $\cos(n\pi x)$. If the uniform state is stable, all the time-dependent coefficients will tend in time to zero. If the uniform state is temporally unstable even in the well-stirred case, but stable to spatial patterning, then the coefficients A_0 and T_0 will grow but the other amplitudes $A_1 - A_\infty$ and $T_1 - T_\infty$ will again tend to zero. If the uniform state becomes unstable to pattern formation, at least some of the higher coefficients will grow. This may all sound rather technical but is really only a generalization of the local stability analysis of chapter 3.

There are two specific advantages to this particular functional form for the perturbations: (i) the space and time dependences have been separated and

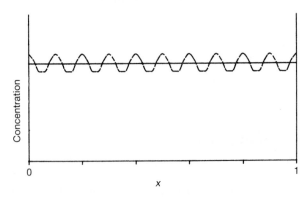

FIG. 10.4.   A typical spatially dependent perturbation about a uniform stationary-state solution.

are, hence, independent; (ii) the spatial dependence is particularly helpful since the second partial derivative of this cosine form is easily found and differs from the original only by a spatially independent factor.

10.3.1. Local stability of uniform state

Let us consider the effect of small perturbations about the uniform stationary state $\bar{\alpha}$, $\bar{\theta}$ given by eqns (10.26) and (10.27). For ease of manipulation we can represent the concentration and temperature rise as

$$\alpha = \bar{\alpha} + \Delta\alpha \qquad (10.28)$$

$$\theta = \bar{\theta} + \Delta\theta \qquad (10.29)$$

where $\Delta\alpha$ and $\Delta\theta$ are the small perturbations given by the sums in eqns (10.26) and (10.27).

If we substitute these expressions for α into eqn (10.10) we obtain

$$\frac{\partial \bar{\alpha}}{\partial \tau} + \frac{\partial \Delta\alpha}{\partial \tau} = \beta\left(\frac{\partial^2 \bar{\alpha}}{\partial x^2} + \frac{\partial^2 \Delta\alpha}{\partial x^2}\right) + \gamma[\mu - \kappa(\bar{\alpha} + \Delta\alpha)\exp(\bar{\theta} + \Delta\theta)]. \quad (10.30)$$

The stationary-state solution α is constant in time and uniform in space, so $\partial\bar{\alpha}/\partial\tau = \partial^2\bar{\alpha}/\partial x^2 = 0$. Our specification that the perturbations $\Delta\alpha$ and $\Delta\theta$ should be small also allows us to linearize the reaction terms, giving

$$\frac{\partial \Delta\alpha}{\partial \tau} = \beta\frac{\partial^2 \Delta\alpha}{\partial x^2} + \gamma[\mu - \kappa\bar{\alpha}e^{\bar{\theta}} - \kappa e^{\bar{\theta}}\Delta\alpha - \kappa\bar{\alpha}e^{\bar{\theta}}\Delta\theta - \cdots]. \quad (10.31)$$

Substituting for $\bar{\alpha}$ and $\bar{\theta}$, the first two terms in the brackets cancel (because that is the uniform stationary-state condition). Proceeding in a similar way for the temperature perturbations, we finally obtain the governing equations

$$\frac{\partial \Delta\alpha}{\partial \tau} = \beta\frac{\partial^2 \Delta\alpha}{\partial x^2} - \gamma\kappa e^{\mu/\kappa}\Delta\alpha - \gamma\mu\Delta\theta - \cdots \qquad (10.32)$$

$$\frac{\partial \Delta\theta}{\partial \tau} = \frac{\partial^2 \Delta\theta}{\partial x^2} + \gamma e^{\mu/\kappa}\Delta\alpha + \gamma\left(\frac{\mu}{\kappa} - 1\right)\Delta\theta + \cdots. \qquad (10.33)$$

These equations are linear in the variables $\Delta\alpha$ and $\Delta\theta$ but still involve partial derivative forms. Comparing (10.28) and (10.29) with (10.26) and (10.27), we see that the perturbations are given by

$$\Delta\alpha = \sum_{n=0}^{\infty} A_n(\tau)\cos(n\pi x) \qquad (10.34)$$

$$\Delta\theta = \sum_{n=0}^{\infty} T_n(\tau)\cos(n\pi x). \qquad (10.35)$$

If we consider just the $n = 1$ terms:

$$\Delta\alpha_1 = A_1(\tau)\cos(\pi x) \qquad (10.36)$$

$$\Delta\theta_1 = T_1(\tau)\cos(\pi x) \qquad (10.37)$$

then the partial derivatives with respect to time are

$$\frac{\partial\Delta\alpha_1}{\partial\tau} = \cos(\pi x)\frac{dA_1}{d\tau} \qquad \frac{\partial\Delta\theta_1}{\partial\tau} = \cos(\pi x)\frac{dT_1}{d\tau} \qquad (10.38)$$

and the second spatial derivatives become

$$\frac{\partial^2\Delta\alpha_1}{\partial x^2} = A_1(\tau)\frac{\partial^2(\cos(\pi x))}{\partial x^2} = -\pi^2 A_1(\tau)\cos(\pi x) \qquad (10.39)$$

$$\frac{\partial^2\Delta\theta_1}{\partial x^2} = T_1(\tau)\frac{\partial^2(\cos(\pi x))}{\partial x^2} = -\pi^2 T_1(\tau)\cos(\pi x). \qquad (10.40)$$

Substituting for $\Delta\alpha_1$, $\partial\Delta\alpha_1/\partial\tau$, $\partial^2\Delta\alpha_1/\partial x^2$, and the equivalent terms in $\Delta\theta_1$ gives

$$\cos(\pi x)\frac{dA_1}{d\tau} = -\beta\pi^2 A_1\cos(\pi x) - \gamma\kappa e^{\mu/\kappa}A_1\cos(\pi x)$$

$$- \gamma\mu T_1\cos(\pi x) \qquad (10.41)$$

$$\cos(\pi x)\frac{dT_1}{d\tau} = -\pi^2 T_1\cos(\pi x) + \gamma e^{\mu/\kappa}A_1\cos(\pi x)$$

$$+ \gamma\left(\frac{\mu}{\kappa} - 1\right)T_1\cos(\pi x). \qquad (10.42)$$

All of the terms in these equations have the factor $\cos(\pi x)$. Cancelling this leaves a pair of linear ordinary differential equations for the amplitudes of the perturbations:

$$\frac{dA_1}{d\tau} = -(\gamma\kappa e^{\mu/\kappa} + \beta\pi^2)A_1 - \gamma\mu T_1 \qquad (10.43)$$

$$\frac{dT_1}{d\tau} = \gamma e^{\mu/\kappa}A_1 + \left[\gamma\left(\frac{\mu}{\kappa} - 1\right) - \pi^2\right]T_1. \qquad (10.44)$$

These are very similar to the equations derived in chapter 3 for the decay or growth of small perturbations in well-stirred systems. Again we can expect exponential growth or decay, depending on the relative magnitudes of the four coefficients in these equations which, in turn, depend on γ, μ, κ, and β.

Similar manipulations can be performed for all the terms in the summations (10.34) and (10.35), i.e. for all the components of different wavelengths.

In general, the time dependence of the perturbation with index n is

$$\frac{dA_n}{d\tau} = -(\gamma\kappa e^{\mu/\kappa} + \beta n^2\pi^2)A_n - \gamma\mu T_n \tag{10.45}$$

$$\frac{dT_n}{d\tau} = \gamma e^{\mu/\kappa}A_n + \left[\gamma\left(\frac{\mu}{\kappa} - 1\right) - n^2\pi^2\right]T_n. \tag{10.46}$$

10.3.2. Time dependence of perturbations

One possible result of our perturbation is that all the modes associated with every possible integer value of n in the above equations will decay. Then $A_n \to 0$, $T_n \to 0$ for all n as $\tau \to \infty$. In this case, the uniform state will be stable to all perturbations.

The decay or possible growth of a given mode will be determined by the eigenvalues of the Jacobian matrix associated with the perturbation equations above: these in turn are determined by the trace and determinant. For a given n the trace has the form

$$\text{tr}(\mathbf{J}) = \gamma(\mu/\kappa - 1 - \kappa e^{\mu/\kappa}) - (1 + \beta)n^2\pi^2. \tag{10.47}$$

We can recognize the first term as the trace of the matrix for the well-stirred system of chapter 4 (let us call this $\text{tr}(\mathbf{U})$) multiplied by the positive quantity γ. We have specified that we are to consider here systems which have a stable stationary state when well stirred, i.e. for which $\text{tr}(\mathbf{U})$ is negative. The additional term associated with diffusion in eqn (10.47) can only make $\text{tr}(\mathbf{J})$ more negative, apparently enhancing the stability. There are no Hopf bifurcations (where $\text{tr}(\mathbf{J}) = 0$) induced by choosing a spatial perturbation with non-zero n.

The determinant of the Jacobian for (10.45) and (10.46) is

$$\det(\mathbf{J}) = \gamma^2\kappa e^{\mu/\kappa} + \gamma[\kappa e^{\mu/\kappa} - \beta(\mu/\kappa - 1)]n^2\pi^2 + \beta n^4\pi^4. \tag{10.48}$$

Again the first term on the right-hand side can be recognized as being proportional to the determinant of the uniform system (now multiplied by γ^2), which was necessarily positive. Equation (10.48) has, however, extra terms which can cause the sign of $\det(\mathbf{J})$ to change, particularly if the ratio of the diffusivities β is large.

If $\det(\mathbf{J})$ is positive for all n, then the amplitudes of all the components of any perturbation will decay back to the spatially uniform stationary state. As mentioned above, $\det(\mathbf{J})$ is positive for $n = 0$, and clearly will always be positive for sufficiently large n when the last term dominates. However, eqn (10.48) is a quadratic in n^2: a completely stable uniform state arises if there are no real solutions to the condition $\det(\mathbf{J}) = 0$. We can write the condition for the determinant to vanish (the dispersion relation) as

$$n^2 = \frac{\beta(\mu/\kappa - 1) - \kappa e^{\mu/\kappa} \pm \{[\beta(\mu/\kappa - 1) - \kappa e^{\mu/\kappa}]^2 - 4\beta\kappa e^{\mu/\kappa}\}^{1/2}}{2\pi^2\beta/\gamma}. \quad (10.49)$$

If β, μ, and κ are chosen suitably, this equation may have two real and positive solutions, n_-^2 and n_+^2 say. The uniform stationary state is then unstable to perturbations which have n half-wavelengths where n is an integer lying in the range

$$n_-^2 \leqslant n \leqslant n_+^2. \quad (10.50)$$

For example, if the various parameters are such that $n_-^2 = 0.2$ and $n_+^2 = 123$, then the uniform state will be unstable to perturbations with $n = 1, 2, 3, \ldots,$ 9, 10, and 11, but not to $n = 12$ or higher.

Clearly n_-^2 and n_+^2 are only real if the discriminant under the square root sign in (10.49) is positive for the experimental conditions (i.e. the values of β, μ, and κ). Thus we require that

$$[\beta(\mu/\kappa - 1) - \kappa e^{\mu/\kappa}]^2 - 4\beta\kappa e^{\mu/\kappa} \geqslant 0. \quad (10.51)$$

The parameters μ and κ are determined by such factors as the initial reactant concentration and the ambient temperature, whereas β is simply the ratio of the mass and thermal diffusivities. We can treat eqn (10.51) as a condition on β (in which it is quadratic) in terms of μ and κ. If the inequality is to be satisfied, giving the possibility of diffusion-driven instability, then β must be larger than a given critical value β_c given by

$$\beta > \beta_c = \frac{\kappa + \mu + 2(\kappa\mu)^{1/2}}{(\mu/\kappa - 1)^2} e^{\mu/\kappa}. \quad (10.52)$$

If the mass diffusion coefficient is sufficiently large compared with the thermal diffusivity, so $\beta > \beta_c$, the range between n_-^2 and n_+^2 will be non-zero. There is another consideration: n can only have discrete integer values, the lowest of which for a non-uniform state is $n = 1$. Thus for observable patterns we must make sure that at least n_+^2 exceeds unity. Equation (10.49) shows that this last requirement puts a lower bound on the size parameter γ: we need $\gamma > \gamma_c$, where

$$\gamma_c = \frac{2\pi^2\beta}{\beta(\mu/\kappa - 1) - \kappa e^{\mu/\kappa} + \{[\beta(\mu/\kappa - 1) - \kappa e^{\mu/\kappa}]^2 - 4\beta\kappa e^{\mu/\kappa}\}^{1/2}}. \quad (10.53)$$

10.3.3. Presentation of results

Even in this simple system we have four parameters to keep track of: μ and κ are in some sense determined by the chemistry, γ and β by the size and diffusion characteristics. In addition we have the wave number n of the

various possible patterned solutions. Because of this large number of parameters there are many different ways of presenting the results. In a given
situation μ, κ, γ, and β may well be determined and we would only need to
find any values of n to which the uniform state might be unstable. The real
purpose of simplified models, however, is to bring out general principles and
trends. Some conclusions can already be drawn.

First, it is possible that diffusional processes can destabilize a uniform
stationary state. We suspect (but have not yet proved) that a destabilized
system will move to a spatially patterned steady state (it could instead
develop into time, and possibly spatially, dependent oscillations). Instability
is favoured by unequal diffusion rates (i.e. large β) in favour of the enhanced
diffusion of the non-autocatalytic species (here A) rather than the
autocatalyst (here ΔT). Similarly, patterned solutions are favoured by large
reaction zones (high γ), and we will see that more modes can be excited as
γ increases. (In biological problems, this suggests greater variety of markings
should appear as animals grow.)

10.3.4. Adjusting size and diffusion

One way of presenting the various results above graphically is to take
particular values for the chemical parameters μ and κ such that the uniform
state is stable. The locus $\det(\mathbf{J}) = 0$ can then be plotted out in the β–γ plane;
there will be different loci for different choices of the wave number n.

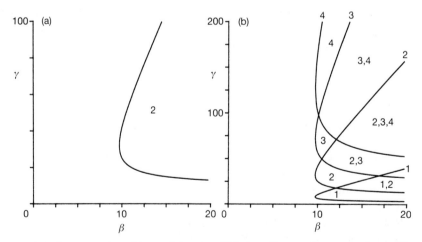

FIG. 10.5.   Representation of the conditions for which the uniform stationary state is unstable to
perturbations with a particular number of half-wavelengths: (a) $n = 2$; (b) the loci for $n = 1, 2, 3$,
and 4, showing how these overlap to create regions in which more than one spatial perturbation
may cause departure from the uniform state.

Figure 10.5(a) shows such a locus for $\mu = 0.5$, $\kappa = 0.1$ and for $n = 2$, so perturbations have the spatial form $\cos(2\pi x)$. To the left of this boundary and below it, the uniform stationary state is stable to this particular perturbation; to the right, for high β and large γ, it is unstable. We can draw similar loci for all integer n, as shown in Fig. 10.5(b), noting in general that the boundary is 'pushed up' the parameter plane by increasing the number of half-wavelengths. The various loci intersect each other, so the regions in which the uniform state is unstable to particular shapes of perturbation overlap. In some regions the uniform state may be only unstable to a single mode; in others it is unstable to many. In the latter case the system cannot move to all the possible choices, and one must dominate. One of the important questions of current interest is the determination of which features determine the particular pattern selected when more than one appears possible.

10.3.5. Pattern selection

The idea of this short subsection is to present a 'rule of thumb' which will allow us to predict which of a given set of possible patterns will be selected. In other words we seek the 'dominant mode'. For any given mode, specified by the number of half-wavelengths n, the growth rate of that component of any given initial perturbation is determined by the magnitude of the corresponding positive eigenvalue

$$\lambda_+^{(n)} = \tfrac{1}{2}\{\mathrm{tr}(\mathbf{J}) + [\mathrm{tr}(\mathbf{J}^2) - 4\det(\mathbf{J})]^{1/2}\}. \tag{10.54}$$

For a given set of parameters (β, μ, κ, and γ) we may ask which mode has the largest positive eigenvalue and postulate that this will be dominant. Treating n as a continuous variable for the moment, we can maximize $\lambda_+^{(n)}$, giving

$$n^2\pi^2 = \frac{\gamma}{\beta - 1}[(1 + \beta)(\mu e^{\mu/\kappa}/\beta)^{1/2} - (\mu/\kappa - 1 + \kappa e^{\mu/\kappa})]. \tag{10.55}$$

Thus for given values of the various parameters the corresponding value for n can be found from this expression. We may expect that the closest integer for this value will give the wave number of the dominant mode.

Boundaries between regions of dominance by different modes correspond to parameter values at which the positive eigenvalues for two adjacent wave numbers n and $(n + 1)$ become equal (and larger than those for all the other modes).

Murray (1982) has confirmed this pattern of behaviour empirically for a variety of two-variable models with zero-flux boundary conditions such as those considered here. In general, the dominant mode increases in wave number n as the size of the reaction zone γ increases, but decreases as the ratio of diffusivities increases—as shown in Fig. 10.6.

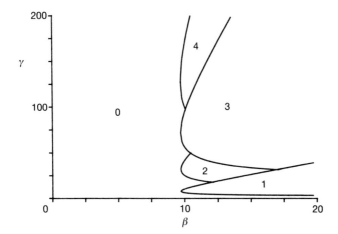

FIG. 10.6. Division of the γ–β parameter plane into regions according to the different dominant spatial forms. For the zero-flux model, an odd number of half-wavelengths appears to be favoured.

10.3.6. Adjusting chemistry

An alternative way of portraying the pattern formation behaviour in systems of the sort under consideration here is to delineate the regions in chemical parameter space (the μ–κ plane) over which the uniform state is unstable to non-uniform perturbations. We have already seen in chapter 4, and in Fig. 10.3, that we can locate the boundary of Hopf instability (where the uniform state is unstable to a uniform perturbation and at which spatially uniform time-dependent oscillations set in). We can use the equations derived in § 10.3.2 to draw similar loci for instability to spatial pattern formation. For this, we can choose a value for the ratio of the diffusivities β and then find the conditions where eqn (10.48), regarded as a quadratic in either γ or n, has two real positive solutions. The latter requires that

$$\kappa e^{\mu/\kappa} - \beta(\mu/\kappa - 1) < 0. \tag{10.56}$$

For the roots to be real, the parameters must also satisfy the condition

$$\kappa^2 e^{2\mu/\kappa} - 2\beta(\mu/\kappa + 1)\kappa e^{\mu/\kappa} + \beta(\mu/\kappa - 1)^2 > 0. \tag{10.57}$$

Recognizing that μ and κ can be related to the uniform stationary-state temperature excess $\theta_{ss} = \mu/\kappa$, the condition for real positive solutions to eqn (10.48) to exist can be expressed as

$$\kappa = \beta e^{-\theta_{ss}}(\theta_{ss}^{1/2} - 1)^2 \tag{10.58}$$

$$\mu = \kappa\theta_{ss} \tag{10.59}$$

with $\theta_{ss} > 1$.

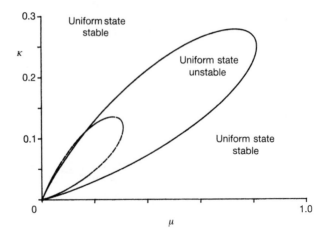

Fig. 10.7. Representation of conditions in the κ–μ parameter plane for instability of the uniform stationary state with respect to spatial perturbations for a system with $\beta = 10$. Also shown (broken curve) is the Hopf bifurcation locus, within which the uniform state is itself unstable. The two loci cross at some point on their upper shores.

Figure 10.7 shows this locus for a system with $\beta = 10$. Also shown, as a broken curve, is the Hopf locus for the well-stirred system. The latter is important, since we must remain outside this region for the uniform system to be stable in the absence of diffusion. Clearly, for this particular choice of β, there is a significant region in which the well-stirred system is stable (and hence the uniform state is stable to uniform perturbations) but unstable to pattern formation.

Changing β merely alters the size of the region of instability to pattern formation, as the values of μ and κ which form the locus are directly proportional to β in eqns (10.58) and (10.59). Thus, increasing β increases the range of μ and κ over which pattern formation is possible. Table 10.1 gives the general coordinates of the locus, which can be used for any β.

The point at which the pattern instability and Hopf bifurcation loci cross, for a system with given β, is easily located by taking eqns (10.21) and (10.58). These give

$$\theta_{ss}^{1/2} = \frac{1 + \beta}{\beta - 1} \tag{10.60}$$

(and also $\theta_{ss} = 1$) from which

$$\kappa = \frac{4\beta}{(\beta - 1)^2} \exp\left(-\frac{(1 + \beta)^2}{(\beta - 1)^2} \right) \quad \mu = \frac{4\beta(1 + \beta)^2}{(\beta - 1)^4} \exp\left(-\frac{(1 + \beta)^2}{(\beta - 1)^2} \right). \tag{10.61}$$

Table 10.1

Location of locus of instability
for pattern formation, from
eqns (10.58) and (10.59)

θ_{ss}	$10^2 \, \kappa/\beta$	$10^2 \, \mu/\beta$
1.0	0	0
1.1	0.0793	0.0872
1.3	0.536	0.696
1.5	1.13	1.69
1.75	1.81	3.17
2.0	2.32	4.64
2.5	2.77	6.93
2.618	2.79	7.28
3.0	2.67	8.00
3.5	2.29	8.02
4.0	1.83	7.33
5.0	1.03	5.15
6.0	5.21	3.13
8.0	0.112	0.897
10.0	0.0212	0.212

Thus the two loci cross at some point provided β is greater than unity. Any inequality in the diffusion coefficients, in favour of the non-autocatalytic species, can support pattern formation over some range of experimental conditions.

We may also note, for the special case $\beta = 1$, that the locus described by eqns (10.58) and (10.59) is exactly that corresponding to the boundary between unstable focus and unstable node for the well-stirred system. This seems to be a general equivalence between the existence of unstable nodal solutions in the well-stirred system and the possibility of diffusion-driven pattern formation in the absence of stirring. We have seen in chapter 5 that unstable nodes are not found in the present model if the full Arrhenius rate law is used and the activation energy is low, i.e. if $E < 4RT_a$. In that case we would also not expect spatial instability.

Returning to the example $\beta = 10$ (and to the exponential approximation), we can choose a set of conditions inside the region of instability but above the Hopf curve, for example $\mu = 0.6$ and $\kappa = 0.2$. We must now determine any restrictions on the size of the system. From eqn (10.53), we can calculate that the size parameter γ must exceed 5.784 if the $n = 1$ pattern is required. In fact we can state for this choice of μ and κ that, for a pattern with n half-wavelengths, we require

$$\gamma \geqslant 5.784n^2. \tag{10.62}$$

10.3.7. Time dependence due to reactant consumption

Before finishing this section, it is interesting to recall that the parameter μ, which is the dimensionless concentration of the reactant P, is not really a constant but varies slowly with time. In fact we have a spatially uniform exponential decrease

$$\mu = \mu_0 e^{-\varepsilon\tau}. \tag{10.63}$$

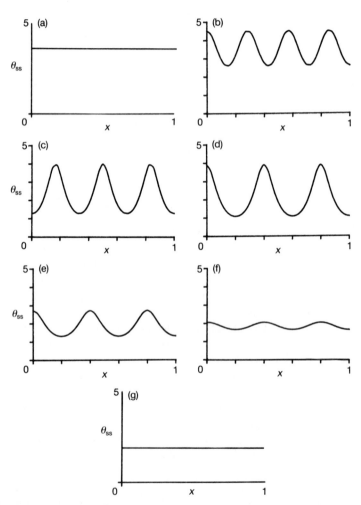

FIG. 10.8. The onset, development, and death of pseudo-stationary-state patterning in the temperature-position profile for the model with precursor decay; $\mu_0 = 1$, $\varepsilon = 0.01$, $\kappa = 0.2$, $\gamma = 550$, and $\beta = 10$: (a) $\tau = 27$, uniform solution still stable; (b) $\tau = 35$, $n = 7$ solution; (c) $\tau = 75$, $n = 6$; (d) $\tau = 85$, $n = 5$; (e) $\tau = 97$, $n = 5$; (f) $\tau = 99$, $n = 5$, but amplitude almost zero; (g) $\tau = 100$, spatially uniform state returns.

We can imagine an experiment starting from a particular point in the μ–κ plane and slowly moving across the diagram, to the left and parallel to the μ axis in Fig. 10.7, as the decreased reactant concentration decays. It is easiest if we consider an example with $\kappa > e^{-2}$, so that we do not traverse the region of Hopf bifurcation at any stage, and most dramatic if we take relatively high values for β and γ so that more modes come into play. We will take a value for μ_0 such that the uniform state is initially stable to spatial perturbations i.e. a point to the right of the region of diffusion-driven instability shown in Fig. 10.7. Thus we take $\mu_0 = 1.0$ with $\kappa = 0.2$ and then choose a dimensionless size and ratio of diffusivities such that $\gamma = 550$ and $\beta = 10$. From eqn (10.62) we can predict that patterns with $n < 10$ may be possible.

With a typically small value for the decay exponent (which is proportional to the rate constant k_0) of $\varepsilon = 0.01$, the concentration α of intermediate A and the temperature rise θ move rapidly from their initial conditions to the appropriate uniform stationary-state values. As time increases and μ decreases, the system eventually crosses the dispersion curve. The uniform state is now unstable to spatial perturbations. The system next moves rapidly away to a non-uniform solution, first settling to a state with seven half-wavelengths ($n = 7$). This bifurcation occurs at $\tau = 28$. This pattern persists, with the amplitude of the waveform increasing in time. Figure 10.8(b) shows the temperature profile at $\tau = 35$. At $\tau = 67$, there is another bifurcation and the profile moves abruptly to one with six half-wavelengths ($n = 6$). We then have a period in which this pattern, shown in Fig. 10.8(c), holds sway. The system then switches to the $n = 5$ pattern at $\tau = 85$. The amplitude of this waveform now decreases in time, as shown through the sequence in Figs 10.8(d–f). As τ approaches 99, the amplitude of the spatial non-uniformity tends to zero. The system passes through the left-hand section of the locus of instability. The uniform state is now stable, and the system settles on to its appropriate flat profile (Fig. 10.8(g)). The final stage then sees this uniform state decay in α and θ to zero, as τ tends to infinity.

10.4. Formation of stable patterns when uniform state is unstable

In the previous section we have taken care to keep well away from parameter values μ and κ for which the uniform stationary state is unstable to Hopf bifurcations. Thus, instabilities have been induced solely by the inequality of the diffusivities. We now wish to look at a different problem and ask whether diffusion processes can have a stabilizing effect. We will be interested in conditions where the uniform state shows time-dependent periodic oscillations, i.e. for which μ and κ lie inside the Hopf locus. We wish to see whether, as an alternative to uniform oscillations, the system can move on to a time-independent, stable, but spatially non-uniform, pattern. In fact the answer to this question is strictly 'no'. However, in a more limited sense a certain stability can be induced: the resulting profiles are always unstable to uniform perturbations, but may be stable to a whole class of other disturbances (Merkin and Needham 1989).

10.4.1. Spatially uniform perturbations

The general equations governing the growth or decay of different components of a disturbance are again given by (10.45) and (10.46). For simplicity we can take the case of equal diffusivities, so $\beta = 1$. Thus, for the nth mode:

$$\frac{\mathrm{d}A_n}{\mathrm{d}\tau} = - (\gamma \kappa e^{\mu/\kappa} + n^2 \pi^2) A_n - \gamma \mu T_n \tag{10.64}$$

$$\frac{\mathrm{d}T_n}{\mathrm{d}\tau} = \gamma e^{\mu/\kappa} A_n + [\gamma(\mu/\kappa - 1) - n^2 \pi^2] T_n. \tag{10.65}$$

For a spatially uniform perturbation ($n = 0$), these become:

$$\frac{\mathrm{d}A_0}{\mathrm{d}\tau} = - \gamma \kappa e^{\mu/\kappa} A_0 - \gamma \mu T_0 \tag{10.66}$$

$$\frac{\mathrm{d}T_0}{\mathrm{d}\tau} = \gamma e^{\mu/\kappa} A_0 + \gamma(\mu/\kappa - 1) T_0. \tag{10.67}$$

Apart from the common factor γ, these are the same as the equations governing growth or decay of small perturbations in the well-stirred system. The eigenvalues $\lambda_{1,2}$ are given by

$$\lambda^2 - \gamma(\mu/\kappa - 1 - \kappa e^{\mu/\kappa})\lambda + \gamma^2 \kappa e^{\mu/\kappa} = 0 \tag{10.68}$$

i.e.

$$\lambda^2 - \gamma(\mathrm{tr}(\mathbf{U}))\lambda + \gamma^2(\det(\mathbf{U})) = 0 \tag{10.69}$$

where $\mathrm{tr}(\mathbf{U})$ and $\det(\mathbf{U})$ are the trace and determinant appropriate to the well-stirred system.

We have specified that the conditions of interest here are those lying within the Hopf bifurcation locus, so $\mathrm{tr}(\mathbf{U})$ will be greater than zero. This means that the eigenvalues appropriate to this uniform state must have positive real parts. The system is unstable to uniform perturbations. We must exclude this $n = 0$ mode from any perturbations in the remainder of this section.

It is convenient to collect here the definitions of $\mathrm{tr}(\mathbf{U})$ and $\det(\mathbf{U})$, for later reference:

$$\mathrm{tr}(\mathbf{U}) = \mu/\kappa - 1 - \kappa e^{\mu/\kappa} > 0 \tag{10.70}$$

$$\det(\mathbf{U}) = \kappa e^{\mu/\kappa} > 0. \tag{10.71}$$

10.4.2. Spatially non-uniform perturbations

For a perturbation with a general non-zero n the eigenvalues $(\omega_{1,2})_n$ of the Jacobian matrix appropriate to eqns (10.64) and (10.65) are given by the roots of

$$\omega_n^2 - (\mathrm{tr}(\mathbf{J}))\omega_n + \det(\mathbf{J}) = 0 \tag{10.72}$$

where

$$\mathrm{tr}(\mathbf{J}) = \gamma(\mathrm{tr}(\mathbf{U})) - 2n^2\pi^2 \tag{10.73}$$

$$\det(\mathbf{J}) = \gamma^2(\det(\mathbf{U})) - \gamma(\mathrm{tr}(\mathbf{U}))n^2\pi^2 + n^4\pi^4. \tag{10.74}$$

The nth component then grows or decays as the real part of ω_n is positive or negative.

A number of qualitative comments can be made immediately. The trace of $\mathbf{J}$ is given by the (positive) trace of the uniform system *minus* the term $2n^2\pi^2$, related to diffusion. The latter can have a stabilizing effect, changing the sign of $\mathrm{tr}(\mathbf{J})$ from positive to negative. This change in sign occurs when

$$n^2\pi^2/\gamma = \tfrac{1}{2}\mathrm{tr}(\mathbf{U}) = \tfrac{1}{2}(\mu/\kappa - 1 - \kappa e^{\mu/\kappa}). \tag{10.75}$$

Provided condition (10.70) is satisfied (i.e. provided the well-stirred system is unstable), this equation has a real solution for the wave number n for given values of μ and κ.

Of course the change in sign of the trace is only important if it leads to a change in the sign of the eigenvalues ω_n. This will be the case, and will only be the case, provided the determinant $\det(\mathbf{J})$ remains positive all the time. However, $\det(\mathbf{J})$ also depends on n and itself becomes zero when

$$n^2\pi^2/\gamma = \tfrac{1}{2}\{\mathrm{tr}(\mathbf{U}) \pm [(\mathrm{tr}(\mathbf{U}))^2 - 4\det(\mathbf{U})]^{1/2}\}. \tag{10.76}$$

This latter condition will have zero or two real solutions depending on the discriminant under the square root sign, and hence on κ. If κ is such that $(\mathrm{tr}(\mathbf{U}))^2 - 4\det(\mathbf{U})$ is always negative, there will be no real roots to eqn (10.76). If, however, the discriminant can become positive, there can be two real roots for n.

We can recognize three different ranges of behaviour. First, if $\kappa > e^{-2}$, the uniform system can never become unstable by Hopf bifurcation: $\mathrm{tr}(\mathbf{U})$ and, hence, $\mathrm{tr}(\mathbf{J})$ are always negative and the determinants can never change sign. All spatially dependent perturbations decay back to the stable uniform state.

Secondly, if κ is smaller but lies in the range

$$\tfrac{1}{2}(3 - \sqrt{5})\exp[-\tfrac{1}{2}(3 + \sqrt{5})] < \kappa < e^{-2} \tag{10.77}$$

then $\mathrm{tr}(\mathbf{U})$ is positive over some range of μ (i.e. Hopf bifurcations occur in the well-stirred system). The uniform oscillations can be 'stabilized' (in the special sense discussed above) when n increases beyond the value given in eqn (10.75).

For these values of κ there are no real conditions for which the determinant of **J** can change sign. This is equivalent to saying that the well-stirred system does not have unstable nodal states for κ in this range.

The third, and most interesting, case corresponds to the smallest values of the dimensionless rate constant:

$$\kappa < \tfrac{1}{2}(3 - \sqrt{5}) \exp[- \tfrac{1}{2}(3 + \sqrt{5})] \approx 0.0279. \qquad (10.78)$$

If we consider the well-stirred system, the stationary state has two Hopf bifurcation points at $\mu^*_{1,2}$, where $\mathrm{tr}(\mathbf{U}) = 0$. In between these there are two values of the dimensionless reactant concentration $\mu'_{1,2}$ where the state changes from unstable focus to unstable node. In between these parameter values we can have $(\mathrm{tr}(\mathbf{U}))^2 - 4\det(\mathbf{U}) > 0$, so there are real roots to eqn (10.76).

We will consider these second and third cases separately.

10.4.3. Well-stirred system has unstable focus

When the dimensionless reaction rate constant lies in the range given by eqn (10.77), the well-stirred system has two Hopf bifurcation points $\mu^*_{1,2}$. Over the range of reactant concentration

$$\mu^*_2 < \mu < \mu^*_1 \qquad (10.79)$$

the uniform state is an unstable focus; nowhere does $(\mathrm{tr}(\mathbf{U}))^2 - 4\det(\mathbf{U})$ become positive.

If we choose any value of μ in the range given by (10.79), $\mathrm{tr}(\mathbf{J})$ will be positive for low n (it is positive for $n = 0$), but we can see from eqn (10.73) that it will become negative if n becomes large.

The locus n–μ at which $\mathrm{tr}(\mathbf{J})$ changes sign for a given κ is shown in Fig. 10.9. This is known as the 'neutral stability curve'. Below the locus, i.e. for low n, the eigenvalues ω_n have positive real parts, so perturbations with these low wave numbers grow in time. The uniform stationary state is unstable to these long-wavelength perturbations, in a similar way in which it is unstable to uniform perturbations ($n = 0$). Above the locus, however, components with larger n have eigenvalues with negative real parts and which, hence, decay. The system is stable to these higher modes.

There is one solution to eqn (10.75) for each value of μ in the range given by (10.79): the maximum in the locus, and hence the highest mode to which the uniform state is unstable, is given by

$$n^2 \pi^2 / \gamma = \tfrac{1}{2} \ln(1/\kappa e^2). \qquad (10.80)$$

The wave number n can only have integer values, so only discrete values of the ordinate $n\pi/\gamma^{1/2}$ on Fig. 10.9 are allowed. The lowest value for n is unity,

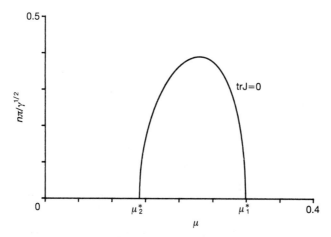

FIG. 10.9. The neutral stability curve in the n–μ plane: below the curve parametrized by $\mathrm{tr}(\mathbf{J}) = 0$ the uniform state is unstable to perturbations of appropriate spatial form (n half-wavelengths).

so non-uniform modes will only come in to play with regard to instability, if

$$\tfrac{1}{2}(\gamma/\pi^2)\ln(1/\kappa e^2) > 1. \tag{10.81}$$

We can think of the reactant concentration and some initial spatial distribution of the intermediate concentration and temperature profiles specifying a point on Fig. 10.9. If we choose a point above the neutral stability curve, then the first response of the system will be for spatial inhomogeneity to disappear. If the value of μ lies outside the range given by (10.79), then the system adjusts to a stable spatially uniform stationary state. If μ lies between μ_2^* and μ_1^*, we may find uniform oscillations.

If, however, we start the system with a given non-uniform distribution, corresponding to $n = 2$ say, and a value for μ such that the initial point lies *beneath* the neutral stability curve, then the spatial amplitudes will not decay. Rather the positive real parts to the eigenvalues will ensure that the perturbation waveform grows. The system may move to a state which is varying both in time and position—a standing-wave solution.

10.4.4. Well-stirred system has unstable node

If the dimensionless rate constant satisfies inequality (10.78), the well-stirred system again has two Hopf bifurcation points μ_1^* and μ_2^*. However, within the range of reactant concentrations between these, the uniform state also changes character from unstable focus to unstable node at μ_1' and μ_2', as shown in Fig. 10.10.

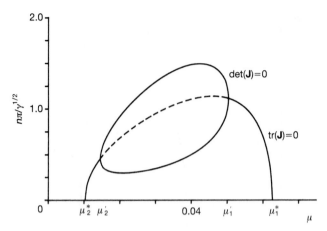

FIG. 10.10. The neutral stability curve for a system with $\kappa < 0.0279$, showing curves paramet-rized by $\mathrm{tr}(\mathbf{J}) = 0$ and $\det(\mathbf{J}) = 0$. Within the latter, non-uniform profiles may be 'stable'.

We can classify the various ranges as follows:

(1) $\mu_1^* < \mu$, uniform state stable, $\mathrm{tr}(\mathbf{U}) < 0$;

(2) $\mu_1' < \mu < \mu_1^*$, uniform state unstable focus, $\mathrm{tr}(\mathbf{U}) > 0$, $(\mathrm{tr}(\mathbf{U}))^2 - 4\det(\mathbf{U}) < 0$;

(3) $\mu_2' < \mu < \mu_1'$, uniform state unstable node, $\mathrm{tr}(\mathbf{U}) > 0$, $(\mathrm{tr}(\mathbf{U}))^2 - 4\det(\mathbf{U}) > 0$;

(4) $\mu_2^* < \mu < \mu_2'$, uniform state unstable focus, $\mathrm{tr}(\mathbf{U}) > 0$, $(\mathrm{tr}(\mathbf{U}))^2 - 4\det(\mathbf{U}) < 0$;

(5) $\mu < \mu_2^*$, uniform state stable, $\mathrm{tr}(\mathbf{U}) < 0$.

In cases (1) and (5) the eigenvalues ω_n have negative real parts for all n, so non-uniform perturbations decay.

For case (2), the eigenvalues ω_n have positive real parts for small n, but are still negative for n sufficiently large. The change in sign (neutral stability) is again given by eqn (10.80). There is only one neutral mode for any given reactant concentration μ in this range. A similar situation holds for case (4). The possible responses in these two situations are similar to those discussed in the previous subsection, i.e. standing-wave solutions.

For case (3), with μ in the range between μ_1' and μ_2', the condition $\mathrm{tr}(\mathbf{J}) = 0$, corresponding to eqn (10.80), is of less importance (and hence the locus is shown as a broken curve in Fig. 10.10). As n increases from zero, so the trace does become less positive, but now the determinant can also change sign. The latter occurs first, in fact, so the eigenvalues ω_n change from two, real, and positive values to two real solutions of opposite sign. The condition for this is given by the lower root of eqn (10.76).

Over a range of n between the lower and upper roots of eqn (10.76) the uniform state appears as a saddle point to perturbations with the appropriate wave numbers, so we expect any such spatial non-uniformity to grow. Only for n larger than the upper root of (10.76) do the eigenvalues become both real and negative.

The neutral stability curve corresponding to the condition $\det(\mathbf{J}) = 0$ gives a closed region, within which we expect the appearance of stable (time-independent) spatially non-uniform profiles.

10.4.5. Presentation of results

For any value of the rate constant κ less than $\frac{1}{2}(3 - \sqrt{5})\exp[-\frac{1}{2}(3 + \sqrt{5})]$, there are two values for the reactant concentration, μ_1' and μ_2', between which some of the spatial modes do not tend to decay. Figure 10.11 shows the curves corresponding to $\det(\mathbf{J}) = 0$ for various values of κ. In each case, there is a closed region of the n–μ parameter plane.

As an example of how these curves should be interpreted, we consider a specific case. Let us take $\kappa = 0.02$ and, for convenience, choose the size of the reaction zone such that $\gamma = 16\pi^2$. The dispersion, or neutral stability curve for this system, is shown separately in Fig. 10.12. The wave number n can only have integer values, so valid modes correspond to the horizontal lines with $n\pi/\gamma^{1/2} = \frac{1}{4}, \frac{1}{2}, \frac{3}{4}$, etc. Only three of these horizontals intersect the

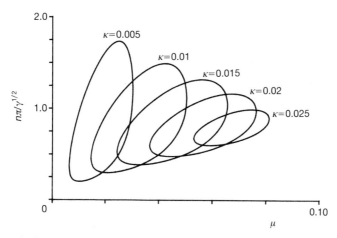

FIG. 10.11. The development of the neutral stability curve for stable pattern formation with the dimensionless rate constant κ.

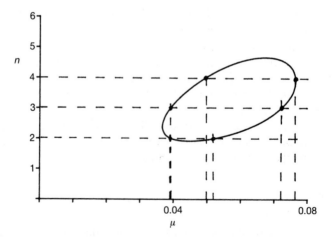

FIG. 10.12. Specific neutral stability curve for $\kappa = 0.02$ and $\gamma = 16\pi^2$, showing the possibility of stabilizing patterns with $n = 2, 3$, or 4 over limit ranges of the precursor concentration μ.

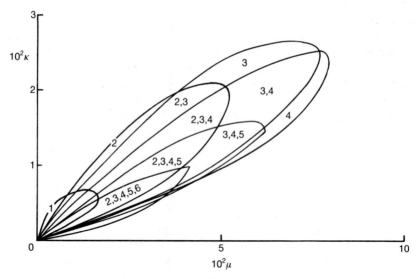

FIG. 10.13. The regions of stability of different spatial forms in the μ–κ parameter plane for a system with $\gamma = 16\pi^2$.

dispersion region: those with $n = 2$, $n = 3$, and $n = 4$. Furthermore, for each of these modes, 'stability' is possible over different ranges of the reactant concentration. Thus for $n = 2$, we require $0.0385 < \mu < 0.052$; for $n = 3$, $0.0389 < \mu < 0.0724$; and for $n = 4$, $0.0495 < \mu < 0.0763$.

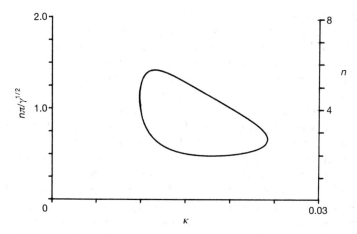

FIG. 10.14. Representation of region for stable spatial stationary state in the n–κ plane for a system with $\mu = 0.05$. The left-hand ordinate gives the general locus for any γ, the right-hand ordinate is appropriate to the specific case $\gamma = 16\pi^2$.

For lower values of κ, the dispersion region is larger and more modes can be excited; e.g. with $\kappa = 0.005$, those with $n = 1$ to $n = 6$ all have non-zero ranges. Figure 10.13 shows how the κ–μ plane is divided into regions in which different modes can be excited, for the above choice of γ. In some regions only one spatial mode shows possible stability, e.g. the region with $n = 3$ towards the top of the diagram. In others, particularly near the centre, many modes may be possible.

The dispersion relation $\det(\mathbf{J}) = 0$ can also be viewed in the n–κ plane, as in Fig. 10.14. This allows us to read off the maximum value of the dimensionless rate constant for which a given mode can be found. The left-hand ordinate corresponds to general values of the group $n\pi/\gamma^{1/2}$, the right-hand ordinate to our specific example $\gamma = 16\pi^2$. Thus, for the latter case, a pattern with $n = 5$ requires $0.01 < \kappa < 0.0148$.

10.5. Concluding remarks

There has been a great deal of analysis and prediction of stable spatial patterns from various model schemes other than that considered here. It is very much harder to find any experimental examples of such stable stationary structures in chemical systems. This may be because the first type considered in § 10.3 (diffusion-driven instabilities) require unequal diffusivities and are particularly favoured when the ratio β is very much larger than unity. With the various intermediate chemical species which are believed to participate in, say, the Belousov–Zhabotinskii reaction and which are of similar molecular mass and size, huge differences in mobility are not likely. In biological systems, the latter type of consideration may not be so true; for instance proteins may be much more sluggish than H^+ ions.

The second mechanism for stable pattern formation, discussed in § 10.4, does not in fact lead to truly stable patterns. These profiles are always unstable to uniform perturbations—or to any perturbations with a uniform component. Thus they may survive for a non-zero period, but eventually we must always expect the uniform oscillatory solution to emerge.

In the most recent years, a new type of reactor has been introduced, designed specifically to allow stationary-state pattern formation. This continuous-flow unstirred reactor (CFUR), also known as a Turing–Prigogine–Nicolis reactor, is essentially the type of reactor discussed above with an added inflow and outflow such as that provided for the CSTR of the previous chapters. The inflow and outflow must be provided in such a way that avoids the setting up of flow currents with consequent spatial mixing: one approach is to carry out the reaction in an inert gel (Noszticzius *et al.* 1987; Tam *et al.* 1988). The reaction–diffusion equations governing such systems would have the form of eqns (10.1)–(10.3) etc., but with the additional terms $k_f(p_0 - p)$ and $k_f(a_0 - a)$ for inflow and outflow in (10.1) and (10.2) respectively. Vastano *et al.* (1988) have considered the behaviour of the simple cubic autocatalytic model of chapters 6 and 8 (without the precursor P) in such a reactor and shown that stable Turing patterns can indeed be formed. Furthermore, by choosing parameter values such that the uniform stationary-state solution (i.e. the CSTR result which will also be a stationary state of the CUFR) is close to having a double-zero eigenvalue (see chapter 8), Pearson and Horsthemke (1989) have found stable non-uniform solutions with a ratio of diffusion coefficient arbitrarily close to unity.

References

Arcuri, P. and Murray, J. D. (1986). Pattern sensitivity to boundary and initial conditions in reaction-diffusion models. *J. Math. Biol.*, **24**, 141–65.

Murray, J. D. (1979). *Lecture notes on nonlinear differential equation models in biology.* Clarendon Press, Oxford.

Murray, J. D. (1982). Parameter space for Turing instability in reaction diffusion mechanisms: a comparison of models. *J. Theor. Biol.*, **98**, 143–63.

Murray, J. D. (1988). How the leopard gets its spots. *Sci. Am.*, **258**, March, 62–9.

Needham, D. J. and Merkin, J. H. (1989). Pattern formation through reaction and diffusion in a simple pooled-chemical system. *Dyn. Stab. Syst.*, **3**,

Noszticzius, Z., Horsthemke, W., McCormick, W. D., Swinney, H. L., and

Tam, W. Y. (1987). Sustained chemical waves in an annular gel reactor: a chemical pinwheel. *Nature*, **329**, 619–20.

Pearson, J. E. and Horsthemke, W. (1989). Turing instabilities with nearly equal diffusion coefficients. *J. Chem. Phys.*, **90**, 1588–99.

Segel, L. A. (ed.) (1980). *Mathematical models in molecular and cellular biology.* Cambridge University Press.

Tam, W. Y., Horsthemke, W., Noszticzius, Z., and Swinney, H. L. (1988). Sustained spiral waves in a continuously unstirred chemical reactor. *J. Chem. Phys.*, **88**, 3395–6.

Turing, A. M. (1952). The chemical basis for morphogenesis. *Phil. Trans. R. Soc.*, **B237**, 37–72.

Vastano, J. A., Pearson, J. E., Horsthemke, W., and Swinney, H. L. (1988). Turing patterns in an open reactor. *J. Chem. Phys.*, **88**, 6175–81.

TRAVELLING WAVES

This chapter deals principally with travelling wavefronts of chemical reactions of uniform speed and constant concentration profile but also considers 'pulses' and periodic wave trains. Isothermal cubic autocatalysis offers deep insights into the phenomenon as a whole. After careful study the reader should be able to:

(1) set up wave equations and cast them into an expressive dimensionless form;

(2) exploit the constancy of wave speed and profile (and recognize its limitations, e.g. to non-accelerating waves);

(3) see how a concentration–distance profile and a gradient profile can be matched through the gradient–concentration relationship;

(4) appreciate the possibility of studying mixed cubic plus quadratic autocatalytic mechanisms;

(5) see how the 'front' and 'tail' of a wave can reflect different parts of a kinetic mechanism.

In the previous chapter we dealt with the existence and spontaneous formation of stable spatially non-uniform patterns. Here, we consider another form of spatial behaviour. We envisage a (long) tube, initially containing a spatially uniform distribution of reactants. The spontaneous reaction converting these reactants to products is exceedingly slow but can be initiated at one end, perhaps by the local addition of some key intermediates (such as an autocatalyst), by local heating, or by a spark (which often serves to achieve both of the former). We are interested to see if and how the reaction can then propagate along the whole length of the tube. If the reactants are gaseous and the process is exothermic, the reaction zone will typically move quickly and we talk of a flame. For reactions in dilute solutions the propagation is typically much slower, and we talk of a chemical wave.

We can further distinguish three types of waves, as illustrated in Fig. 11.1: fronts, pulses, and periodic wave trains. A 'front', like a flame generally converts initial reactants to final products, so the compositions ahead of and behind the wave are quite different. With a 'pulse', some intermediate may be produced by the first front, but this is then converted back (or to some further product) by the second (recovery) wave. Thus the concentration of the

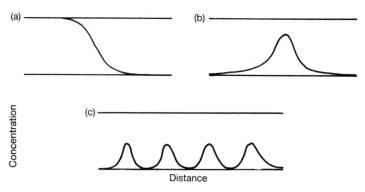

FIG. 11.1.   Three types of travelling wave: (a) a front, (b) a pulse, and (c) a train.

intermediate is the same ahead of and behind the pulse. Finally 'periodic wave trains' are comprised of a series of fronts or pulses which are continuously being initiated at some point or 'centre' and which follow each other along tube.

In this chapter we will concentrate on fronts and pulses, and we will illustrate these with the isothermal autocatalytic models seen previously. We start with the single cubic autocatalytic process

$$A + 2B \rightarrow 3B \qquad \text{rate} = k_1 ab^2. \qquad (11.1)$$

This is a particularly suitable choice for the mathematical modelling. We will imagine that our tube initially contains only species A, so $b = 0$ and the reaction rate is everywhere zero. Initiation is brought about by the addition of some B locally, as indicated in Fig. 11.2(a). In the initiation region, reaction begins to occur. As the concentration of the autocatalyst increases, B begins to diffuse ahead into the fresh reactant, where further reaction is initiated (Fig. 11.2(b)). Behind this diffusing front, all the reactant is eventually consumed, leaving a region of product B. Eventually this reaction and diffusion process may lead to a front moving into the region of unused reactant at a constant velocity (Fig. 11.2(c)).

At any given time, however, there is still fresh reactant far enough ahead of the front and into which no B has yet diffused. In this region, the reaction rate is still zero and so the system must wait until the leading edge of the front arrives. This latter point is particularly useful. The mathematical analysis is most easily achieved in the limit of considering an infinitely long tube (so the wave can settle to a steady velocity in much the same way as previous chapters have dealt with systems approaching stationary states in infinite time). However, unless the reaction rate ahead of the wave is exactly zero, the conversion of A to B will not wait an infinite time to start, and reaction can become complete before the wave arrives.

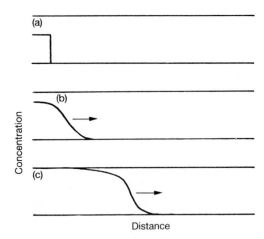

Fɪɢ. 11.2.   Schematic representation of the development of a constant velocity, constant wave-form travelling front: (a) localized initiation caused by introducing some autocatalyst at one 'end'; (b) the autocatalyst begins to diffuse into the fresh reactant ahead, meanwhile reaction begins near the original initiation site; (c) the reaction approaches completion behind the freely propagating wavefront.

11.1. Travelling wavefronts with cubic autocatalysis

We consider an infinitely long thin tube ($-\infty < r < +\infty$) along which the concentrations of A and B vary. The sides of the tube are impermeable, and we imagine that there are no concentration gradients along the axes perpendicular to the r axis (so we have a one-dimensional wave). A general picture of the system at some time after initiation is that corresponding to Fig. 11.2(c).

11.1.1. Setting up the model: wave equations

If we concentrate on a thin slice of the concentration profiles, then the rate of change of concentration in time in that slice is given by the net rate of diffusion in and the chemical rate of removal by reaction. The former is given by a term involving the second partial derivative with respect to space and the latter by the term $k_1 ab^2$. Thus, for the reactant A, we have the governing equation

$$\frac{\partial a}{\partial t} = D_A \frac{\partial^2 a}{\partial r^2} - k_1 ab^2. \tag{11.2}$$

Similarly for the autocatalyst

$$\frac{\partial b}{\partial t} = D_B \frac{\partial^2 b}{\partial r^2} + k_1 ab^2. \tag{11.3}$$

We need to specify the boundary conditions: at one end of the tube ($r \to +\infty$, say) no reaction has occurred, so $a = a_0$, $b = 0$; at the other end reaction is complete, with $a = 0$ and $b = a_0$. Thus the boundary conditions are

$$
\begin{aligned}
a = 0 \qquad b = a_0 \qquad &\text{at} \qquad x = -\infty \\
a = a_0 \qquad b = 0 \qquad &\text{at} \qquad x = +\infty.
\end{aligned}
\tag{11.4}
$$

We will also assume that A and B are similar chemical species, with the same value for their diffusion coefficients, $D_A = D_B = D$ (this is only a special case in the sense that the mathematics is significantly easier).

If we add eqns (11.2) and (11.3), the kinetic terms cancel, giving

$$\frac{\partial(a + b)}{\partial t} = D \frac{\partial^2(a + b)}{\partial r^2}. \tag{11.5}$$

This is an equation in the total concentration of chemicals. One solution of eqn (11.5) which satisfies the boundary conditions is that the sum of the concentrations $a + b$ should be equal to a constant everywhere and at all times. This constant is determined by the reaction stoichiometry and gives the sensible relationship

$$a + b = a_0 \qquad \text{all } r \text{ and } t. \tag{11.6}$$

If this result is used to substitute for a in (11.3) we obtain a single equation for the one independent concentration, b:

$$\frac{\partial b}{\partial t} = D \frac{\partial^2 b}{\partial r^2} + k_1 b^2(a_0 - b). \tag{11.7}$$

This is sometimes known as the cubic Fisher equation and has been applied variously in chemical kinetics, population biology, and to the spread of infectious diseases or advantageous genes.

11.1.2. Dimensionless forms

The following dimensionless groups reduce eqn (11.7) to its simplest form:

concentration $\beta = b/a_0$ (11.8)

time $\tau = k_1 a_0^2 t$ (11.9)

distance $x = (k_1 a_0^2/D)^{1/2} r.$ (11.10)

Note that the diffusion coefficient, which has units of m^2s^{-1}, has been used to provide a length scale. With these, we have

$$\frac{\partial \beta}{\partial \tau} = \frac{\partial^2 \beta}{\partial x^2} + \beta^2(1 - \beta) \qquad (11.11)$$

with boundary conditions

$$\beta = 1 \quad \text{at} \quad x = -\infty \qquad \text{and} \qquad \beta = 0 \quad \text{at} \quad x = +\infty. \quad (11.12)$$

11.1.3. General approach to travelling-wave solutions

The behaviour we are expecting to emerge from this physico-chemical model is that of a steady wave of reaction moving from left to right in Fig. 11.2 into the region of unreacted A. By steady, we really mean that the wavefront should maintain its shape as it moves with a constant speed. It is this shape and speed which we seek to determine (and express in terms of the rate constant, diffusion coefficient, etc.).

First, then, we assume that there actually is a steady wave and that it has a constant velocity $c = dx/d\tau$. (We can identify a particular point on the wavefront, e.g. the point at which the reaction is half-way to completion, $\alpha = \beta = 0.5$, and c is then the speed at which this point is moving to the right.)

A useful technique which can be applied in cases where constant velocity solutions arise is that of changing from a fixed coordinate system (the present x coordinate has an origin whose position is fixed in space) to travelling-wave coordinates. With the latter, the origin moves from left to right with the front at the same constant velocity c; we are constantly adjusting our frame of reference so that within the frame the reaction front appears stationary. This new coordinate z is defined in terms of x and τ by

$$z = x - c\tau. \qquad (11.13)$$

We must now write our reaction–diffusion equation in terms of the z coordinate. For this we need the identities

$$\frac{\partial \beta}{\partial \tau} = -c\frac{d\beta}{dz} \qquad \frac{\partial^2 \beta}{\partial x^2} = \frac{d^2 \beta}{dz^2}. \qquad (11.14)$$

Substitution of these forms into eqn (11.11) yields

$$\frac{d^2 \beta}{dz^2} + c\frac{d\beta}{dz} + \beta^2(1 - \beta) = 0 \qquad (11.15)$$

with the boundary conditions

$$\beta = 0 \text{ at } z = +\infty \qquad \text{and} \qquad \beta = 1 \text{ at } z = -\infty. \qquad (11.16)$$

Equation (11.15) is a highly significant development from the original equations. In stages, we have come from two equations to one and have now manipulated that partial differential form in x and τ into an *ordinary* differential equation in z.

11.1.4. Solving the equation

We are looking for the solution of a non-linear (cubic) second-order ordinary differential equation (11.15), subject to boundary conditions (11.16). The solution we are looking for gives the concentration of autocatalyst as a function of the coordinate $\beta(z)$. Any such solution will have the form shown in Fig. 11.2, with x replaced by z. As well as noticing that β must tend to certain values (1 and 0) at $z = \pm \infty$, it can be seen that the concentration gradient $d\beta/dz$ must tend to zero at both these limits. In between, for finite z and for $0 < \beta < 1$, the concentration gradient is negative, $d\beta/dz < 0$. If we choose to write this gradient in terms of β, then we must have

$$\frac{d\beta}{dz} = 0 \quad \text{for} \quad \beta = 0, 1 \qquad \frac{d\beta}{dz} < 0 \quad \text{for} \quad 0 < \beta < 1. \tag{11.17}$$

A suitable form is the simple parabolic relationship

$$\frac{d\beta}{dz} = -k\beta(1 - \beta) \tag{11.18}$$

where k is some constant we hope to determine together with the speed c. (This should be possible since we have a second-order differential equation and two boundary conditions.)

Differentiation of eqn (11.18) with respect to z gives

$$\frac{d^2\beta}{dz^2} = -k(1 - 2\beta)\frac{d\beta}{dz} = k^2\beta(1 - \beta)(1 - 2\beta). \tag{11.19}$$

We now have expressions for the two derivatives, which can be substituted into (11.15) to give

$$k^2\beta(1 - \beta)(1 - 2\beta) - ck\beta(1 - \beta) + \beta^2(1 - \beta) = 0. \tag{11.20}$$

Cancellation of the common factor $\beta(1 - \beta)$ leaves the relationship

$$k^2(1 - 2\beta) - ck + \beta = 0. \tag{11.21}$$

The two undetermined constants c and k may now be chosen so that the equality holds for all values of β. Equation (11.21) can be rewritten as

$$(1 - 2k^2)\beta + k(k - c) = 0 \tag{11.22}$$

which will hold for all β provided that

$$1 - 2k^2 = 0 \quad \text{and} \quad k - c = 0 \Rightarrow k = c = 1/\sqrt{2} \tag{11.23}$$

which determines k and c as required.

11.1.5. Velocity of wavefront

In dimensionless terms, the wave travels with a steady velocity

$$c = \frac{\mathrm{d}x}{\mathrm{d}\tau} = \frac{1}{\sqrt{2}}. \tag{11.24}$$

Returning to the original rate constants, diffusion coefficient, etc., this gives the speed as

$$\frac{\mathrm{d}r}{\mathrm{d}t} = \frac{1}{\sqrt{2}}(Dk_1 a_0^2)^{1/2}. \tag{11.25}$$

For example, if we take as typical values $k_1 = 2.5 \times 10^9 \, \text{mol}^{-2} \, \text{dm}^6 \, \text{s}^{-1}$ (as in Chapter 2) and $D = 2 \times 10^{-5} \, \text{cm}^2 \, \text{s}^{-1}$, then the wavefront moves with a velocity of approximately $-9.5 \, \text{cm} \, \text{min}^{-1}$.

We may compare the speed of this wave, which will take approximately 6 s to move 1 cm, with the corresponding diffusion of B in the absence of reaction. In this latter case, the diffusion time is simply of the order r^2/D, so for $r = 1$ cm we have $t_{\mathrm{diff}} \approx 10^5$ s. The diffusion-only wave would propagate more than four orders of magnitude slower than the reaction front.

The velocity of the front increases linearly with the initial reactant concentration and with the square root of the rate constant and the diffusion coefficient. Fast chemistry and high diffusion both increase the speed of the wave.

11.1.6. Wave profile

The gradient of the wave at any point is given by eqn (11.18), with $k = 1/\sqrt{2}$ from (11.23). The actual profile can thus be obtained from a simple integration, giving

$$\beta(z) = \frac{\exp[-(1/\sqrt{2})(z + z_0)]}{1 + \exp[-(1/\sqrt{2})(z + z_0)]} \tag{11.26}$$

where z_0 is a constant of integration. For any finite value of z_0, eqn (11.26) satisfies boundary conditions (11.16), with $\beta(z) \to 1$ as $z \to -\infty$ and $\beta(z) \to 0$ as $z \to +\infty$.

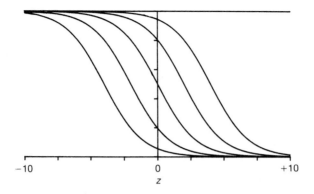

FIG. 11.3. The reaction–diffusion wave for cubic autocatalysis in wave-fixed coordinates. The five fronts correspond to $z_0 = -4, -2, 0, 2,$ and 4.

Equation (11.26) corresponds to a whole family of solutions, wavefronts moving from left to right with velocity $c = 1/\sqrt{2}$ and each front differing from the others only by a slight displacement along the z axis, as shown in Fig. 11.3. Each possible profile has the same shape, with a inflection point occurring at 50 per cent reaction ($\beta = 0.5$) at the coordinate location $z = -z_0$. Without loss of generality, as they say in all the best mathematical texts, we can take any profile as representative of the wave, and the simplest choice is that with $z_0 = 0$. The inflection then occurs at the origin of our travelling-wave coordinate frame.

The equation for the profile in terms of the original physical quantities k_1, D, etc. is now

$$b(r, t) = a_0 \exp\left[\frac{-k_1 a_0^2}{\sqrt{2}}\left(\frac{r}{(k_1 a_0^2 D)^{1/2}} - \frac{t}{\sqrt{2}}\right)\right]\Bigg/$$
$$\times \left\{1 + \exp\left[\frac{-k_1 a_0^2}{\sqrt{2}}\left(\frac{r}{(k_1 a_0^2 D)^{1/2}} - \frac{t}{\sqrt{2}}\right)\right]\right\}. \quad (11.27)$$

11.1.7. Sharpness of wavefront

As a measure of the sharpness of the wavefront, we can consider two quantities: (i) the magnitude of the maximum (negative) gradient, $-(db/dr)_{max}$, which occurs at $b = \frac{1}{2}a_0$ and (ii) the distance Δr between the points corresponding to 25 per cent and 75 per cent conversion (see Fig. 11.4)

In dimensionless terms and with travelling coordinates, these two quantities are simply

$$-(d\beta/dz)_{max} = 1/\sqrt{32} \qquad \Delta z = \sqrt{2} \ln 9. \quad (11.28)$$

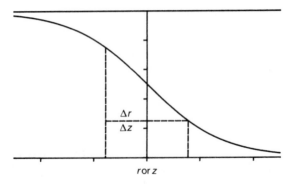

FIG. 11.4. Possible measures of the 'sharpness' of the reaction front: the text considers the maximum gradient appropriate to $\beta = \frac{1}{2}$ at $z = 0$ and also the distance between the locations for 25 per cent and 75 per cent conversion as illustrated.

Sharp fronts correspond to high gradients and small distances Δr (and either one of these two conditions implies the other). The results given in (11.28) also imply the following dependences:

$$ - (db/dr)_{max} \propto 1/\Delta r \propto (k_1 a_0^2 /D)^{1/2}. \qquad (11.29) $$

so sharpness is favoured by fast chemical kinetics. High diffusion rates, on the other hand, tend to smear out the wavefront (although, as shown above, this tends also to increase the front speed).

Thus, with the simple cubic autocatalytic rate law, we have been able to find an analytical expression for the time and space dependence of a steady reaction–diffusion wave and make various quantitative and qualitative comments about the behaviour of the wave in terms of the kinetic and diffusion parameters. We now turn to the apparently simpler kinetics of a quadratic autocatalysis, hoping for similar rewards.

11.2. Travelling waves with quadratic autocatalysis

In the previous section, full analytical solutions for constant velocity travelling waves in reaction with cubic autocatalysis were obtained. We might hope that the equivalent system with quadratic autocatalysis

$$ A + B \rightarrow 2B \qquad rate = k_q ab \qquad (11.30) $$

which gives rise to a lower-order non-linearity might also yield simple solutions. Unfortunately, the quadratic Fisher equation, as it is best known, shows a great deal of subtlety, some of which we discuss below. There is, at first, even no simple result for the velocity, since an infinite number of possibilities appear. Some firm conclusions can, however, be drawn and this is a particularly important system in practical situations, demanding our attention in this chapter.

11.2.1. Governing equations

Proceeding as in the previous section, the appropriate dimensionless reaction–diffusion equation can be written in terms of the wavefront coordinates as

$$\frac{d^2\beta}{dz^2} + c\frac{d\beta}{dz} + \beta(1 - \beta) = 0 \tag{11.31}$$

with

$$\beta = 1 \quad \text{at} \quad z = -\infty \qquad \beta = 0 \quad \text{at} \quad z = +\infty \tag{11.32}$$

and

$$d\beta/dz = 0 \quad \text{at} \quad z = \pm\infty \tag{11.33}$$

(i.e. when $\beta = 0$, 1). In the previous section, condition (11.33) suggested a parabolic dependence of the gradient on the concentration, as given by eqn (11.18). With the cubic non-linearity this led to self-consistent expressions for k, c, and the wave profile. A similar substitution with the present quadratic term, however, leads to the conclusion $k = c = 0$. No constant velocity wave with such a parabolic relationship between $d\beta/dz$ and β can, therefore, exist.

11.2.2. Phase plane analysis

A general systematic technique applicable to second-order differential equations, of which (11.31) is a particular example, is that of phase plane analysis. We have seen this approach before (chapter 3) in the context of systems with two first-order equations. These two cases are, however, equivalent. We can replace eqn (11.31) by two first-order equations by introducing a new variable g, which is simply the derivative of the concentration with respect to z. Thus

$$\frac{d\beta}{dz} = g \tag{11.34}$$

$$\frac{dg}{dz}\left(= \frac{d^2\beta}{dz^2} \right) = -cg - \beta(1 - \beta). \tag{11.35}$$

Dividing one equation by the other gives the phase plane equation

$$\frac{dg}{d\beta} = \frac{-[cg + \beta(1 - \beta)]}{g}. \tag{11.36}$$

This determines how the gradient varies with the concentration, but the integral has no closed form analytical solution.

The singular points of this system, where $d\beta/dz = dg/dz = 0$, are $(\beta, g) = (1, 0)$ and $(0, 0)$. In fact these are the points corresponding to the boundary conditions (11.32) and (11.33). Thus the reaction wave can be represented by a trajectory passing through the β–g phase plane, originating from one singular point $(1, 0)$ and entering the other $(0, 0)$, as shown in Fig. 11.5.

11.2.3. Minimum velocity

The local stability and character of the singular points can be determined by the usual analysis of the eigenvalues of the Jacobian matrix

$$\mathbf{J} = \begin{pmatrix} 0 & 1 \\ -(1 - 2\beta) & -c \end{pmatrix}_{(1,\,0)\text{ or }(0,\,0)} \tag{11.37}$$

with

$$\lambda^2 + c\lambda + (1 - 2\beta) = 0. \tag{11.38}$$

For the initial point ($\beta = 1$), the two eigenvalues are real and have opposite sign: this is a saddle point, and the system moves away when given a push by the diffusion process. The final point, which is the origin in the phase plane, has eigenvalues

$$\lambda_\pm = \tfrac{1}{2}[-c \pm (c^2 - 4)^{1/2}]. \tag{11.39}$$

This is thus a stable node or a stable focus, depending on the size of the wave velocity c. For $c \geqslant 2$, the eigenvalues are real; for lower wave velocities with

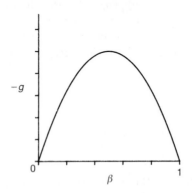

FIG. 11.5.   Phase plane representation of a travelling wavefront for quadratic autocatalysis. The trajectory emerges from the initial singularity at $\beta = 1$, $g = 0$ and tends to the final state $\beta = 0$, $g = 0$.

$c < 2$, the eigenvalues $\lambda_\pm$ form a complex pair. In the latter case, the trajectories would spiral into the origin. This spiralling would lead to the concentration β becoming negative at some stages of the reaction. Physically this cannot occur and so no steady chemical wave can propagate with these lowest velocities. We therefore have one requirement of the wave speed:

$$c \geqslant c_{min} = 2. \tag{11.40}$$

Apparently, the wave can propagate with any velocity, provided that velocity exceeds a certain minimum.

11.2.4. Stability of minimum velocity wave

Although the previous discussion appears to permit any wave velocity from the spectrum in eqn (11.40), computed solutions and appropriate experimental evidence suggest that such systems adjust quickly so that they travel with the minimum speed allowed. This indicates that the wave with velocity c_{min} is somehow a stable solution, whereas those with higher speeds are unstable.

The evaluation of stability for travelling-wave solutions is by no means a simple process and will not even be attempted here. The result that systems governed by quadratic Fisher equations tend to pick up their minimum permitted velocity will be used later.

11.3. Mixed cubic and quadratic autocatalysis

The reaction between iodate and arsenite ions appears to have contributions from both cubic and quadratic autocatalysis (the autocatalyst is the product, iodide ion). In the previous sections we have treated these two rate laws separately and by different methods. Both methods can be applied to the system in which these processes occur simultaneously, yielding results which, despite not being consistent at first sight, can be resolved by the idea of stability.

If both reactions (11.1) and (11.30) occur, the dimensionless reaction–diffusion equation for the concentration of B in terms of the travelling–wave coordinate z can be written as

$$\frac{d^2\beta}{dz^2} + c\frac{d\beta}{dz} + \beta(1 - \beta)(\beta + q) = 0. \tag{11.41}$$

Here the quantity q is a measure of the relative rates of the quadratic and cubic terms

$$q = k_q/k_1 a_0. \tag{11.42}$$

The other dimensionless terms, like those in eqns (11.8)–(11.10) are based on the cubic reaction timescale $t_{ch} = 1/k_1 a_0^2$: the limit $q \to 0$ corresponds to the cubic contribution becoming dominant, $q \to \infty$ to a quadratic regime.

Equation (11.41) is a cubic equation. The kinetic term $\beta(1 - \beta)(\beta + q)$ has three roots: $\beta = 0$, the initial condition; $\beta = 1$, complete reaction; $\beta = -q$, an unphysical state with negative concentrations. We wish, therefore, to find a constant velocity (steady) wave connecting $\beta = 0$ at $z = +\infty$ to $\beta = 1$ at $z = -\infty$ (with $d\beta/dz = 0$ at both boundaries). Following the procedure in § 11.1.4, we obtain an explicit prediction of the wave velocity:

$$c = (1/\sqrt{2})(1 + 2q). \tag{11.43}$$

In the limit of pure cubic autocatalysis ($q \to 0$), we regain the previous result, $c \to 1/\sqrt{2}$. When cubic and quadratic processes combine, the speed of the wave increases.

It is not apparent that eqn (11.43) goes over to the quadratic result $c \geqslant c_{min} = 2$ as q tends to infinity: there are additional considerations to be taken into account, as we show below, but even then we shall see that this result does not in fact have the correct limiting form.

Equation (11.41) can also be treated by the phase plane methods of § 11.2.2 and 11.2.3. Again the singularity at $z = -\infty$, with $\beta = 1$, $d\beta/dz = 0$, is a saddle point, whilst the origin with $\beta = d\beta/dz = 0$ is a stable node or a stable focus depending on the velocity c. Arguing against focal character on the grounds that β cannot become negative, we again find a minimum velocity requirement:

$$c \geqslant c_{min} = 2q^{1/2}. \tag{11.44}$$

Clearly these two approaches give different predictions, as eqns (11.43) and (11.44) are quite different functional forms. We have seen that (11.43) goes over to the pure cubic result as q tends to zero; eqn (11.44) does not have the same limiting value.

If we consider the opposite extreme, where quadratic autocatalysis dominates, we should also change our characteristic timescale. So far we have based the wave velocity c on the cubic chemical time $t_{ch} = 1/k_1 a_0^2$:

$$c = \frac{dx}{d\tau} = \frac{1}{(Dk_1 a_0^2)^{1/2}} \frac{dr}{dt}. \tag{11.45}$$

The quadratic Fisher result of the previous section is based on the quadratic chemical timescale $t_{ch} = 1/k_q a_0$: if we represent the wave velocity measured in these terms as κ, defined by

$$\kappa = \frac{1}{(Dk_q a_0)^{1/2}} \frac{dr}{dt} \tag{11.46}$$

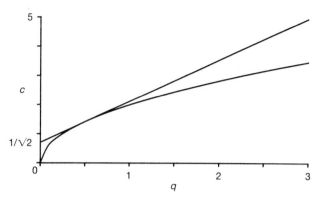

FIG. 11.6. The two possible dependences of wave velocity on the ratio of quadratic to cubic contributions to the autocatalytic reaction: the straight line corresponds to eqn (11.43), the cubic result; the parabola to eqn (11.44), the quadratic form. The two loci touch tangentially at $q = \frac{1}{2}$, for which $c = \sqrt{2}$.

then we have the relationship

$$\kappa = c/q^{1/2} \tag{11.47}$$

and so eqn (11.44) really does correspond to the correct limiting velocity $\kappa = 2$ at large q.

The dependences of wave velocity on the ratio of cubic to quadratic processes, described by eqns (11.43) and (11.44) are shown in Fig. 11.6. The 'quadratic' curve always lies below the 'cubic' result, but the two are tangential at $q = 1/2$, $c = \sqrt{2}$. Numerical computations reveal the following detail concerning the stability of the respective solutions. For $q < \frac{1}{2}$, where cubic dominates quadratic, it is the velocity determined by eqn (11.43) which is selected by the system. Above $q = \frac{1}{2}$, however, the quadratic character takes over, and the minimal velocity described by eqn (11.44) emerges dominant.

11.4. Autocatalysis with decay

In the various situations we have seen before, allowing a finite decay rate for the catalyst B has had significant results. The concentrations of A and B are then decoupled and this has allowed oscillations, isolas, and mushrooms. In the present case of reaction–diffusion waves, the uncoupling is again an important step upwards in complexity, sufficiently so as to prevent any completely general form of analysis.

We may still begin by arguing intuitively. In the case of quadratic autocatalysis, the chemistry involves local competition between a production of B which depends on the product of concentrations $\alpha\beta$ and a removal of

B by means of the step

$$B \rightarrow C \qquad rate = k_2 b \qquad (11.48)$$

which, in dimensionless terms, has a rate $\kappa_2 \beta$ (here $\kappa_2 = k_2/k_q a_0$). As α is always less than or equal to unity, we may guess that autocatalysis can only compete successfully, and hence waves can only propagate, if also $\kappa_2 \leqslant 1$. This prediction appears to be borne out by analysis and computation. If κ_2 is less than unity, a steady wave can be set up: in the leading front there is conversion of A to B (and hence to some C); in a following tail, or recovery front, B is converted to C. Thus the system shows a travelling 'pulse', as shown in Fig. 11.7.

We may comment further on the importance of the condition on the dimensionless decay rate κ_2. This term is the ratio of decay and autocatalytic rate constants, and the condition $\kappa_2 \leqslant 1$ becomes, in terms of the physical rate constants,

$$k_2 \leqslant k_q a_0. \qquad (11.49)$$

The analogy between autocatalysis and branched-chain radical reactions has already been mentioned in Chapter 1. For such a system, k_2 represents the rate constant for radical termination, whilst the term $k_q a_0$ gives a first-order form for the branching rate constant. Condition (11.49) then requires that the branching rate must exceed the termination rate. Branched-chain reactions are often discussed in terms of the difference in the branching and termination rate, i.e. the net branching factor $\phi = k_q a_0 - k_2$. Thus the condition for a travelling-wave pulse to propagate in the above system is equivalent to requiring that the net branching factor should be positive.

An estimate of the steady pulse velocity in this model can also be obtained. In a similar way to that found in §11.2, the system has a stable minimum velocity c_{min} which depends on the value of the decay rate constant:

$$c_{min} = 2(1 - \kappa_2)^{1/2} \qquad (11.50)$$

again showing that the wave can be expected to fail if $\kappa_2 > 1$.

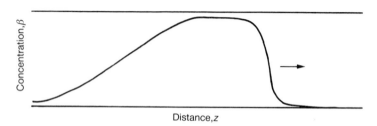

FIG. 11.7.   A travelling pulse, comprising a leading front and a recovery wave, for cubic autocatalysis with decay.

More challenging is the behaviour of a system with cubic autocatalysis coupled with decay. Here a term $\alpha\beta^2$ must compete with the removal $\kappa_2\beta$. For $\alpha \leqslant 1$ and $\beta \leqslant 1$ it is not immediately obvious that any wave solutions will exist at all. Such considerations also suggest that wave formation will be favoured by particularly high autocatalyst concentrations at the point of initiation, so $\beta > 1$ in a localized region. Thus we may expect some threshold for the initiation.

Again the scenario we envisage is similar to that shown qualitatively in Fig. 11.7: we expect our best chance of such behaviour if the decay rate is small, i.e. $\kappa_2 \ll 1$. The reaction wave has a leading front moving with a steady velocity c_1, through which most of the conversion of A to B occurs. After this front, the dimensionless concentration of A is almost zero and that of B is almost unity. At some distance, the first front is followed by a recovery wave, possibly more diffuse, in which A is completely removed and the autocatalyst also decays. The velocity of the recovery wave is c_2. If c_1 exceeds c_2, the first front will move away from the second, so the pulse will increase in width; if $c_1 = c_2$, the pulse will move with a constant shape; if, however, c_2 exceeds c_1, we can expect the second wave to catch the first, in which case propagation may fail.

In the approximate treatment which follows we consider the two parts of the wave separately and see that the leading front can be considered as a cubic Fisher wave and the recovery front by a quadratic form.

11.4.1. First reaction front

We assume that a steady wavefront with velocity c_1 exists. The system must be governed now by two travelling coordinate equations for the two concentrations of A and B. These are

$$\frac{d^2\alpha}{dz^2} + c_1 \frac{d\alpha}{dz} - \alpha\beta^2 = 0 \tag{11.51}$$

$$\frac{d^2\beta}{dz^2} + c_1 \frac{d\beta}{dz} + \alpha\beta^2 - \kappa_2\beta = 0. \tag{11.52}$$

Ahead of the reaction front (at $z \to +\infty$), no conversion has occurred, so the boundary condition is

$$\alpha = 1, \beta = 0 \quad \text{with} \quad \frac{d\alpha}{dz} = \frac{d\beta}{dz} = 0 \quad \text{at } z \to +\infty. \tag{11.53}$$

We do not know the exact conditions after the wavefront has passed. These must emerge from the analysis.

The first important assumption to be made is that the autocatalyst should decay only slowly, so $\kappa_2 \ll 1$. This will lead to a distinct separation between the two wavefronts and hence to a region in which the concentration gradients are relatively small.

Coupled wave equations are not easy to handle. It will help greatly if we can somehow find some sort of approximate relationship between α and β, which will allow us to substitute for α, say, and reduce the system to a single equation. This relationship need only apply during the first wavefront. What should we take? The behaviour of this model in a CSTR suggests one such form. By analogy with the stationary-state relationship, eqn (6.50), we try the form

$$\beta = \frac{1 - \alpha}{1 + \kappa_2} \Rightarrow \alpha = 1 - (1 + \kappa_2)\beta. \qquad (11.54)$$

Our approach is now to substitute for α into eqn (11.52). This gives

$$\frac{d^2\beta}{dz^2} + c_1 \frac{d\beta}{dz} - \beta[(1 + \kappa_2)\beta^2 - \beta + \kappa_2] = 0. \qquad (11.55)$$

This is a cubic Fisher equation. The final term corresponds to the chemical kinetics, and we need to establish when this term becomes zero. One solution is $\beta = 0$, which matches the boundary condition as $z \to +\infty$. The other two zeros in the kinetic term arise when

$$\beta_\pm = \frac{1 + [1 - 4\kappa_2(1 + \kappa_2)]^{1/2}}{2(1 + \kappa_2)}. \qquad (11.56)$$

These non-zero roots are only real if κ_2 is sufficiently small, such that

$$\kappa_2 < \tfrac{1}{2}(\sqrt{2} - 1) \qquad (11.57)$$

so we have a first estimate on the decay rate constant for possible wave propagation.

For small κ_2, the upper root (which we may denote β_+), is given by

$$\beta_+ \approx 1 - 2\kappa_2. \qquad (11.58)$$

In the absence of decay ($\kappa_2 = 0$), β_+ tends to unity and then corresponds to the final, complete conversion state of §11.1. Thus decay of the autocatalyst shifts this singular point in the phase plane (Fig. 11.8), along the β axis, to lower concentrations.

The lower root β_- is given by

$$\beta_- \approx \kappa_2. \qquad (11.59)$$

This singularity moves away from the origin in the phase plane as the decay of autocatalyst is included.

The corresponding values of the concentration of the reactant A can be obtained from eqn (11.54), and in the limit of small κ_2 the above forms give

$$a_+ \approx \kappa_2 \qquad\qquad a_- \approx 1 - \kappa_2. \qquad (11.60)$$

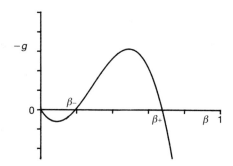

FIG. 11.8.   Influence of the decay reaction on the phase plane for cubic autocatalysis: the pulse will emerge from its initial singularity β_+ and tend to a state of incomplete conversion β_-.

The wave equation can be written as

$$\frac{d^2\beta}{dz^2} + c_1\frac{d\beta}{dz} - (1 + \kappa_2)\beta(\beta - \beta_+)(\beta - \beta_-) = 0. \qquad (11.61)$$

The β_+ state corresponds to almost complete conversion of A, which we expect to follow the first front. Thus we seek a wave connecting $\beta = 0$ to $\beta = \beta_+$.

The appropriate boundary conditions after the front are

$$d\beta/dz \approx 0 \text{ for } \beta = \beta_+. \qquad (11.62)$$

Following the procedure of § 11.1.4, we assume a gradient varying with the concentration in a parabolic manner, with

$$d\beta/dz = -k\beta(\beta_+ - \beta) \qquad (11.63)$$

where k is a constant to be determined. The same process of differentiation, substitution, and matching of coefficients then leads to the result

$$k = -[\tfrac{1}{2}(1 + \kappa_2)]^{1/2} \qquad (11.64)$$

and a wave velocity c_1

$$c_1 = [\tfrac{1}{2}(1 + \kappa_2)]^{1/2}(b_+ - 2a_+) \qquad (11.65)$$

and so for small κ_2 we have

$$c_1 \approx \frac{1}{\sqrt{2}}(1 - \tfrac{3}{2}\kappa_2). \qquad (11.66)$$

As the autocatalyst decay rate tends to zero, this speed approaches the previous results (§ 11.1), $c_1 = 1/\sqrt{2}$. Thus the decay reaction slows this front down as well as reducing the extent of conversion through the front.

11.4.2. Recovery front

The second front in the pulse, moving with velocity c_2, sees the conversion of B to C. Thus β must fall from β_+ to zero. There may also be a final decay in the reactant concentration from α_+ to zero. However, α_+ is already small, so we will make the approximation that $\alpha = \alpha_+ \approx \kappa_2$ throughout the whole of this front. Assuming constant α means that again the system is reduced to one governing equation. Substituting $\alpha = \kappa_2$ into eqn (11.52) gives a quadratic wave equation

$$\frac{d^2\beta}{dz^2} + c_2 \frac{d\beta}{dz} - \kappa_2 \beta(1 - \beta) = 0 \qquad (11.67)$$

which has the form of a quadratic Fisher equation.

The boundary condition after the front (at $z = -\infty$) is simply $\beta = d\beta/dz = 0$. Ahead of the wave we have $\beta = \beta_+$ and $d\beta/dz = 0$ at some (unknown) z. In order to fix these pre-front conditions at $+\infty$, we must rescale our distance coordinate. If we introduce a new travelling wave coordinate ξ defined by

$$\xi = z/\kappa_2 \qquad (11.68)$$

then for vanishingly low decay rates with $\kappa_2 \to 0$, the boundary ahead of the wave (finite, positive z) is scaled to $\xi = +\infty$.

Writing eqn (11.67) in terms of derivatives with respect to ξ, we obtain

$$\frac{d^2\beta}{d\xi^2} + C_2 \frac{d\beta}{d\xi} - \kappa_2^3 \beta(1 - \beta) = 0 \qquad (11.69)$$

where $C_2 = \kappa_2 c_2$ is the wave velocity in the new coordinate, $d\xi/d\tau$. The boundary conditions for this equation become

$$\beta = 0 \text{ at } \xi = -\infty \qquad \beta = \beta_+ \approx 1 \text{ at } \xi = +\infty. \qquad (11.70)$$

This form is now the same as that of § 11.2. Equation (11.69) permits any wave velocity above a minimum $C_{2,\,min}$ given by

$$C_{2,\,min} = 2\kappa_2^{3/2}. \qquad (11.71)$$

We expect this minimum velocity to be the stable velocity and hence that which is established. In terms of the previous coordinate z and velocity c_2 this give

$$c_2 = 2\kappa_2^{1/2}. \qquad (11.72)$$

Because we have been dealing with slowly decaying catalysts, $\kappa_2 \ll 1$, this recovery wave may move quite slowly. In the extreme limit $\kappa_2 \to 0$, the recovery wave fails to propagate as no B is converted to C.

Equations (11.66) and (11.72) can only be treated as leading-order expressions at best. They do, however, convey the spirit that for sufficiently small decay rates, a two-front pulse of reaction may propagate through the tube of reactant producing an overall conversion of A to C through the intermediate B. Comparing the two expressions, we can also see that if κ_2 increases to approximately $\frac{1}{8}$ the two velocities will become comparable, and for larger decay rates the second wave may catch up the first.

11.5. Travelling waves in two and three dimensions

We have only considered the simplest possible geometry—that of a plane wavefront in an infinitely long tube. Much current interest in chemical and biochemical wave propagation is concerned with the forms that arise in 'real' geometries. In two dimensions, such as for very thin layers of reaction medium, these include the target patterns, such as those shown in Fig. 1.17(c). A single circular wavefront is clearly to be expected in such an experimental arrangement if a travelling front of the type discussed above is initiated from some point. The successive concentric rings of the target indicate repeated initiation in some form at the 'centre' of the rings. If the target waves are broken, the cleaved ends curl into spiral wave patterns. Spirals also arise spontaneously in 'excitable' media (see chapter 5). In three dimensions, one form of travelling wave is known as a scroll. Details of these various forms can be found in the papers by Tyson and Keener and by Winfree listed at the end of this chapter.

References

Fisher, R. A. (1937). The wave of advance of advantageous genes. *Ann. Eugen*, **7**, 355–69.

Gray, P., Showalter, K., and Scott, S. K. (1987). Propagating reaction–diffusion fronts with cubic autocatalysis: the effects of reversibility. *J. Chim. Phys.*, **84**, 1329–33.

Hanna, A., Saul, A., and Showalter, K. (1982). Detailed studies of propagating fronts in the iodate oxidation of arsenous acid. *J. Am. Chem. Soc.*, **104**, 3838–44.

Kolmogorov, A. N., Petrovskii, I. G., and Piskounov, N. S. (1937). A study of the diffusion equation associated with the growth in the amount of a substance and its application to a biological problem. *Bull. Univ. Moscow Ser. Int., Sect.*, A1, 1.

Luther, R. (1906). Propagation of chemical reactions in space. *Z. Elektrochem.*, **12**, 596.

Luther, R. (1987). *J. Chem. Educ.*, **64**, 740–2 (trans. R. Arnold, K. Showalter, and J. J. Tyson).

Murray, J. D. (1979). *Lecture notes on nonlinear differential equation models in biology.* Clarendon Press, Oxford.

Ortoleva, P. and Schmidt, S. L. (1985). The structure and variety of chemical waves. In *Oscillations and traveling waves in chemical systems*, (ed. R. J. Field and M. Burger), ch. 10, pp. 333–417. Wiley, New York.

Saul, A. and Showalter, K. (1985). Propagating reaction-diffusion fronts. In *Oscillations and traveling waves in chemical systems*, (ed. R. J. Field and M. Burger), ch. 11, pp. 419–39. Wiley, New York.

Segal, L. A. (ed.) (1980). *Mathematical models in molecular and cellular biology.* Cambridge University Press.

Showalter, K. (1987). Chemical waves. In *Kinetics of nonhomogeneous processes*, (ed. G. R. Freeman). Wiley, New York.

Tyson, J. J. (1985). A quantitative account of oscillations, bistability and traveling waves in the Belousov–Zhabotinskii reaction. In *Oscillations and traveling waves in chemical systems*, (ed. R. J. Field and M. Burger), ch. 3, pp. 93–143. Wiley, New York.

Tyson, J. J. and Keener, J. P. (1988). Singular perturbation theory of traveling waves in excitable media (a review). *Physica*, D **32**, 327–61. (December)

Winfree, A. T. (1974). Rotating chemical reactions. *Sci. Am.* **230**, (December) 82–95.

Winfree, A. T. (1980). *The geometry of biological time.* Springer, New York.

Winfree, A. T. (1985). Organizing centers for chemical waves in two and three dimensions. In *Oscillations and traveling waves in chemical systems*, (ed. R. J. Field and M. Burger), ch. 12, pp. 441–71. Wiley, New York.

Ye, Q.-X. and Wang, M.-X. (1987). Traveling wave front solutions of Noyes-Field system for Belousov–Zhabotinskii reaction. *Nonlinear Anal. Theor. Methods. Appl.*, **11**, 1289–302.

Zeldovich, Ya. B., Barenblatt, G. I., Librovich, V. B., and Makhviladze, G. M. (1985). *The mathematical theory of combustion and explosions.* Consultants Bureau, New York.

HETEROGENEOUS REACTIONS

A full study of heterogeneous reactions lies beyond the scope of this chapter but some broad principles are both important and can help with the perspective. Several different representative circumstances are examined and these include:

(1) non-linearities introduced by activated adsorption, where an activation energy depending on the degree of surface coverage can give rise to multiple stationary states;

(2) non-linearities introduced by the participation of one or more unoccupied surface sites which can do the same;

(3) in particular, when surface reaction requires the participation of two unoccupied sites, strong analogies to cubic isothermal autocatalysis emerge, with multiple stationary states and oscillatory behaviour;

(4) non-linearities introduced or enhanced by competitive chemisorption illustrate how 'poisoning' can underlie oscillatory behaviour.

The previous chapters have dealt at some length with the wide range of non-linear kinetic behaviour which can arise when there is autocatalysis, either because of chemical intermediates or through self-heating. Reactions involving more conventional catalysts, such as heterogeneous surfaces or enzymes, are also known as a good source of multiple stationary states and oscillations. Many simple oxidation processes, including those of H_2, CO, and NH_3, show ignition and extinction (hysteresis loops) and periodic reaction.

Here, we examine a number of simple models which have been variously proposed to account for, or at least mimic, such responses.

12.1. Multiplicity in adsorption reaction isotherms

We begin with the simplest model scheme for a heterogeneously catalysed reaction, with Langmuir–Hinshelwood kinetics. A reactant P is adsorbed, reversibly, onto a surface S. There it may react to give a product C. which is immediately and irreversibly desorbed:

$$P + S \rightleftharpoons P - S \tag{12.1}$$

$$P - S \rightarrow C + S. \tag{12.2}$$

Such a mechanism is also used for the simplest model of enzyme kinetics—the Michaelis–Menten model.

If the pressure of reactant P above the surface is maintained constant, the rate equation for the concentration of adsorbed reactant is

$$\frac{d[P - S]}{dt} = k_1[P][S] - k_{-1}[P - S] - k_2[P - S]. \tag{12.3}$$

The total concentration of surface sites $[S]_0$, occupied or unoccupied and expressed in terms of moles per unit area, will remain constant, so $[P - S]$ and $[S]$ are related at all times by the stoichiometric relationship

$$[P - S] + [S] = [S]_0. \tag{12.4}$$

With this, eqn (12.3) can be rewritten as

$$\frac{d[P - S]}{dt} = k_1[P]([S]_0 - [P - S]) - (k_{-1} + k_2)[P - S]. \tag{12.5}$$

It is usual to work in terms of the fractional coverage of the surface θ_p given by

$$\theta_p = [P - S]/[S]_0 = 1 - \theta_s \tag{12.6}$$

where $\theta_s = [S]/[S]_0$. If we also denote the group of rate constants $(k_{-1} + k_2)/k_1$ by K' and the time by $\tau' = k_1 t$, eqn (12.5) becomes

$$\frac{d\theta_p}{d\tau'} = [P](1 - \theta_p) - K'\theta_p. \tag{12.7}$$

Equation (12.7) is not quite dimensionless because of the units of the reactant concentration (usually this would be expressed as a pressure). Dividing throughout by a standard pressure p^0, denoting the partial pressure of the reactant thus obtained by p, we have

$$\frac{d\theta_p}{d\tau} = p(1 - \theta_p) - K\theta_p \tag{12.8}$$

where

$$K = \frac{k_{-1} + k_2}{k_1 p^0}. \tag{12.9}$$

For enzyme reactions K is the traditional Michaelis constant. For a heterogeneous surface on which adsorption and desorption but no reaction occurs $(k_2 = 0)$ K is simply an equilibrium constant for adsorption. (Actually we are not being as economical as we could be in this non-dimensionalization of the equations. We could have divided throughout by K' instead of introducing the pressure scale p^0, and eqn (12.9) would then have read

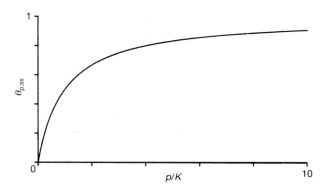

Fig. 12.1. The Langmuir–Hinshelwood adsorption isotherm, showing the fractional coverage of the catalyst surface as a function of the partial pressure of p in the gas phase.

$d\theta_p/d\tau = p(1 - \theta_p) - \theta_p$ with $p = [P]/K'$ and $\tau = K'\tau'$. The equation then has one variable θ_p dependent on one parameter p, with no unfolding parameter to alter the dependence of one on the other. We will, however, allow ourselves some luxury here and continue with eqn (12.9) as given above.)

The stationary-state extent of surface coverage for a given partial pressure p is obtained by setting $d\theta/d\tau$ equal to zero. This gives

$$\theta_{p, ss} = \frac{p}{p + K}. \tag{12.10}$$

Figure 12.1 shows $\theta_{p, ss}$ as a function of p/K. For any given value of the latter the extent of surface coverage is uniquely determined: there is no multiplicity. At low reactant pressures the coverage increases almost linearly with p/K; at high pressures the system approaches complete coverage, with θ_p tending to unity as p/K tends to infinity.

The rate of reaction, defined in terms of the production of C, is given by $dc/dt = k_2[S]_0\theta_p$. Denoting $k_2[S]_0$ by V_{max}, we then have

$$\frac{dc}{dt} = \frac{V_{max}\, p}{p + K}. \tag{12.11}$$

The dependence of the rate of the reactant pressure has a similar form to that shown in Fig. 12.1, with a limit of V_{max} as $p \to \infty$. Again, multiple solutions are not possible for any values of p/K.

12.2. Adsorption and desorption with interactions

The above analysis makes the implicit assumption that molecules of P adsorbed on the surface do not interact with each other in any attractive or repulsive way. Thus the rate constants for adsorption (k_1) and desorption (k_{-1}) are independent of the surface coverage. We now consider a situation where there are interactions, so that the heat of adsorption varies with the extent of coverage in some linear manner

$$\Delta H = \Delta H_0 - \alpha' \theta_p \tag{12.12}$$

where ΔH_0 is the heat of adsorption on a clean surface. If α' is positive, the heat of adsorption will become more negative as the coverage increases. This may be reflected as a commensurate increase in the activation energy for desorption. Thus, k_{-1} will depend on θ_p in a way which we can represent as

$$k_{-1} = (k_{-1})_0 e^{-\alpha \theta_p} \tag{12.13}$$

where $\alpha = \alpha'/RT$. Even in the absence of the reaction step (12.2), this model can lead to multiplicity of stationary-state extents of coverage.

The rate equation equivalent to (12.7) can be written as

$$\frac{d\theta_p}{dt} = k_1 p(1 - \theta_p) - (k_{-1})_0 \theta_p e^{-\alpha \theta_p} - k_2 \theta_p. \tag{12.14}$$

The stationary-state relationship between the partial pressure p and the extent of coverage is then

$$p = \frac{(k_{-1})_0 \theta_p e^{-\alpha \theta_p} + k_2 \theta_p}{k_1 (1 - \theta_p)}. \tag{12.15}$$

For the special case $k_2 = 0$, this simplifies to

$$p = \frac{K \theta_p e^{-\alpha \theta_p}}{1 - \theta_p} \tag{12.16}$$

where $K = (k_{-1})_0/k_1$. The form of the stationary-state locus described by eqn (12.16) depends on α. For weak interactions, α is small and the behaviour is not significantly different from that of the previous section ($\alpha = 0$). The extent of surface coverage increases monotonically with the partial pressure, tending to unity as $p/K \to \infty$, as shown in Fig. 12.2(a).

For larger attractive interactions, however, the stationary-state locus becomes S-shaped with a hysteresis loop (Fig. 12.2(b)). The critical value of α at which the hysteresis loop first appears can be determined using the singularity theory equations of §7.3.1. The appropriate form of the stationary-state condition F is

$$F = p(1 - \theta_p) - K \theta_p e^{-\alpha \theta_p} = 0. \tag{12.17}$$

We then find the derivatives F_{θ_p} and $F_{\theta_p \theta_p}$ and equate these to zero along with F. These requirements are satisfied for

$$\theta_p = \tfrac{1}{2} \qquad p/K = e^{-2} \qquad \text{and} \qquad \alpha = 4. \tag{12.18}$$

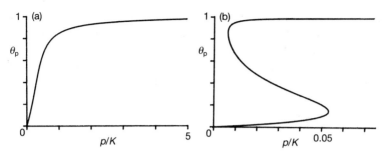

FIG. 12.2. The stationary-state fractional coverage of the surface for adsorption with coverage-dependent parameters and no reaction: (a) $\alpha = 2$, showing unique solution for all partial pressures; (b) $\alpha = 4.5$, typical of all $\alpha > 4$, showing multiplicity at low pressures.

Thus, multiple stationary-state solutions are a feature of this model with $\alpha > 4$. The turning points in the locus can be located from the condition $dp/d\theta_p = 0$ from eqn (12.16), which gives

$$\theta_p = \frac{\alpha \pm (\alpha^2 - 4\alpha)^{1/2}}{2\alpha} \qquad (12.19)$$

again showing the need for $\alpha > 4$.

The interpretation of the loci for systems with such large attractive interactions is that as the reactant pressure is increased and the extent of coverage increases with it, there is a point of sudden 'condensation' at which something approaching a monolayer of adsorbed reactant is formed.

If reaction occurs so k_2 is not zero, a similar situation arises, although the jump to high surface coverage will also then favour a jump to high reaction rates (which then tends to decrease the concentration of P on the surface). The stationary-state condition from eqn (12.14) can be written in a fully dimensionless form as

$$F = p(1 - \theta_p) - K\theta_p(K_2 + e^{-\alpha\theta_p}) = 0 \qquad (12.20)$$

where $K = k_{-1}/k_1$ as above and $K_2 = k_2/k_{-1}$.

The condition for the appearance of a hysteresis loop is again most easily obtained using the singularity theory recipe, from which we find the relationship

$$\alpha_H = \frac{4}{1 - e^2 K_2}. \qquad (12.21)$$

If the attractive interaction for a given system is such that α is greater than α_H, as defined above, then the surface coverage and the reaction rate curves will show multiplicity over some range of reactant partial pressures. We may now talk of ignition and extinction, since the jump to high surface coverages also corresponds to a jump to high rates of production of C.

Equation (12.21) shows that the influence of the reaction is to make multiplicity less favoured. The higher the value of the reaction rate constant k_2 compared with the desorption step k_{-1}, the greater the attractive interaction α_H needed. If k_2 increases so that K_2 exceed e^{-2}, multiple stationary states will not occur for any finite value of α.

12.3. Multiplicity from reaction

Non-linearities arising from non-reactive interactions between adsorbed species will not be our main concern. In this section we return to variations of the Langmuir–Hinshelwood model, so the adsorption and desorption processes are not dependent on the surface coverage. We are now interested in establishing which properties of the chemical reaction step (12.2) may lead to multiplicity of stationary states. In particular we will investigate situations where the reaction step requires the involvement of additional vacant sites. Thus the reaction step can be represented in the general form

$$P - S + nS \rightarrow (n + 1)S + C \qquad \text{rate} = k_2[P - S][S]^n \qquad (12.22)$$

where n additional vacant sites are involved in the reaction step.

12.3.1. Involvement of vacant sites

The particular form for the reaction rate law invoked above has been justified in a number of ways. One interpretation is that a second adsorbed reactant may also be involved in the final step, but that the adsorption and desorption of this species occurs on a much faster timescale than that of P or of the reaction step. Thus, if this second reactant is denoted R, which may be polyatomic and adsorb onto n surface sites, the kinetics become

$$P + S \rightleftharpoons P - S \qquad (12.23)$$

$$R + nS \rightleftharpoons R - S_n \qquad (12.24)$$

with the reaction step

$$P - S + R - S_n \rightarrow (n + 1)S + C. \qquad (12.25)$$

However, if the forward and reverse processes in step (12.24) are sufficiently rapid that the adsorption and desorption of R is kept virtually in a state of equilibrium, then (12.24) and (12.25) can be replaced by a form such as (12.22).

An overall rate of the form of (12.22) can also arise if the initial adsorption of P on to one site is followed by rearrangement and perhaps further dissociation prior to reaction, e.g.

$$P + S \rightleftharpoons P - S \qquad (12.26)$$

$$P - S + 2S \rightarrow B - S + C - S_2 \qquad (12.27)$$

with then two rapid desorption steps

$$B - S \rightarrow B + S \tag{12.28}$$

$$C - S_2 \rightarrow C + 2S. \tag{12.29}$$

12.3.2. Multiplicity of stationary states

Returning now to the condensed form of the model, with eqn (12.22), the dimensionless rate equation for the surface coverage with P has the form

$$\frac{d\theta_p}{d\tau} = p(1 - \theta_p) - K\theta_p - KK_2\theta_p(1 - \theta_p)^n \tag{12.30}$$

where $K = k_{-1}/k_1$ and $K_2 = k_2[S]_0^n/k_{-1}$. These definitions are similar to those used in the previous section, but the dimensionless decay rate constant K_2 now also involves the total concentration of surface sites $[S]_0$.

The stationary-state solutions of (12.30) satisfy

$$F = p(1 - \theta_p) - K\theta_p - KK_2\theta_p(1 - \theta_p)^n = 0 \tag{12.31}$$

and hence we have

$$\frac{p}{K} = \frac{\theta_p[1 + K_2(1 - \theta_p)^n]}{1 - \theta_p}. \tag{12.32}$$

Singularity theory can be used again to find the conditions at which a hysteresis loop first appears in the $\theta_p - (p/K)$ locus. Setting $F = F_{\theta_p} = F_{\theta_p\theta_p} = 0$, some algebra yields

$$\theta_p = \frac{2}{n + 1} \qquad K_2 = \left(\frac{n + 1}{n - 1}\right)^{n+1} \qquad \frac{p}{K} = \frac{4n}{(n - 1)^2}. \tag{12.33}$$

These forms show that multiple stationary states will be possible, provided n is greater than unity. As n is the number of vacant sites being recruited into the reaction step, it probably ought to be an integer. Thus multiplicity with simple Langmuir–Hinshelwood adsorption requires at least two vacant sites to be involved in the reaction.

12.4. Cubic catalysis for reaction rate

The reaction requiring two vacant sites for the conversion of $P - S$ to the product C is the simplest form of the above scheme which will display multiplicity. The reaction scheme can be written as

$$P + S \rightleftharpoons P - S \qquad k_1, k_{-1} \tag{12.34}$$

$$P - S + 2S \rightarrow 3S + C \qquad k_2. \tag{12.35}$$

The second step here can clearly be related to the cubic autocatalysis seen in earlier chapters (with S replacing B).

With a slight change of notation from the previous section, the dimensionless rate equation becomes

$$\frac{d\theta_p}{d\tau} = p(1 - \theta_p) - K\theta_p - \kappa_2\theta_p(1 - \theta_p)^2 \tag{12.36}$$

so that $\kappa_2 = k_2/k_1$ appears instead of KK_2.

Using the singularity analysis again, the condition for the appearance or disappearance of a hysteresis loop is given in terms of the two unfolding parameters K and κ_2 as

$$K = \tfrac{1}{27}\kappa_2. \tag{12.37}$$

If the desorption rate constant K is less than the value given in eqn (12.37), the stationary-state locus (whether for extent of conversion or for reaction rate as functions of the reactant partial pressure p) displays multiplicity, as shown in Fig. 12.3. As the desorption rate increases, the range of multiple solutions decreases and vanishes when (12.37) is satisfied. For larger desorption rates, the reaction rate and extent of coverage increase monotonically with partial pressure.

A special case for this system is that where no desorption occurs at all $(K = 0)$. Under such conditions, any non-zero κ_2 gives rise to multiplicity. The stationary-state equation is

$$p(1 - \theta_p) = \kappa_2\theta_p(1 - \theta_p)^2. \tag{12.38}$$

One solution of this is that corresponding to complete surface coverage $\theta_p = 1$, and this exists for all partial pressures and all values of κ_2. The other

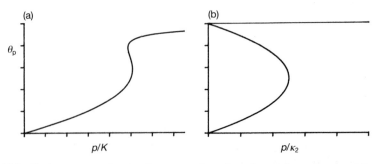

FIG. 12.3. Stationary-state fractional coverage for adsorption and reaction involving two vacant sites: (a) $\kappa_2/K = 36$ showing multiplicity; (b) multiplicity in absence of desorption; now one solution corresponding to a fully covered surface exists for all reactant pressures.

solutions are given by a quadratic

$$\theta_p = \frac{1}{2}\left[1 \pm \left(1 - \frac{4p}{\kappa_2}\right)^{1/2}\right] \tag{12.39}$$

so that these are real for $p < \frac{1}{4}\kappa_2$ (the point of ignition). With no desorption there is no extinction.

The condition for multiplicity, equivalent to eqn (12.37), for general n is

$$K = \left(\frac{n-1}{n+1}\right)^{n+1}\kappa_2. \tag{12.40}$$

12.5. Effect of catalyst poisons

We now take the simple cubic model above and allow for the competitive adsorption and desorption of a second species Q. Thus the model becomes

$$P + S \rightleftharpoons P - S \qquad k_2, k_{-1} \tag{12.41}$$

$$P - S + 2S \rightarrow 3S + C \qquad k_2 \tag{12.42}$$

$$Q + S \rightleftharpoons Q - S \qquad k_3, k_{-3}. \tag{12.43}$$

One feature of this extension is that there are now two different adsorbed species whose concentrations are independent. In terms of the fractional surface coverages, we have the condition

$$\theta_s = 1 - \theta_p - \theta_q \tag{12.44}$$

where θ_s is the fraction of the surface uncovered.

The rate equations for the surface concentrations of the two species can be written as

$$\frac{d\theta_p}{d\tau} = p(1 - \theta_p - \theta_q) - K\theta_p - \kappa_2\theta_p(1 - \theta_p - \theta_q)^2 \tag{12.45}$$

$$\frac{d\theta_q}{d\tau} = \kappa_3 q(1 - \theta_p - \theta_q) - \kappa_{-3}\theta_q \tag{12.46}$$

where $\kappa_3 = k_3/k_1$ and $\kappa_{-3} = k_{-3}/k_1$.

12.5.1. Stationary states and multiplicity

The stationary-state solutions are obtained by setting $d\theta_p/d\tau = d\theta_q/d\tau = 0$. These give

$$\theta_q = \frac{K_3(1 - \theta_p)}{1 + K_3} \tag{12.47}$$

where $K_3 = \kappa_3 q/\kappa_{-3}$ is an adsorption constant for Q. The concentration of P on the surface is then given by the roots of the cubic equation

$$\kappa_2 \theta_p^3 - 2\kappa_2 \theta_p^2 + [(1 + K_3)p + (1 + K_3)^2 K + \kappa_2]\theta_p - (1 + K_3)p = 0.$$
$$(12.48)$$

Using the singularity theory approach, we find that the stationary-state locus will display a hysteresis loop provided that

$$K \leqslant \frac{1}{27} \frac{\kappa_2}{(1 + K_3)^2}.$$
$$(12.49)$$

Again then, we see that if there is no desorption of the reactant P, i.e. if $K = 0$, the system will display multiplicity, irrespective of the poison Q.

12.5.2. Oscillations

The reaction processes on the surface are now determined by two coupled equations (12.45) and (12.46). There are two independent concentrations and a cubic non-linearity, hence we expect that Hopf bifurcation may also be a feature of this model.

For simplicity, we will consider the case without reactant desorption, $K = 0$, in which analytical expressions for the stationary-state concentrations can be obtained. With $K = 0$, the stationary-state condition requires

$$p(1 - \theta_p - \theta_q) = \kappa_2 \theta_p (1 - \theta_p - \theta_q)^2.$$
$$(12.50)$$

One solution of this $(1 - \theta_p - \theta_q) = 0$, which from eqn (12.47) then implies $\theta_p = 1$ and $\theta_q = 0$, i.e. the surface is completely covered with P.

The remaining possible solutions are given by the roots of a quadratic equation obtained by substituting from (12.47) into (12.50), yielding

$$\theta_p = \frac{1}{2}\left[1 \pm \left(1 - \frac{4p}{\kappa_2}(1 + K_3)\right)^{1/2}\right].$$
$$(12.51)$$

Thus, the stationary-state locus is similar in form to that shown in Fig. 12.3(b) with the two branches described by eqn (12.51) existing for

$$p \leqslant \frac{1}{4} \frac{\kappa_2}{(1 + K_3)}.$$
$$(12.52)$$

The Hopf bifurcation analysis proceeds as described previously, the required condition being that the trace of the Jacobian matrix corresponding to eqns (12.45) and (12.46) should become equal to zero for some stationary-state concentration given by the lower root from (12.51). (The solution with the upper root corresponds to the middle branch of stationary states for which the determinant of the Jacobian is always negative; it is a saddle point and the vanishing of the trace does not satisfy the Hopf criteria.)

The resulting expression for the partial pressure of the reactant p has the solution

$$\frac{p^*}{\kappa_{-3}} = \frac{2(1 + K_3)^2(2 + K_3) + (\kappa_2/\kappa_{-3}) + \{(\kappa_2/\kappa_{-3})[(\kappa_2/\kappa_{-3}) - 4(1 + K_3)(2 + K_3)]\}^{1/2}}{2(2 + K_3)^2}.$$

(12.53)

The stationary state is unstable for $p > p^*$. Thus there will be a point of Hopf bifurcation provided the rate constants for reaction and the adsorption and desorption of the poison satisfy the condition

$$\frac{\kappa_2}{\kappa_{-3}} > 4(1 + K_3)(2 + K_3).$$

(12.54)

This requirement is favoured by high values for the reaction rate κ_2 compared with the desorption of the poison κ_{-3}. Small values of κ_{-3}, however, also tend to increase $K_3 = \kappa_3/\kappa_{-3}$ increasing the right-hand side of (12.54).

As an example we may consider a system with $K_3 = 9$ and $\kappa_2/\kappa_{-3} = 500$. Then $(p^*/\kappa_{-3}) = 11.9$. This point lies on the lowest branch of stationary-state solutions which exists up to $p/\kappa_{-3} = 12.5$. Thus, if $\kappa_{-3} = 0.002$, we take $\kappa_3 = 0.018$ and $\kappa_2 = 1$. The Hopf bifurcation point then occurs for $p = p^* = 0.0237$, with the saddle–node turning point at $p = 0.025$. For this

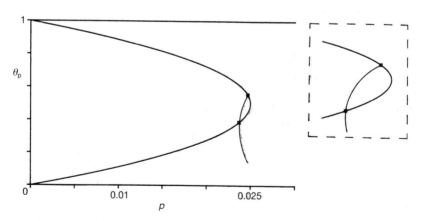

FIG. 12.4. Stationary-state solutions and limit cycles for surface reaction model in presence of catalyst poison: $K_3 = 9$, $\kappa_2 = 1$, $\kappa_3 = 0.018$. There is a Hopf bifurcation on the lowest branch $p = 0.0237$. The resulting stable limit cycle grows as the dimensionless partial pressure increases and forms a homoclinic orbit when $p = 0.0247$ (see inset). The saddle–node bifurcation point is at $p = 0.025$.

Table 12.1

Growth of oscillations for heterogeneous catalysis model with poisoning: A = amplitude, τ_p = period

p	A_{θ_p}	A_{θ_q}	τ_p
0.023 80	0.051 32	0.044 63	1221
0.0240	0.101 73	0.088 19	1299
0.0242	0.152 99	0.131 85	1404
0.0245	0.250 55	0.211 79	1656
0.0246	0.2972	0.248 06	1815

exemplary case also, the Hopf bifurcation is supercritical: a stable limit cycle emerges, as shown in Fig. 12.4. As p increases beyond p^*, the limit cycle survives and grows. Sustained oscillations in the surface coverages are possible over the range $p^* \leqslant p < 0.0247$. At the upper limit of this range, the limit cycle is destroyed by homoclinic orbit formation, with the oscillatory period becoming infinite. The variation in oscillatory amplitude and period with the dimensionless partial pressure of P is given in Table 12.1. For $0.0247 < p < 0.025$, the lowest stationary state is unstable, with no stable limit cycle, so the system must tend to the state with $\theta_p = 1$.

12.6. Competitive chemisorption of reactants

The simple model just discussed shows multistability even when the system is 'clean' but requires the involvement of a poison for oscillations. One reason for this is that the latter is needed to provide a second independent surface concentration, so we then have a two-variable system. It was mentioned in §12.3.1 that implicit in the rate law used above may be the adsorption of a second reactant which participates in the reaction step. The latter did not provide a second concentration variable there since its adsorption and desorption processes were assumed to be on a very much faster (instantaneous) timescale.

In this section we turn to a model where the adsorption and desorption of two reactants occur on similar timescales. The adsorption is competitive, i.e. both reactants are adsorbed on to the same surface sites. Again, a number of vacant sites will be involved in the reaction step. The model is

$$P + S \rightleftharpoons P - S \qquad k_1, k_{-1} \qquad (12.55)$$

$$R + S \rightleftharpoons R - S \qquad k_2, k_{-2} \qquad (12.56)$$

$$P - S + R - S + 2S \rightarrow 4S + products \qquad k_3. \qquad (12.57)$$

The rate equations for the surface coverages θ_p and θ_r can be written in the dimensionless form

$$\frac{d\theta_p}{d\tau} = p(1 - \theta_p - \theta_r) - \kappa_1 \theta_p - \theta_p \theta_r (1 - \theta_p - \theta_r)^2 \qquad (12.58)$$

$$\frac{d\theta_r}{d\tau} = r(1 - \theta_p - \theta_r) - \kappa_2 \theta_r - \theta_p \theta_r (1 - \theta_p - \theta_r)^2 \qquad (12.59)$$

where

$$p = \frac{k_1 P}{k_3} \qquad r = \frac{k_2 R}{k_3} \qquad \kappa_1 = \frac{k_{-1}}{k_3} \qquad \kappa_2 = \frac{k_{-2}}{k_3}. \qquad (12.60)$$

Although this set of equations has four parameters, it will be enough to demonstrate that multiple stationary states and sustained oscillatory responses are possible to consider variations in the dimensionless reactant partial pressures p and r, with fixed values for the desorption rate constants $\kappa_1 = 0.001$ and $\kappa_2 = 0.002$.

12.6.1. Stationary states and multiplicity

At a stationary state, the following relationships hold:

$$\theta_r = \frac{(p - r) - (p - r + \kappa_1)\theta_p}{p - r - \kappa_2} \qquad 1 - \theta_p - \theta_r = \frac{(\kappa_1 + \kappa_2)\theta_p - \kappa_2}{p - r - \kappa_2}.$$
$$(12.61)$$

Substituting these into eqn (12.58), the stationary-state condition then yields a quartic in θ_p:

$$F(\theta_p) = p(p - r - \kappa_2)^2 [(\kappa_1 + \kappa_2)\theta_p - \kappa_2] - \kappa_1(p - r - \kappa_2)^3 \theta_p$$
$$- \theta_p^3 [(p - r) - (p - r + \kappa_1)\theta_p][(\kappa_1 + \kappa_2)\theta_p - \kappa_2]^2 = 0. \qquad (12.62)$$

There may be either one or three solutions to this lying in the range $0 \leqslant \theta_p \leqslant 1$ for any given set of positive parameters.

Taking $\kappa_1 = 0.001$ and $\kappa_2 = 0.002$, the boundaries of the region in the partial pressure $(p-r)$ parameter plane in which multiple solutions are found can be located by solving the equation with the extra condition $F_{\theta_p} = 0$. This roughly triangular region, shown in Fig. 12.5, has two cusp points C and C', and a tip N. The coordinates of the cusp points, which satisfy $F = F_{\theta_p} = F_{\theta_p \theta_p} = 0$, for the above choices of κ_1 and κ_2 are $(p = 0.021\,442, r = 0.018\,746)$ for C and $(p = 0.020\,133, r = 0.030\,296)$ for C', whilst the tip N lies at $(p = 0.037\,975, r = 0.039\,159)$.

The bifurcation diagrams in which the stationary-state coverage is plotted as a function of one of the partial pressures, r say, have some surprising and

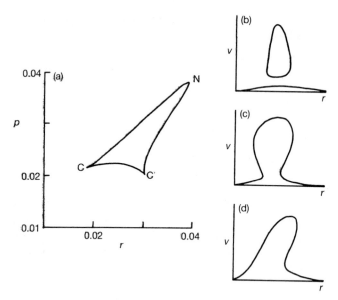

FIG. 12.5. (a) The region of multiple stationary-state behaviour for the Takoudis–Schmidt–Aris model of surface reaction, with $\kappa_1 = 10^{-3}$ and $\kappa_2 = 2 \times 10^{-3}$; (b), (c), and (d) show how the stationary-state reaction rate varies with the gas-phase pressure of reactant R for different values of p, giving isola, mushroom, and single hysteresis loop respectively. (Adapted and reprinted with permission from McKarnin, M. A. *et al.* (1988). *Proc. R. Soc.*, **A415**, 363–87.)

potentially misleading complexities. The behaviour of the system is slightly easier to discuss in terms of the stationary-state reaction rate v as a function of r, where

$$v = \theta_p \theta_r (1 - \theta_p - \theta_r)^2 \qquad (12.63)$$

with θ_p and θ_r taking their appropriate stationary-state values.

If we choose a high value for the partial pressure of reactant P, $p > 0.037\,975$, then the system does not cross the region of multiplicity as r is varied. The reaction rate varies monotonically with r. A similar situation occurs if the fixed partial pressure of P is too small, such that $p < 0.020\,133$. However, with any value of p in the range $0.020\,133 < p < 0.037\,975$, there are multiple stationary-state reaction rates over some range or ranges of the partial pressure of R.

Three different patterns of multistability appear possible from Fig. 12.5. For a cut across the diagram such as that represented by (b), the rate curve shows an isola as r passes through the region of multistability. If p just exceeds the value corresponding to the cusp point C, then a horizontal cut across the diagram such as (c) intersects the boundary four times: the stationary-state reaction rate locus has four turning points and gives a mushroom pattern. Finally, for $0.020\,133 < p < 0.021\,442$, the traverse (d) shows a single hysteresis loop. Qualitatively similar responses emerge if the partial pressure of R is held constant and that of P varied during an experiment.

The occurrence of multiple stationary states of course also indicates the existence of points of 'ignition' and 'extinction'. In the present model, mushroom and isola patterns arise, indicating for instance that extinction of the reaction can be achieved either by decreasing the partial pressure of one of the reactants too far, or by increasing it. Here we see the effect of the competitive adsorption. If the partial pressure of R is low, the surface will have a high coverage with the other reactant P, so θ_p may be close to unity, but θ_r will be small. The low value of θ_r makes the reaction rate small. Increasing the pressure of R then increases θ_r at the expense of θ_p, but the effect on the reaction rate of the increase in the former more than compensates for the decrease in θ_p. (The maximum rate possible will occur with $\theta_p = \theta_r = \frac{1}{4}$, with θ_s, the fractional coverage of free sites, equal to $\frac{1}{2}$, so P – S, R – S, and S are in their stoichiometric proportions.) Eventually the conditions for ignition may be met, and the rate v will jump to a high value. If r is now increased further, it will begin to displace sufficient P from the surface to cause the rate to decrease again. The reactant in excess can act as a poison, choking the surface of the minor reactant and eventually killing off the reaction.

12.6.2 Hopf bifurcations

Again we have a two-variable system, so we can look for points of Hopf bifurcation in terms of the trace and determinant of the Jacobian matrix evaluated for the stationary-state solutions. Thus we seek conditions such that

$$\text{tr}(\mathbf{J}) = 0 \quad \text{with} \quad \det(\mathbf{J}) > 0. \tag{12.64}$$

With the values of the dimensionless desorption rate constants used above, $\kappa_1 = 0.001$ and $\kappa_2 = 0.002$, condition (12.64) describes two curves in the p–r parameter plane. These are shown in Fig. 12.6, which also gives their relative location with respect to the loci of turning points (i.e. where $\det(\mathbf{J}) = 0$) which mark the boundaries of the region of multiplicity.

The two lines of Hopf bifurcations begin and end at points which lie on the boundaries of multistability. These points, labelled K, L, M, and N, correspond to parameter combinations for which there is a stationary state with two zero eigenvalues (the double-zero eigenvalue degeneracy of §8.4.1). The respective locations of these points are: K, ($p = 0.021\,970$, $r = 0.020\,491$); L, (0.029\,730, 0.029\,133); M, (0.020\,146, 0.030\,287); N, (0.037\,975, 0.039\,159).

Two other points are marked, one along each Hopf curve. These are the degenerate bifurcation points at which the emerging limit cycle changes from stable (supercritical) to unstable (subcritical). These have the locations

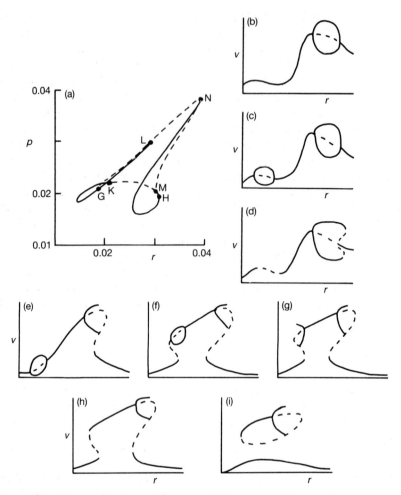

FIG. 12.6. (a) Hopf bifurcation loci for the Takoudis–Schmidt–Aris model with $\kappa_1 = 10^{-3}$ and $\kappa_2 = 2 \times 10^{-3}$. Also shown (broken curves) are the saddle–node boundaries from Fig. 12.6. (b)–(i) The eight qualitative arrangements of Hopf and saddle–node bifurcation points. (Adapted and reprinted with permission from McKarnin, M. A. *et al.* (1988). *Proc. R. Soc.*, **A415**, 363–87.)

$G = (0.020\,668,\ 0.018\,330)$ and $H = (0.019\,308,\ 0.030\,686)$. Along the segments GK and HN the Hopf bifurcation is supercritical; for GL and HM the bifurcations are subcritical.

As an example, consider the horizontal cut obtained by varying r with $p = 0.017$. This cut does not traverse the region of multiplicity, so the reaction rate varies monotonically with r. Nor do we intersect the curve of Hopf points LK. We do, however, encounter the other Hopf curve, NM. The stationary state loses stability at the first crossing, Fig. 12.6(b). Here there is a supercritical Hopf bifurcation and a stable limit cycle emerges, growing as r increases and surrounding the unstable state. Sustained oscillation in the surface coverages and in the reaction rate now prevail. After some further increase in r, the limit cycle begins to decrease in size again. It shrinks to zero at the second supercritical bifurcation as we cross the Hopf curve a second time and re-emerge into the realm of a stable stationary state.

With $p = 0.019$, the traverse cuts both Hopf curves, so the stationary-state locus has four Hopf bifurcation points, as shown in Fig. 12.6(c), each one supercritical. There are two separate ranges of the partial pressure of R over which a stable limit cycle and hence sustained oscillations occur.

Next, consider the case with $p = 0.020\,14$. The traverse across Fig. 12.6(a) as r is varied now also cuts the region of multistability. It passes above the cusp point C' (see Fig. 12.5), giving rise to two turning points in the stationary-state locus, but below the double-zero eigenvalue point M. There are still four intersections with the Hopf curve, so there are four points of Hopf bifurcation. The Hopf point at highest r is now a subcritical bifurcation. The dependence of the reaction rate on r for this system is shown in Fig. 12.6(d).

The double-zero eigenvalue points, such as M, represent the coalescence of Hopf bifurcation and stationary-state turning points. As mentioned above, they thus represent the points at which the Hopf bifurcation loci begin and end. They also have other significance. Such points correspond to the beginning or end of loci of homoclinic orbits. For the present model, with the given choices of κ_1 and κ_2, there are two curves of homoclinic orbit points, one connecting M to N, the other K to L, as shown schematically in Fig. 12.7.

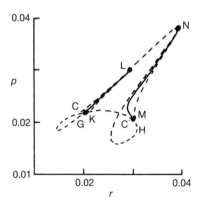

FIG. 12.7.   Curves of homoclinic orbits for Takoudis–Schmidt–Aris model with $\kappa_1 = 10^{-3}$ and $\kappa_2 = 2 \times 10^{-3}$: see text for details. (Adapted and reprinted with permission from McKarnin, M. A. *et al.* (1988). *Proc. R. Soc.*, **A415**, 363–87.)

Thus, if we take $p = 0.021$, the traverse with varying r cuts across the region of multiplicity above point M. There are only three intersections with Hopf curves; the fourth such intersection has been replaced by an intersection with the homoclinic orbit curve. Thus the stationary-state bifurcation diagram, Fig. 12.6(e), shows three Hopf bifurcation points and one region of multi-stability (with two turning points). Within the region of multiple stationary states, the limit cycle which emerged from the third Hopf point is destroyed through homoclinic orbit formation.

The continuation of such experiments, varying r with higher and higher values of p, will see first the appearance of a second region of multistability at low r above the cusp point C (Fig. 12.6(f)). Then will come the formation of a homoclinic orbit instead of the first Hopf bifurcation point (Fig. 12.6(g)), when p exceeds K. The mushroom will change to an isola, as described previously. Such an isola has two Hopf bifurcation points, as shown in Fig. 12.6(h). In between these points the stationary state is stable. Both emerging limit cycles terminate in homoclinic orbits. Above L, the lower Hopf point disappears after coalescing with the extinction turning point. The system then has just one Hopf bifurcation, at high r, at which a stable limit cycle is born and grows as r increases further but soon terminates in a homoclinic orbit (Fig. 12.6(i)).

12.7. The oxidation of carbon monoxide on platinum

The various models above come tantalizingly close to offering a sensible prototype scheme for one of the most extensively studied heterogeneous oscillators—the oxidation of carbon monoxide on platinum or palladium surfaces. This reaction shows oscillatory behaviour with a variety of forms for the catalyst—in particular for platinum or palladium supported on either γ-alumina (Mukesh et al. 1983; Onken and Wicke 1986) or xeolites (Jaeger et al. 1985) or simply with a platinum wire (Scott and Watts 1981; Galwey et al. 1985; Scott et al. 1989). In each of these cases, multiple stationary-state behaviour is readily observed within certain limits of concentration and operating condition. Oscillatory behaviour, on the other hand, tends not to be found with pure reactants unless some inhibitor or poison is added. Also, Ertl and coworkers (1989) have found oscillatory responses with single platinum crystals at operating pressures well below atmospheric—often in the ultra-high vacuum.

In addition to 'kinetic' interpretations such as those mentioned in the previous sections, three different types of explanation have been proposed. Scott and Watts (1981) have suggested a vital involvement of boundary layer films, allowing for considerable depletion of the bulk gas-phase concentrations of the reactants close to the catalyst. In the work on single crystals, Ertl found oscillations with Pt(100) and (110) planes but not with the (111) form. The explanation offered for the behaviour under these conditions is that a phase transition is important. At low CO coverage, a Pt(100) surface exists with a quasi-hexagonal arrangement of surface atoms, differing from the square arrangement appropriate to the bulk. This 'hex' form becomes unstable with respect to the normal (1 × 1) form as the CO coverage increases. Importantly, the two forms also differ considerably in their ability to absorb oxygen. The low-coverage hex form has a very low sticking coefficient for oxygen and hence virtually no catalytic activity. The (1 × 1) form is more efficient at absorbing oxygen dissociatively. There are clearly similarities in this argument to the idea of coverage-dependent kinetic parameters as discussed in §12.2.

In many other cases it is not at all clear that these exothermic reactions are operated in such a way that the system can remain isothermal. Self-heating and hence thermal feedback routes can be expected to have a strong autocatalytic effect on the reaction, perhaps in addition to chemical mechanisms. Recent modelling invoking cellular automata (Jaeger *et al.* 1985) has been to some extent successful at matching qualitatively many of the rather exotic responses which have been observed experimentally.

References

Bykov, V. I. and Yablonskii, G. S. (1981). Simplest model of catalytic oscillator. *React. Kinet. Catal. Lett.*, **16**, 377–81.

Cameron, P., Scott, R. P., and Watts, P. (1986). The oxidation of carbon monoxide on a platinum catalyst at atmospheric pressure. *J. Chem. Soc. Faraday Trans I*, **82**, 1389–403.

Eigenberger, G. (1978). Kinetic instabilities in heterogeneously catalyzed reactions I. *Chem. Eng. Sci.*, **33**, 1255–61.

Eigenberger, G. (1978). Kinetic instabilities in heterogeneously catalyzed reactions II. *Chem. Eng. Sci.*, **33**, 1263–8.

Ertl, G. (1989). The oscillatory catalytic oxidation of carbon monoxide on platinum surfaces. In *Spatial inhomogeneities and transient behaviour in chemical kinetics*, (ed. P. Gray, G. Nicolis, F. Baras, P. Borckmans, and S. K. Scott), ch. 37, pp. 563–76. Manchester University Press.

Galwey, A. K., Gray, P., Griffiths, J. F., and Hasko, S. M. (1985). Surface retexturing of Pt wires during the catalytic oxidation of CO. *Nature*, **313**, 668–71.

Imbihl, R., Cox, M. P., and Ertl, G. (1985). Kinetic oscillations in the catalytic CO oxidation on Pt(100): theory. *J. Chem. Phys.*, **83**, 1578–87.

Imbihl, R., Cox, M. P., and Ertl, G. (1986). Kinetic oscillations in the catalytic CO oxidation on Pt(100): experiments. *J. Chem. Phys.*, **84**, 3519–34.

Jaeger, N. I., Möller, K., and Plath, P. J. (1985). The development of a model for the cooperative behaviour of palladium crystalites during the catalytic oxidation of CO. *Ber. Bunsenges. Phys. Chem.*, **89**, 633–7.

McKarnin, M. A., Aris, R., and Schmidt, L. D. (1988). Autonomous bifurcations of a simple biomolecular surface-reaction model. *Proc. R. Soc.*, **A415**, 363–87.

McKarnin, M. A., Schmidt, L. D., and Aris, R. (1988). Forced oscillations of a self-oscillating bimolecular surface reaction model. *Proc. R. Soc.*, **A417**, 363–88.

Möller, P., Wetzl, K., Eiswirth, M., and Ertl, G. (1986). Kinetic oscillations in the catalytic CO oxidation on Pt(100): computer simulations. *J. Chem. Phys.*, **85**, 5328–36.

Mukesh, D., Kenney, C. N., and Morton, W. (1983). Concentration oscillations of carbon monoxide, oxygen and 1-butene over a platinum supported catalyst. *Chem. Eng. Sci.*, **38**, 69–77.

Onken, H. U. and Wicke, E. (1986). Statistical fluctuations of temperature and conversion at the catalytic CO oxidation in an adiabatic packed bed reactor. *Ber. Bunsenges. Phys. Chem.*, **90**, 976–81.

Schwankner, R. J., Eiswirth, M., Möller, P., Wetzl, K., and Ertl, G. (1987). Kinetic oscillations in the catalytic CO oxidation on Pt(100): periodic perturbations. *J. Chem. Phys.*, **87**, 742–9.

Scott, R. P. and Watts, P. (1981). Kinetic considerations of mass transport in heterogeneous, gas–solid catalytic reactions. *J. Phys. E: Sci. Instrum.*, **14**, 1009–13.

Scott, S. K., Griffiths, J. F., and Galwey, A. K. (1989). In *Spatial inhomogeneities and transient behaviour in chemical kinetics*, (ed. P. Gray, G. Nicolis, F. Baras, P. Borckmans, and S. K. Scott), ch. 38, pp. 577–90. Manchester University Press.

Takoudis, C. G., Schmidt, L. D., and Aris, R. (1981). Multiple steady states in reaction controlled surface catalysed reactions. *Chem. Eng. Sci.*, **36**, 377–86.

Takoudis, C. G., Schmidt, L. D., and Aris, R. (1981). Isothermal sustained oscillations in a very simple surface reaction. *Surf. Sci.*, **105**, 325–33.

Takoudis, C. G., Schmidt, L. D., and Aris, R. (1981). Isothermal oscillations in surface reactions with coverage independent parameters. *Chem. Eng. Sci.*, **37**, 69–76.

COMPLEX OSCILLATIONS AND
CHEMICAL CHAOS

We have concentrated on two-variable examples hitherto because so many of the key features of oscillatory behaviour can be vividly portrayed in these terms. However, real systems all have more numerous chemical constituents and may give more complex responses than those illustrated previously. The inclusion of a third variable or varying one of the parameters of the system in time takes us to new complexities, and these are now examined. After a careful study the reader should be able to:

(1) see the connections between iterative mapping procedures and period-doubling sequences and the relationships between bifurcations like the Feigenbaum ratio;

(2) appreciate the distinction between chaotic and random behaviour;

(3) see how periodic forcing of a parameter (e.g. inflow concentration) causes limit cycles to buckle and distort and to entrain natural oscillations to the forcing period;

(4) represent responsiveness by Arnol'd tongues (or resonance horns);

(5) use the Poincaré map to illustrate the stability of limit cycles and to quantify this by the Floquet multiplier;

(6) recognize the versatility of autonomous three-variable systems as elementary exemplars of chemical chaos whether by isothermal or thermal feedback.

In previous chapters we have dealt only with systems which have one or two independent concentrations. This has been sufficient for a wide range of intricate behaviour. Even with just a single independent concentration (one variable), reactions may show multiple stationary states and travelling waves. Oscillations are, however, not possible. To understand the latter point we can think in terms of the phase plane or, more correctly for a one-dimensional system, the phase line (Fig. 13.1(a)). As the concentration varies in time, so the system moves along this line. Stationary-state solutions are points on the line: the arrows indicate the direction of motion along the line, as time increases, towards stable states and away from unstable ones. Figure 13.1(b) shows this motion and phase line behaviour represented in terms of some potential, with stable states a minima and an unstable (saddle) solution as a maximum.

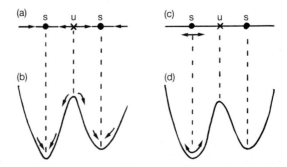

FIG. 13.1. (a) The phase line for a one-variable system showing three stationary points—two stable (s) and one unstable (u); (b) representation of the potential associated with points along the phase line showing the stationary points as extrema; (c), (d) the disallowed motions on the phase line or potential curve which would correspond to oscillatory behaviour, but also to crossing the singular point.

For such phase space representations with one variable, or indeed with any number of independent concentrations, there are a number of rules which the trajectories must obey. In particular, trajectories cannot cross themselves, except at singular points (the stationary states) or if they form closed orbits (such as limit cycles or some other forms we will introduce later). Also, the trajectories cannot pass over singular points. The first of these rules is perhaps most easily shown for two-dimensional systems, where we have a two-dimensional phase plane. Let us assume that the rate equations for the two independent concentrations (or concentration and temperature), x and y, can be written in the form

$$dx/dt = f(x, y, p) \tag{13.1}$$

$$dy/dt = g(x, y, p) \tag{13.2}$$

where f and g are functions and p represents the various parameters which may occur in the equations. (For instance, for the pool chemical model with cubic autocatalysis of chapters 2 and 3 we have $x = \alpha$, $y = \beta$, $f = \mu - \alpha\beta^2 - \kappa_u\alpha$, $g = \alpha\beta^2 + \kappa_u\alpha - \beta$, and $p = (\mu, \kappa_u)$.)

The phase plane, Fig. 13.2, is a plot of x as a function of y, and as x and y vary in time they draw out a trajectory curve across the plane. The gradient of the trajectory at any point is simply the derivative dx/dy given by

$$dx/dy = f(x, y, p)/g(x, y, p). \tag{13.3}$$

Any given point in the phase plane corresponds to a particular value of x and a particular value of y. For any such particular pair (x, y) both the rates f and g also have particular values, i.e. given an x and a y (and of course explicit values for the parameters p) the functions f and g are uniquely determined.

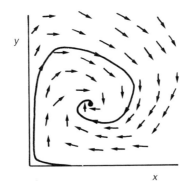

FIG. 13.2. The phase plane for a two-variable system showing a particular trajectory and the surrounding vector field. In this case, trajectories are all directed towards the stable singular point.

Because f and g are uniquely determined, the gradient dx/dy is uniquely determined at every given point. Thus, although the gradient varies in magnitude and sign at each different point on the plane, there is only one possible value for the gradient at any given point. The arrows on the phase plane in Fig. 13.2 show how the gradient 'vectors' may be arranged across the plane.

Trajectories which cross would have a crossing point at which there would be two different gradients, i.e. would require some (x, y) pair at which f/g could have two different values, and this does not occur.

There is only one exception to this property of uniquely determined gradients. If we have an x and a y such that both f and g are simultaneously zero, dz/dy will be singular and not determined. Of course, conditions where the two time derivatives vanish simultaneously correspond to stationary-state solutions. Thus whole sets of trajectories may, and indeed do, tend towards (or grow away from) stationary-state singularities. (We may also note that where we have used the term bifurcation in previous chapters we have been referring to situations in which there is some qualitative change in the way the various gradients or vectors in the phase plane are oriented. Thus there are qualitative changes when new singularities (stationary states) appear, or when a closed trajectory (limit cycle) emerges. We can thus talk more formally about 'bifurcations of the vector field' associated with the rate expressions f, g, etc.)

To disallow the crossing of a stationary-state singularity we need only note that once a trajectory reaches such a point all time derivatives (rates of change) are simultaneously zero, so there is no driving force for further evolution of the concentrations.

Returning to our one-dimensional phase line, any oscillatory motion, such as a rocking motion in one of the potential minima, would correspond to motion backwards and forwards along the line, repeatedly traversing the singular point (Figs 13.1(c) and (d)). In fact, the rules above even disallow any damped settling down on to the stationary state. Such a one-variable system cannot even act like a damped pendulum: it must move towards one of the stationary states, where it settles without overshoot (the approach, as we have seen, is generally exponential and so takes infinite time). In passing we may note that the equation of motion of a pendulum (Newton's second law) involves the second derivative d^2x/dt^2, making a two-variable system (x and dx/dt).

As we have seen above, when a system (mechanical or chemical) has two independent variables, the phase space is a two-dimensional plane. Stationary states are still singular points (Fig. 13.3(a)). The corresponding 'potential' must now be shown as a three-dimensional plot, such as Fig. 13.3(b): stable stationary states correspond to local minima, unstable states to local maxima. Oscillatory behaviour is now permissible, since the trajectory may orbit the now unstable stationary state, settling on to a limit cycle which surrounds the singularity without ever crossing it (Fig. 13.3(c)). We can think of the stationary state becoming unstable by rising from a local minimum to a local maximum, creating a surrounding moat—a closed connected path of equal minima—within which the concentrations circulate (Fig. 13.3(d)). In fact, the flattening out of the local minimum to create this moat is a representation of a Hopf bifurcation.

Two-variable systems are, however, still somewhat constrained. Only simple limit cycles are possible—so, once established, each oscillation is

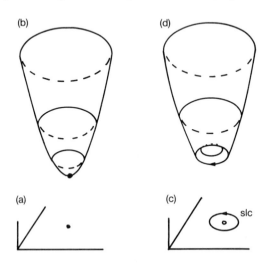

FIG. 13.3.   (a), (b) Phase plane and quasi-three-dimensional representation of a stable singular point for a two-dimensional system. (c), (d) Equivalent forms for a stable limit cycle surrounding an unstable singular point.

exactly the same in amplitude and period as the previous one. More complex forms, which would repeat say every two cycles are not possible, because the trajectory along the cycle would cross itself at one point at least.

If we have more than two dimensions to work in, however, this constraint can be obviated. The apparent crossing can be avoided by lifting the cycle out of the plane, so the trajectory passes over, rather than through itself, as shown in Fig. 13.4(a). Thus oscillations which repeat every two maxima are possible (Fig. 13.4(b)). Indeed, once we have left two dimensions, any number of such overpasses can be allowed, so very complex periodicities can be sustained for systems with three or more independent concentrations. At the extreme, a trajectory—which is a line in the volume—can wander for ever without quite retracing its steps or actually crossing, drawing out some attracting surface on to which the system moves but which has no periodicity. Nor need this 'strange attractor' be a simple two-dimensional surface: it may have a final dimension which is non-integer—a fractal structure.

In general we may expect real chemical systems to have many interacting species. For instance, the stoichiometrically simple H_2–O_2 reaction involves the participation of H, OH, O, HO_2, H_2O, and H_2O_2 as well as the reactants themselves and a contribution from self-heating. The number of independent concentrations will actually be fewer than the number of species. For instance, in the H_2–O_2 system there are two conservation relations such that the total concentrations of H and O atoms in all their various forms must not change during the reaction. Nevertheless, chemical systems will only rarely have just one or two variables. This does not mean that every reaction must show complex oscillations, or even simple 'two-variable' oscillations. In many cases the behaviour of the reaction will be dominated by the dynamics of just a small subset of the participants, with the others adjusting on a very much faster timescale to stay in 'transient equilibrium' with the former. Thus, if only two concentrations call the tune, the system will behave like a two-variable scheme, possibly exhibiting simple oscillations and multistability, but no higher features. We can talk of a two-dimensional 'centre manifold' or surface

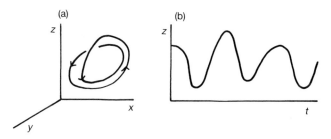

FIG. 13.4.   (a) Phase space representation of a period-2 limit cycle for a three-variable system; (b) the corresponding evolution of one variable in time.

on to which the rapidly varying concentrations move quickly and then across which the remaining motion of the slower variables occurs.

It would be hoping for too much that all reactions should be no more than one- or two-dimensional over all of the experimental conditions attainable. Thus we will look at some ways of modelling the complex responses of higher-order systems in the next few sections. Three different scenarios will be discussed: mappings, forced systems, and autonomous systems. The first of these will be introduced by means of a 'game' which shows the various extensions beyond stationary state and period-1 behaviour. We again make use of our 'cubic autocatalysis' form for the rules of the mapping. This mapping process is of interest pedagogically, but we make no claim that this scheme is in any way easily related to any particular set of chemical rate equations.

A way of transforming a two-variable system to one of higher order is to make one of the parameters in the system a function of time. Thus with a CSTR we might vary the pumping rate (and hence alter the residence time) in a time-dependent and perhaps oscillatory manner. The interaction of the original chemical non-linearity and the imposed forcing shows similar patterns to that displayed by the map. Finally, chemical systems with three or more independent concentrations may drive themselves, of their own free will so to speak, to the heights of complexity.

It is important to stress, and we hope to show in the next few sections, that none of these complex responses, not even chaos, are in any way random events. They are the resulting output from explicit differential or difference equations and are thus completely determinate. Probability does not enter into the discussion, and there is a remarkable structure underlying the sometimes bewildering array of different waveforms and the transitions between them.

13.1. Mappings

The 'game' we wish to use in this section can be played on a non-programmable calculator and proceeds as follows. We choose a particular starting value x_0 for our variable x, lying in the range $0 < x < 1$. Now from x_0 we calculate a new value x_1 and from x_1 we find and x_2, from x_2 and x_3 and so on, according to the rule

$$x_{n+1} = Ax_n(1 - x_n)^2 \tag{13.4}$$

where A is some constant. This is an example of a 'cubic map'. As an example we may have $A = 2$ and choose $x_0 = \frac{1}{2}$: the resulting sequence is then $x_1 = 0.25$, $x_2 = 0.281\,25$, $x_3 = 0.290\,588$, $x_4 = 0.292\,485\,8$, etc. For $n \geqslant 10$ we find $x_n = 0.292\,893\,2$, so the sequence is converging to a 'stationary-state' solution. The same stationary-state value is obtained from any initial condition within the range specified above.

13.1.1. 'Stationary states' of the map

We can easily determine the stationary-state values of x for the present map. We require simply that $x_{n+1} = x_n$, i.e. we must solve

$$x_n = Ax_n(1 - x_n)^2. \tag{13.5}$$

One solution is $x_n = 0$. Non-zero solutions are also possible, with

$$x^n = 1 \pm A^{-1/2}. \tag{13.6}$$

For x_n to lie between 0 and 1, we require $A > 1$ and take the lower root of eqn (13.6). For our example above, $A = 2$, so $x_{n,ss} = 1 - 1/\sqrt{2} = 0.292\,893\,2$, as we discovered by evaluating the sequence.

13.1.2. Pictorial representation of the map

The map described by eqn (13.4) can be represented graphically, as shown in Fig. 13.5. The cubic curve has a maximum of $\frac{4}{27}A$ at $x_n = \frac{1}{3}$ and a zero minimum at $x_n = 1$. The locations of the stationary states are given by the interactions of the mapping curve with the straight line $x_{n+1} = x_n$. There is one at the origin, one at high x_n, and one lying in the range $0 < x_{n,ss} < 1$ given by the lower root of eqn (13.6).

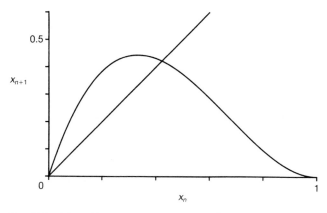

FIG. 13.5.   The cubic map $x_{n+1} = Ax_n(1 - x_n)^2$, specifically for $A = 3$.

If we are to require that x_n (and hence x_{n+1}) should remain in the range between zero and unity, we have the following restriction on the parameter A:

$$1 \leqslant A \leqslant 27/4. \tag{13.7}$$

We have seen that for $A = 2$ the system evolves to the stationary state, i.e. we have a stable solution. How does the response change as we vary A? A stationary-state solution exists across the whole range, but is it always stable?

13.1.3. Stability of stationary state: bifurcations to periodic solutions

As the parameter A is varied, so the point of intersection of the mapping curve and the straight line in Fig. 13.5 changes. The gradient of the cubic curve at the point of intersection also varies during this process. It turns out that the stationary state represented by the intersection is stable provided the value of the gradient at the point is greater than -1. The condition for the equivalent of a Hopf bifurcation in the map is thus

$$f(x_{n,\,\mathrm{ss}}) = x_{n,\,\mathrm{ss}} \qquad \partial f(x_{n,\,\mathrm{ss}})/\partial x_n = -1 \tag{13.8}$$

where $f(x_n) = ax_n(1 - x_n)^2$ is the functional form of the mapping. The first equality in eqn (13.8) is the stationary-state condition, the second the requirement on the gradient. Using the present cubic form, we have

$$\partial f(x_{n,\,\mathrm{ss}})/\partial x_n = A(1 - x_n)(1 - 3x_n) \tag{13.9}$$

and substituting the lower root from eqn (13.6), the condition on the gradient becomes

$$A - 4A^{1/2} + 3 = -1. \tag{13.10}$$

'Hopf' bifurcation thus occurs when $x_{n,\,\mathrm{ss}} = \frac{1}{2}$ at A^*, where

$$A^* = 4. \tag{13.11}$$

As an example, let us consider $A = 4.5$. The stationary-state solution given by eqn (3.6) is $x_{n,\,\mathrm{ss}} = 0.5286$. If we start with any other initial value for x in the range between zero and unity, however, we soon find that the sequence settles to an alternation between $x = \frac{1}{3}$ and $\frac{2}{3}$.

With this simple pattern of alternation, the value of x is repeated every second iteration; i.e. we have that

$$x_{n+2} = f(f(x_n)). \tag{13.12}$$

Using our simple cubic form for $f(x_n)$, a double application of the mapping is equivalent to a higher-order polynomial form:

$$x_{n+2} = A[Ax_n(1 - x_n)^2][1 - Ax_n(1 - x_n)^2]^2. \tag{13.13}$$

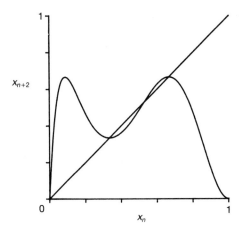

FIG. 13.6.   The second iteration, cubic map $x_{n+2}(x_n)$, appropriate to eqn (13.13).

Figure 13.6 shows this double mapping, $x_{n+2}(x_n)$, with $A = 4.5$. This is typical for any A in the range $4 < A < 5$. There are three intersections, with $x_{n+2} = x_n$, over the range of x between zero and unity. One is at $x = 0.5286$, the (unstable) stationary-state solution, and the other two are at $x = \frac{1}{3}$ and $x = \frac{2}{3}$, the 'limit cycle' solutions between which x_n oscillates.

For the present value of the parameter $A = 4.5$, the gradient of the double mapping described by eqn (13.13) is greater than -1 at the two 'oscillatory' intersections, reflecting the stability of the simple alternating solution. As A increases, however, so the gradient at these points decreases, and may also reach -1. The system then bifurcates again to a solution which repeats every four iterations. This occurs at $A = 5$. With $A = 5.121\,122$, the eventual oscillatory sequence is $x_n = x_{n+4} = 0.758\,685$, $x_{n+1} = 0.226\,254$, $x_{n+2} = 0.693\,678$, $x_{n+3} = 0.333\,33$.

Table 13.1

Location of period-doubling bifurcations for cubic map

Period	A	$(A_n - A_{n-1})/(A_{n+1} - A_n)$
1	4.0000	
2	5.0000	4.2537
4	5.235 09	4.577 39
8	5.286 449	4.649 14
16	5.297 496	4.673 01
32	5.299 86	4.653 54
64	5.300 368	4.668 29
128	5.300 477	4.669 01
256	5.300 500	

The period-4 oscillations become unstable and give way, in turn, to period-8 at $A = 5.23509$. There are further period doublings, as listed in Table 13.1. Examples of the oscillatory time series for some of the different periodicities are given in Fig. 13.7.

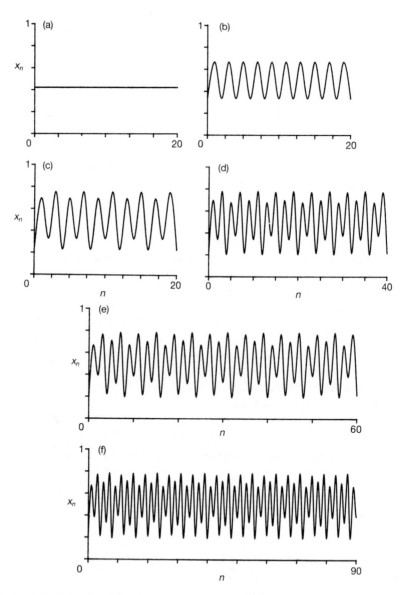

FIG. 13.7.   Periodicities $x_n(n)$ of the cubic map for different values of the parameter A: (a) $A = 3$, period-1; (b) $A = 4.5$, period-2; (c) $A = 5.1$, period-4; (d) $A = 5.25$, period-8; (e) $A = 5.290$, period-16; (f) $A = 5.298$, period-32.

The successive period-doubling bifurcations get closer and closer together as the number of iterations per complete cycle increases. For the highest periodicities, the range of A over which they are stable becomes vanishingly small. The final column of Table 13.1 shows the inverse of the ratio of the ranges for successive periodicities. As well as revealing the steady decrease, there is another point to this list. The value of the entry in this column is tending to the value 4.669 20 . . . . We would find the same value for any particular form we might take for the function $f(x_n)$. It is a universal constant for such mappings, known as the Feigenbaum number.

We could continue the table with periods of 2^n for any n beyond 8. The bifurcations would occur as A increases, but would also tend to a finite limit as n tends to infinity. For the above model this limit is 5.300 506, so all of the periodicities $2^9, 2^{10}, \ldots$, etc. must be squeezed in between 5.300 500 and this value.

What happens for values of A above the limit of the period-doubling sequence? We have an upper limit on A of 27/4, so we still have the range $5.300 506 < A < 6.75$ to investigate. One particularly important form of behaviour, found just above the convergence limit, is that of aperiodicity or 'chaos'. Now the sequence x_n gives a different value at each step, never repeating itself no matter how many iterations we make.

Figure 13.8 gives an indication of how the behaviour varies with A across the whole range $1 < A < 27/4$. It shows the values of $x_n(A)$ for 1000 iterations (after initial transient have died away). For any given value in the range of stationary-state stability, $1 < A < 4$, we see that each x_n is the same for all n, so the curve is single valued. At $A = 4$, the response bifurcates, splitting into two branches: across the range $4 < A < 5$, the iterations alternate between the two branches for odd and even n. Further period doublings then ensue.

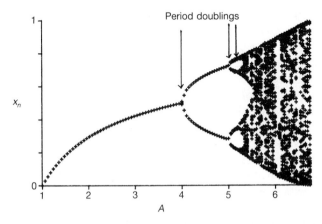

FIG. 13.8.   Variation of the iterations x_n ($n = 1$–100) with one parameter A. The first period-doubling sequence, beginning at $A = 4$, is clearly visible and 'windows' of intermittence can be seen at higher A.

For $A > 5.300\,506$, x_n is different at each stage of the iteration and we could plot an infinite number of points for each A without repetition. Even so we would not fill the whole x_n range between 0 and 1—there are systematic gaps in between the 'line' formed by the infinite number of points. At larger A, there are occasional ranges where the behaviour becomes quite simple again. For instance at $A = 6.055\,678$, a period-3 pattern emerges. This is followed by a period-doubling sequence, with period-6, period-12, etc. This sequence also converges, to $A \approx 6.088$, reproducing the Feigenbaum number as it does so. Other periodic 'windows' can be detected—one based on period-5 exists for $A \approx 6.28$—but the range of A for such periodicities can become extremely small.

13.1.4. Sensitivity to initial conditions in chaotic regime

The title 'chaos' is an unfortunate misnomer. As shown above, there is considerable structure displayed in the onset and interruption of 'chaotic' behaviour. Even within the chaotic regime the value of x_n evolves according to completely defined rules—one value explicitly determines the next, with 100 per cent certainty. There is no randomness, no element of chance uncertainty or irregularity. If we know the rules and can measure a given starting condition exactly, even a chaotic pattern can be predicted exactly.

In real systems, however, we know that we can never achieve an exact measurement of all the concentrations, simultaneously at any given time. There is always some finite uncertainty in our initial conditions (and possibly in our parameter values). In many situations this is not so important. If two systems or numerical integrations begin very close together, they often remain close together for all ensuing time. If the system is approaching a stable stationary state or a simple limit cycle, the convergence on to the attractor will usually cause a decrease in the difference between the two traces. When such a situation holds we can make reproducible measurements of the sort on which science relies.

For chaotic behaviour, with an underlying 'strange attractor', this is not the case. Now trajectories tend to diverge. Two different initial conditions, no matter how close together, do not have similar histories. They begin to move apart, the initial separation growing exponentially, and eventually (and often quite quickly) become completely uncorrelated. Under these circumstances we lose all predictive power. Having performed one integration, we can say how a second starting from exactly the same point will behave, but can say nothing about how any other starting point will evolve. (An initial point, starting *very* close to the first, may follow a trajectory that is not too different for some time, but we have no way of estimating that time either.)

13.1.5. Uncertainty in parameters

In the previous subsection, some of the consequences and limitations on predictability arising from inexact knowledge of initial conditions have been mentioned. We should also recognize that in practice we will also have imprecise knowledge of the values of our parameters. Thus, on a good day, we may be able to measure our ambient temperature to $\pm 0.5\,\text{K}$ and the reactant gas pressure to ± 0.1 per cent. Small uncertainties such as these will lead to smaller or larger uncertainties in our parameters. Furthermore, if we cannot measure these factors more precisely than this, we probably cannot control them to maintain absolutely constant values to any greater extent. Now, if a particular type of response is found over a wide range of parameter values, this may not be too important a limitation: there may be no qualitative change and the quantitative knock-on effect may be smaller than the precision of our measurements.

With the various period-doubling sequences seen above, however, the parameter ranges over which some of the higher-order periodicities exist can become very small. If we cannot control our conditions within these ranges, we may not be able to sustain these patterns.

13.1.6. Interpretations of simple maps

As we have already commented, mappings of the type discussed above are not in any way easily related to a given set of reaction rate equations. Such mappings have, however, been used for chemical systems in a slightly different way. A quadratic map has been used to help interpret the oscillatory behaviour observed in the Belousov–Zhabotinskii reaction in a CSTR. There, the variable x_n is not a concentration but the amplitude of a given oscillation. Thus the map correlates the amplitude of one peak in terms of the amplitude of the previous excursion.

With this identification, the stable 'stationary-state' behaviour (found for the cubic model with $1 < A < 4$) corresponds to oscillations for which each amplitude is exactly the same as the previous one, i.e. to period-1 oscillatory behaviour. The first bifurcation ($A = 4$ above) would then give an oscillation with one large and one smaller peak, i.e. a period-2 waveform. The period doubling then continues in the same general way as described above. The B–Z reaction (chapter 14) shows a very convincing sequence, reproducing the Feigenbaum number within experimental error.

Mapping techniques, and the associated bifurcation analyses, are also of great importance when applied with the Poincaré map described in the appendix to chapter 5. These are used to establish local stability, and changes in that stability, for periodic solutions—oscillations are reduced to fixed points in a way similar to that just described in terms of the amplitude. A simple, stable, period-1 limit cycle appears as either a stable node or a stable focus solution of the map; a bifurcation of that solution gives a period-2 etc.

13.2. Periodic forcing of oscillatory systems

The idea of imposing a periodic perturbation to chemical systems already exhibiting oscillations probably grew from academic interest. Some justification is readily found: in experiments, peristaltic pumps and oven temperature controllers can cause such features; in complex situations such as living systems, different oscillatory processes are often coupled, perhaps to a pulsed bloodstream supply of reagents. More recently, heterogeneously catalysed reactions have been investigated with periodic inflow of reactants, often revealing quite substantial increases in the overall efficiency of the conversion process. Oscillatory forcing has been used in other disciplines to probe the underlying behaviour of linear systems, and there may be extensions which will provide very sensitive tests of mechanisms proposed for non-linear chemical systems.

Three model kinetic schemes have been studied relatively intensively with periodic forcing: the first-order non-isothermal CSTR of chapter 7; the Brusselator model, which is closely related to the cubic autocatalysis of chapters 2 and 3; and the surface reaction model discussed in § 12.6. We will use the last of these to introduce some of the general features.

13.2.1. Forced chemical systems: rate equations

The reaction model taken from §12.6 is

$$P + S \rightleftharpoons P - S \qquad\qquad k_1, k_{-1} \qquad (13.14)$$

$$R + S \rightleftharpoons R - S \qquad\qquad k_2, k_{-2} \qquad (13.15)$$

$$P - S + R - S + 2S \rightarrow 4S + \text{products} \qquad k_3. \qquad (13.16)$$

The appropriate dimensionless rate equations for the surface coverages θ_p and θ_r are

$$\frac{d\theta_p}{d\tau} = p(1 - \theta_p - \theta_r) - \kappa_1 \theta_p - \theta_p \theta_r (1 - \theta_p - \theta_r)^2 \qquad (13.17)$$

$$\frac{d\theta_r}{d\tau} = r(1 - \theta_p - \theta_r) - \kappa_2 \theta_r - \theta_p \theta_r (1 - \theta_p - \theta_r)^2 \qquad (13.18)$$

where

$$p = \frac{k_1 P}{k_3} \qquad r = \frac{k_2 R}{k_3} \qquad \kappa_1 = \frac{k_{-1}}{k_3} \qquad \kappa_2 = \frac{k_{-2}}{k_3}. \qquad (13.19)$$

Of particular interest are the parameters p and r, which are the dimensionless partial pressures of the reactant gases flowing over the catalyst surface. If we wish to force this system, a conceptually simple way is to vary one of these concentrations periodically about some mean value. We will follow McKarnin et al. (1988), who imposed a cosine variation on the partial pressure of R:

$$r = r_0 + r_f \cos(\omega \tau). \qquad (13.20)$$

Here, r_0 is the mean value of the dimensionless partial pressure, r_f is the amplitude of the forcing, so r varies between $r_0 \pm r_f$, and ω is the dimensionless forcing frequency (the forcing peiod τ_f is given by $2\pi/\omega$).

The rate equations can now be written as

$$\frac{d\theta_p}{d\tau} = p(1 - \theta_p - \theta_r) - \kappa_1 \theta_p - \theta_p \theta_r (1 - \theta_p - \theta_r)^2 \qquad (13.21)$$

$$\frac{d\theta_r}{d\tau} = [r_0 + r_f \cos(\omega \tau)](1 - \theta_p - \theta_r) - \kappa_2 \theta_r - \theta_p \theta_r (1 - \theta_p - \theta_r)^2. \qquad (13.22)$$

We now have a total of six parameters: four from the autonomous system (p, r_0, and the desorption rate constants κ_1 and κ_2) and two from the forcing (r_f and ω). The main point of interest here is the influence of the imposed forcing on the natural oscillations. Thus, we will take just one set of the autonomous parameters and then vary r_f and ω. Specifically, we take $p = 0.019$, $r_0 = 0.028$, $\kappa_1 = 0.001$, and $\kappa_2 = 0.002$. For these values the unforced model has a unique unstable stationary state surrounded by a stable limit cycle. The natural oscillation of the system has a period $\tau_0 = 911.98$, corresponding to a natural frequency of $\omega_0 = 0.006\,889\,6$.

13.2.2. Influence of forcing

We shall first investigate the effect of imposing a periodic perturbation at twice the frequency of the natural oscillations, i.e. taking $\omega = 2\omega_0 = 0.013\,779\,2$, for various forcing amplitudes. The change in the response on increasing r_f is displayed through the sequence in Figs 13.9(a–h). Figure 13.9(a) corresponds to the autonomous unforced system with $r_f = 0$: 10 complete cycles are shown, with a period of 911.98 as described above. The θ_p–θ_r phase plane has a simple limit cycle.

With a small forcing amplitude, $r_f = 0.002$ in Fig. 13.9(b) and $r_f = 0.003$ in 13.9(c), the oscillatory traces show only a slight deformation, with a shoulder

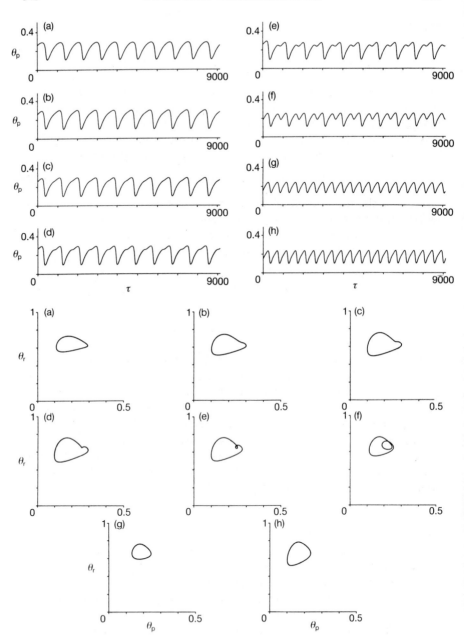

FIG. 13.9. The forced Takoudis–Schmidt–Aris model with a forcing frequency twice that of the natural oscillation: (a) zero-amplitude forcing (autonomous oscillation and limit cycle); (b) $r_f = 0.002$; (c) $r_f = 0.003$; (d) $r_f = 0.004$; (e) $r_f = 0.005$; (f) $r_f = 0.006$; (g) $r_f = 0.007$; (h) $r_f = 0.01$. Traces show the time series $\theta_p(\tau)$ over 10 natural periods (or 20 forcing periods) and the associated limit cycle in the θ_p–θ_r plane.

appearing before the maximum in $\theta_p(\tau)$. The motion in the phase plane shows a buckling of the limit cycle.

As r_f is increased through the range 0.004 to 0.006, the shoulder develops into a second maximum and minimum, so the oscillatory trace now shows one large and one small excursion per full cycle. The trajectory in the phase plane now has a second loop, which grows as the forcing amplitude increases. There are still 10 full cycles in each of Figs 13.9(d–f), with a period still of 911.98.

For a slightly higher forcing amplitude, $r_f = 0.007$, the large–small alternation has disappeared. Each pulse has the same size, so there are now 20 identical cycles in Fig. 13.9(g). The oscillatory period is now 455.99, exactly that of the forcing and half that of the natural oscillations. The phase plane shows a simple limit cycle again corresponding to this fully entrained behaviour. Full entrainment with the forcing frequency is found for all higher r_f, with the oscillatory amplitude simply increasing with the forcing amplitude.

13.2.3. Motion on a torus: quasi-periodicity and phase locking

Figure 13.10 shows a representation of the phase plane behaviour appropriate to small-amplitude forcing. There are two basic cycles which make up the full motion: first, there is the natural limit cycle, corresponding for example to Fig. 13.9(a) around which the unforced system moves; secondly, there is a small cycle, perpendicular to the limit cycle, corresponding to the periodic forcing term. The overall motion, obtained as the small cycle is swept around the large one, gives a torus and the 'buckled' limit cycle oscillations at low r_f in Fig. 13.9 draw out a path over the surface of such a torus.

In the previous subsection, the forcing frequency was exactly twice the natural oscillatory frequency. Thus the motion around one oscillation gives exactly two circuits of the forcing cycle for one revolution of the natural limit cycle. The full oscillation of the forced system has the same period as the autonomous cycle and twice the forcing period. The concentrations θ_p and θ_r return to exactly the same point at the top of the cycle, and subsequent oscillatory cycles follow the same close path across the toroidal surface. This is known as 'phase locking' or resonance. We can expect such locking, with a closed loop on the torus, whenever the ratio of the natural and forcing

FIG. 13.10.   Cooperation of two different oscillatory modes leading to a phase space trajectory on the surface of a torus.

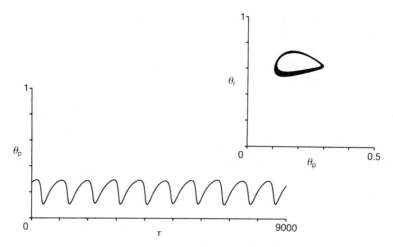

FIG. 13.11. A quasi-periodic trace with $\omega/\omega_0 \approx 1.7$: the attractor in the phase space (inset) has broadened into a torus.

frequencies is given by some rational number. Thus with $\omega/\omega_0 = m/n$, we will find m forcing cycles for n revolutions of the natural limit cycle.

If the quotient ω/ω_0 is irrational, the path across the toroidal surface will return to a different point on the completion of each cycle. Eventually the trajectory will pass over every point on the surface of the torus without ever forming a closed loop. This is 'quasi-periodicity', and an example is shown in Fig. 13.11. The corresponding concentration histories do not necessarily give complex waveforms, as can be seen from the figure. However, the period of the oscillations is neither simply that of the natural cycle nor just that of the forcing term, but involves both.

13.2.4. The stroboscopic map

A technique for distinguishing between phase-locked and quasi-periodic responses, and which is particularly useful when m and n are large numbers, is that of the stroboscopic map. This is essentially a special case of the Poincaré map discussed in the appendix of chapter 5. Instead of taking the whole time series $\theta_p(\tau)$, for all τ, we ask only for the value of this concentration at the end of each forcing period. Thus at times $\tau = 2k\pi/\omega$, with $k = 1$, 2, etc., we measure the surface concentrations of one of our species. If the system is phase locked on to a closed path with $\omega/\omega_0 = m/n$, then the stroboscopic map will show the measured values moving in a sequence between m points, as in Fig. 13.12(a). If the system is quasi-periodic, the iterates of θ_p will never repeat and, eventually, will draw out a closed cycle (Fig. 13.12(b)) in the

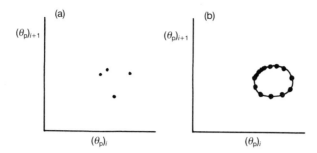

FIG. 13.12.   Stroboscopic map plotting the value of a given variable at the end of each forcing period against its value at the end of the previous period: (d) phase-locked response, giving a finite number of discrete points; (b) quasi-periodic response—the points will eventually fill the complete closed curve in the plane.

stroboscopic plane. Thus the mapping reduces the dimension of the attractor: the closed path (phase locking) on the toroidal surface is reduced to m points, and the full coverage (quasi-periodicity) of the surface to a closed curve. There is great current activity in establishing the bifurcations which can occur on the stroboscopic map and then relating them to the behaviour of the forced system, and some recent references are given at the end of this chapter.

13.2.5. The excitation diagram

In Figure 13.9, we considered a sequence over which the amplitude of the forcing perturbation was increased for one particular forcing frequency. Figure 13.13 shows a different cross-section of the forcing parameter plane, taking $r_f = 0.002$ and varying the forcing frequency. With $\omega/\omega_0 = 1.7$ (Fig. 13.13(a)), the motion appears on visual inspection to repeat every three excursions. However, the period of this cycle is neither that of the natural oscillation nor that of the forcing frequency: it is a quasi-periodic trace. Similarly, with $\omega/\omega_0 = 1.8$ (Fig. 13.13(b)), the motion is quasi-periodic— more believably so. For $\omega/\omega_0 = 1.9$, however, the oscillation becomes phase locked on to the 2/1 resonance, as shown in Fig. 13.13(c); the concentration repeats itself exactly after two forcing periods. The system remains phase locked with $\omega/\omega_0 = 2$, as seen previously.

This example shows that for forcing amplitudes which are not particularly small, phase locking to a given resonance can occur even when the ratio of forcing and natural frequencies ω/ω_0 departs from the exact value required. Figure 13.14 shows how the forcing amplitude–forcing frequency parameter plane divides into regions corresponding to the different possible resonances. The regions indicated in this 'excitation diagram' are known as resonance horns or Arnol'd tongues. These grow out from the forcing frequency axis as

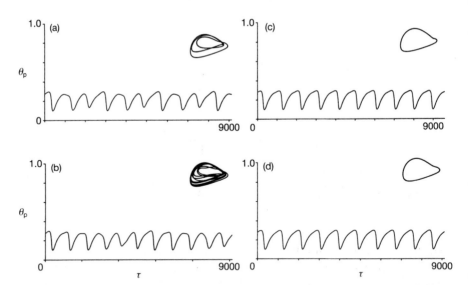

FIG. 13.13. Phase locking appearing from quasi-periodicity as the forcing frequency is brought close to a simple integer, $r_f = 0.002$: (a) $\omega/\omega_0 = 1.7$; (b) $\omega/\omega_0 = 1.8$; (c) $\omega/\omega_0 = 1.9$; (d) $\omega/\omega_0 = 2$.

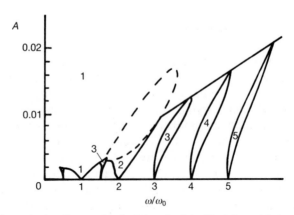

FIG. 13.14. The excitation diagram for the Takoudis–Schmidt–Aris model showing resonance horns (Arnol'd tongues) emerging from integer quotients of forcing and natural frequencies. For details of the behaviour in the closed broken curve see Fig. 13.16. (Reproduced with permission from McKarnin, M. A. *et al.* (1988). *Proc. R. Soc.*, **A417**, 363–88.)

V-shaped regions, but in general close up again at sufficiently large amplitude. The number marked in each of the tongues indicates how many forcing periods are required for one full oscillation or, equivalently, how many points appear in the stroboscopic map.

In theory, there is a resonance horn emerging from every rational number along the abscissa. We have not attempted to show more than a few, just those which are most important in the sense that the corresponding oscillations are relatively easily obtained and exist of a reasonable range of frequencies.

In between the resonance horns are regions of the parameter plane for which the response is quasi-periodic. Note that it is even possible for the frequencies to have a simple ratio and yet for the system to lie outside the corresponding resonance horn if the amplitude is raised. Figure 13.15 shows two time series for forcing with $\omega/\omega_0 = 10/1$. At low forcing amplitude, $r_f = 0.005$, we have phase locking and a simple if rather crumpled limit cycle. With $r_f = 0.01$, however, the response is quasi-periodic: a few cycles are shown and demonstrate quite well how the trajectory begins to wind around the torus.

For sufficiently large forcing amplitudes the oscillation becomes completely entrained, with a period exactly equal to one forcing period, whatever that value of ω/ω_0. The entrainment may arise from a phase-locked response—as seen previously in Fig. 13.9—or from a quasi-periodic pattern. The boundary for full entrainment appears as an almost straight line with positive slope of $\omega/\omega_0 > 1$ and negative slope for $\omega/\omega_0 < 1$.

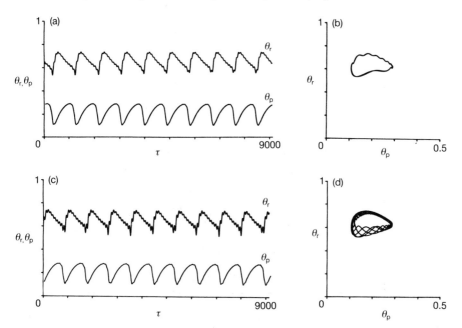

FIG. 13.15. Onset of quasi-periodicity from phase locking as forcing amplitude is increased: (a), (b) $r_f = 0.005$, showing a phase-locked response; (c), (d) $r_f = 0.01$, showing quasi-periodicity with a torus growing in the phase plane.

13.2.6. Period doubling

Also drawn on the excitation diagram, at the top of the 2/1 resonance horn, is a closed region, marked period doubling (Fig. 13.14). This lies across, and interrupts, the line corresponding to the onset of full entrainment. For the present model and with the particular values chosen for the autonomous parameters p, r, κ_1, and κ_2, the response here is simply an oscillation with twice the forcing period. Thus, if we enter this region from a state of full entrainment, we get a period doubling. An example is seen in going from (g) to (f) in Fig. 13.9.

As with most of the features described so far, the existence of a period-doubling region is common amongst forced chemical models. Frequently there are further period doublings within the region, giving a set of concentric 'ellipses', as indicated schematically in Fig. 13.16 (which is loosely based on the forced Brusselator model studied by Kai and Tomita). At the centre of this nesting may lie a non-zero region of chaotic behaviour. Thus we can obtain a period-doubling cascade, culminating in aperiodicity and which obeys the Feigenbaum convergence mentioned in §13.1.3. We can even continue or traverse through the parameter plane and emerge from chaos, through a 'period-halving' sequence, to the reassurance of simple entrainment.

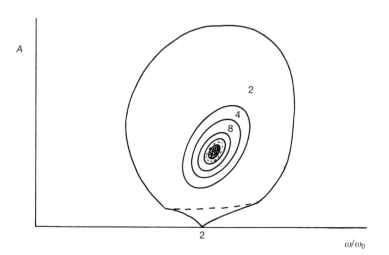

FIG. 13.16. A typical set of concentric period-doubling loci at higher forcing amplitudes, leading ultimately to a region of chaotic behaviour.

13.3. Complex oscillations and chaos in autonomous systems

The previous section considered a non-autonomous reaction system with the time appearing explicitly in the right-hand side of the rate equations through the forcing term $\cos(\omega\tau)$. The model had only two concentration variables, complex oscillations being possible because the phase of the forcing term acts as a third variable. We now turn to fully autonomous systems, i.e. those in which any driving force for complex periodicities must be generated by the concentrations and/or temperature variations themselves. Clearly, the simplest extension for two variables is just to introduce a third, so we will analyse a number of three-variable schemes. First, some general comments about local stability, Hopf bifurcation, and higher bifurcations in multivariable systems can be made.

13.3.1. Local stability analysis and Hopf bifurcation for n variables

The local stability analysis for a two-dimensional system has been given in some detail in chapter 3 and used frequently in later sections. The same general principles apply to any number of variables. Provided the perturbations imposed upon the stationary state are infinitesimal, the growth or decay for an n variable system will be governed by the sum of n exponential terms of the form $e^{\lambda_i\tau}$ with $i = 1, n$.

The eigenvalues λ_i are obtained from what is now an $n \times n$ Jacobian matrix. If all the λ_i have negative real parts, the exponential terms will all decay in time, so the stationary state is stable. On the other hand, if we change the values of the parameters and one real eigenvalue then passes through zero, becoming positive, the stationary state becomes an unstable saddle point (in general we then have a saddle–node bifurcation). Alternatively, instability may arise from a complex pair of eigenvalues having real parts passing through zero: this constitutes a Hopf bifurcation for the n-variable system. In each case, the bifurcation requires that all other eigenvalues should still have negative real parts. The eigenvalue or eigenvalue pair which change sign first is known as the 'principal eigenvalue' (or 'principal pair of eigenvalues'). Subsequent changes of sign amongst the other λ_i will generally have no physical consequence since the system will already have moved away from the unstable stationary state.

In the remaining sections of this chapter we will deal with three-variable systems: saddle–node bifurcation will correspond to a principal real eigenvalue becoming positive, with the remaining eigenvalues being either both real and negative or a complex pair with negative real parts. Hopf bifurcation occurs when a principal pair has a real part equal to zero (and thus becoming purely imaginary at that point) with a real negative eigenvalue. The next subsection gives the appropriate formulae for the Hopf condition in three-variable systems, and then we discuss the idea of local stability and bifurcation from the corresponding limit cycles.

13.3.2. Hopf bifurcation for three-variable systems

The reaction rate equations which describe the evolution of three concentration variables x, y, and z can be written in general terms as

$$dx/dt = f(x, y, z, \mathbf{p}) \tag{13.23}$$

$$dy/dt = g(x, y, z, \mathbf{p}) \tag{13.24}$$

$$dz/dt = h(x, y, z, \mathbf{p}) \tag{13.25}$$

where $\mathbf{p}$ represents the parameters involved in the system. The 3×3 Jacobian matrix $\mathbf{J}$ has the form

$$\mathbf{J} = \begin{pmatrix} j_{11} & j_{12} & j_{13} \\ j_{21} & j_{22} & j_{23} \\ j_{31} & j_{32} & j_{33} \end{pmatrix} \tag{13.26}$$

where $j_{11} = \partial f / \partial x$ etc. The components j_{ik} are evaluated for a given stationary state with particular values of the parameters. The three eigenvalues are then given by the roots of the cubic equation

$$\lambda^3 + b\lambda^2 + c\lambda + d = 0 \tag{13.27}$$

where the coefficients b, c, and d are

$$b = -\operatorname{tr}(\mathbf{J}) = -(j_{11} + j_{22} + j_{33}) \tag{13.28}$$

$$c = \begin{vmatrix} j_{11} & j_{12} \\ j_{21} & j_{22} \end{vmatrix} + \begin{vmatrix} j_{22} & j_{23} \\ j_{32} & j_{33} \end{vmatrix} + \begin{vmatrix} j_{11} & j_{13} \\ j_{31} & j_{33} \end{vmatrix}$$

$$= j_{11}j_{22} + j_{11}j_{33} + j_{22}j_{33} - j_{12}j_{21} - j_{13}j_{31} - j_{23}j_{32} \tag{13.29}$$

$$d = -\det(\mathbf{J}) = -(j_{11}j_{22}j_{33} + j_{21}j_{13}j_{32} + j_{31}j_{12}j_{23}$$

$$- j_{11}j_{23}j_{32} - j_{22}j_{31}j_{13} - j_{33}j_{12}j_{21}). \tag{13.30}$$

(With luck, and at least with simple models, some of the coefficients j_{ik} will be zero, so that the forms for c and d may simplify quite dramatically.)

As described above, the first condition on the eigenvalues for a Hopf bifurcation in a three-variable scheme is that the principal pair should be purely imaginary and the third should be real and negative. For this to be the case, the cubic equation (13.27) must be factorizable in the form

$$(\lambda^2 + \omega_0^2)(\lambda + a) = 0 \qquad a > 0. \tag{13.31}$$

Then the eigenvalues are

$$\lambda_{1,2} = \pm i\omega_0 \qquad \lambda_3 = -a. \tag{13.32}$$

Comparing eqns (13.27) and (13.31) we find the conditions

$$bc - d = 0 \qquad b > 0 \tag{13.33}$$

and the relationships

$$a\omega_0^2 = d \qquad \omega_0 = c^{1/2}. \tag{13.34}$$

The last two equalities, along with $a > 0$, specify that the determinant of the Jacobian (d) must be negative, whilst the sum of the minors c must be positive. Also, with $b > 0$, Hopf bifurcation occurs for $\mathrm{tr}(\mathbf{J}) < 0$, not when the trace changes sign.

Equation (13.33), with the addition requirement on the sign of b, is the first part of the Hopf condition. There are other non-degeneracy conditions, similar to those which hold for two-variable systems. The real part of the principal pair must actually pass through zero (transcritically) and the various coefficients of the higher-order terms such as β_2 in the Hopf expansions must also be non-zero (see §§3.4.1, 5.1.1, 5.1.2, and 8.4). If all of these are satisfied, however, then the behaviour of a three-variable system at a Hopf point so defined is much the same as that seen before. With a supercritical bifurcation, a stable limit cycle emerges around the unstable stationary state; at a subcritical bifurcation an unstable limit cycle emerges and grows when the stationary state is stable.

Let us imagine a scenario for which a supercritical Hopf bifurcation occurs as one of the parameters, μ say, is increased. For $\mu < \mu^*$, the stationary state is locally stable. At μ^* there is a Hopf bifurcation: the stationary state loses stability and a stable limit cycle emerges. The limit cycle grows as μ increases above μ^*. It is quite possible for there to be further bifurcations in the system if we continue to vary μ. With three variables we might expect to have period-doubling sequences or transitions to quasi-periodicity such as those seen with the forced oscillator of the previous section. Such bifurcations, however, will not be signified by any change in the local stability of the stationary state. These are bifurcations from the oscillatory solution, and so we must test the local stability of the limit cycle. We now consider how to do this.

13.3.3. Local stability of limit cycle

The philosophy of the test for local stability of a limit cycle is much the same as that applied to stationary-state solutions, namely will a small perturbation decay or grow? Even for the simplest chemical models such as those considered here, we do not generally obtain analytical expressions for the limit cycle or the corresponding oscillations over all of their course. Rather, these have to be determined by numerical computation. By similar measure, the local stability can then only be determined numerically, but the principles which apply are analogous to those of stationary-state local stability (chapter 3) and, in part, the Hopf theory (chapter 5).

As the motion around the limit cycle is periodic, we can only talk of a perturbation decaying or growing if we compare successive measurements made at the same point on the cycle. Thus we impose an initial perturbation Δx_0 and observe its temporal evolution at the end of each successive circuit of the limit cycle. For a system with n independent variables, the perturbation Δx is an n-component vector. If the oscillatory period is given by T_p, then the perturbation at the end of the first cycle can be represented in the form

$$\Delta x(T_p) = J(T_p)\Delta x_0 \tag{13.35}$$

where J is a (time-dependent) $n \times n$ matrix, evaluated at time $t = T_p$. More strictly, the 'Jacobian' has the form

$$J(T_p) = \exp(BT_p) \tag{13.36}$$

where B is also an $n \times n$ matrix.

After m periods, the perturbation will be given by

$$\Delta x(mT_p) = [J(T_p)]^m \Delta x_0. \tag{13.37}$$

If matrix $J(T_p)$ is in some sense 'smaller than unity', then the magnitude of the perturbation will in the same sense decrease on each successive cycle. This implies local stability as the system returns to the limit cycle. On the other hand, if $J(T_p)$ is 'larger than unity', there will be a growth in the perturbation over each cycle, corresponding to instability of the limit cycle.

The 'size' of the matrix as it operates on the perturbation vector is directly related to the eigenvalues of J (or of B). The eigenvalues of J are known as the Floquet multipliers μ_i; the eigenvalues of B are the Floquet exponents β_i. In general the former are easier to evaluate, although we should identify the parameter β_2 introduced in chapter 5 with the Hopf bifurcation formula as a Floquet exponent for the emerging limit cycle (then $\beta_2 < 0$ implies stability, $\beta_2 > 0$ gives instability, and $\beta_2 = 0$ corresponds to a 'bifurcation' between these two cases).

An n-variable system has n associated Floquet multipliers: one of these is always equal to $+1$. This latter point arises because if our initial perturbation is in a direction exactly along the limit cycle, then it will neither decay nor grow—there will just be a shift in phase. The remaining multipliers may be real or occur as conjugate complex pairs. The values of the multipliers can be represented by points on the complex plane, as shown in Fig. 13.17.

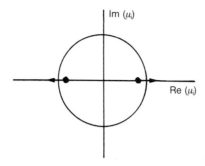

FIG. 13.17.   Floquet multipliers lying within the unit circle, indicating a stable periodic motion: if the CFM leaves the unit circle through -1 a period doubling occurs; if it goes out through $+1$ there is a saddle–node bifurcation with the disappearance of the periodic solution.

Also drawn on the diagram is the unit circle: a Floquet multiplier lying inside the circle has magnitude less than unity; one lying outside has magnitude greater than unity. If the parameters of the system are such that all the Floquet multipliers μ_i lie within the unit circle, $\mathbf{J}(T_p)$ is 'less than unity' in the sense described above, and so the limit cycle is locally stable. Instability occurs when the multiplier with the largest modulus (the 'critical Floquet multiplier' or CFM) crosses the unit circle. The way in which the CFM leaves the unit circle is also significant.

If the CFM, μ_2 say, is real and crosses the unit circle along the real axis at $\mathrm{Re}(\mu_2) = -1$, then the limit cycle becomes unstable by means of a period-doubling bifurcation. A new periodic solution (period $= 2T_p$) appears. On the other hand, if the real CFM leaves the unit circle at $\mathrm{Re}(\mu_2) = +1$, there is a saddle–node bifurcation (i.e. a turning point) in the locus describing the growth in amplitude of the limit cycle. We have seen examples of this bifurcation with the non-isothermal pool chemical system of chapter 4. When the full Arrhenius form for the temperature dependence of the reaction rate constant is retained, the upper Hopf bifurcation may be subcritical. The emerging unstable limit cycle grows as the reactant concentration is increased, but at some point merges with the stable limit cycle born at the lower (supercritical) Hopf bifurcation. This merging of the two different limit cycles gives a saddle–node point in Fig. 4.9 in the same way as does the merging of two stationary-state branches.

If the CFM is formed by a complex pair, the unit circle can be crossed at some point (actually at two points simultaneously) off the real axis. Generally this will correspond to the bifurcation from a stable limit cycle to a quasi-periodic motion on a torus. In special cases, however, where the crossing point corresponds to the kth complex root of -1, the limit cycle bifurcates to a phase-locked cycle (closed loop on the torus) corresponding to period-k resonance such as that described in §13.3.

This concludes the mathematical preliminaries needed for our discussion of systems with three or more independent concentrations. The different dynamical responses which arise from the various bifurcations listed above will now be exemplified through a number of model schemes, each with three variables.

13.4. Autonomous three-variable schemes

The specific models we will analyse in this section are an isothermal autocatalytic scheme due to Hudson and Rossler (1984), a non-isothermal CSTR in which two exothermic reactions are taking place, and, briefly, an extension of the model of chapter 2, in which autocatalysis and temperature effects contribute together. In the first of these, chaotic behaviour has been 'designed' in much the same way that oscillations were obtained from multiplicity with the heterogeneous catalysis model of §12.5.2. In the second, the analysis is firmly based on the critical Floquet multiplier as described above, and complex periodic and aperiodic responses are observed about a unique (and unstable) stationary state. The third scheme has coexisting multiple stationary states and higher-order periodicities.

13.4.1. Chemical chaos in an isothermal oscillator

In chapter 12 we discussed a model for a surface-catalysed reaction which displayed multiple stationary states. By adding an extra variable, in the form of a catalyst poison which simply takes place in a reversible but competitive adsorption process, oscillatory behaviour is induced. Hudson and Rossler have used similar principles to suggest a route to 'designer chaos' which might be applicable to families of chemical systems. They took a two-variable scheme which displays a Hopf bifurcation and, thus, a periodic (limit cycle) response. To this is added a third variable whose role is to switch the system between oscillatory and non-oscillatory phases.

The two-variable 'subscheme' is

$$P \rightarrow A \qquad \text{rate} = k_1 p \qquad\qquad (13.38)$$

$$Q \rightarrow B \qquad \text{rate} = k_2 q \qquad\qquad (13.39)$$

$$A + B \rightarrow 2B \qquad \text{rate} = k_3 ab \qquad\qquad (13.40)$$

$$B \rightarrow R \qquad \text{rate} = k_4 b/(b + K). \qquad\qquad (13.41)$$

Thus P and Q are converted through a sequence of irreversible steps to a final product R. There are two intermediate species, A and B, formed by the slow decay from the precursor reactants, and which are involved in an autocatalytic step. The concentrations of A and B are the variables in this system; those of P and Q decay only 'slowly' and exponentially and are considered to be constant in the spirit of the pool chemical approximation used in chapters 2–5. The decay of B to the product R is not a simple step. Instead it is imagined to follow a saturating rate law, as often found with Langmuir–Hinshelwood or Michaelis–Menten kinetics. The rate of this step does not increase linearly with the concentration of B, but approaches a limiting value of k_4 at high concentration. This second non-linearity is vital if the two-variable subscheme is to show oscillatory behaviour because the autocatalysis in step (13.40) is only quadratic.

The additional reaction to be included with the above subscheme is the reversible interconversion of A to a third intermediate C:

$$A \rightleftharpoons C \qquad \begin{cases} \text{forward rate} = k_5 a \\ \text{reverse rate} = k_{-5} c. \end{cases} \qquad (13.42)$$

Even without this additional step, the scheme is capable of supporting Hopf bifurcation as the various rate constants and concentrations of P and Q are varied. In order to introduce the chaotic behaviour of the full model, however, it is more convenient to imagine a situation where the reversible step (13.42) is included but the concentration of species C is somehow maintained constant rather than allowed to vary in response to A.

For given values of the rate constants $k_1 - k_{-5}$ and K, the behaviour of this reduced system is determined by the constant value of c. If the concentration of C is maintained at a sufficiently low value, the unique stationary-state solution (a_{ss}, b_{ss}) is locally stable. As c is increased, however, there may be a Hopf bifurcation, at $c = c^*$ say. For $c > c^*$, then, the stationary state will be unstable and surrounded by a stable limit cycle whose amplitude grows in some non-linear fashion with c. This bifurcation behaviour is illustrated in Fig. 13.18.

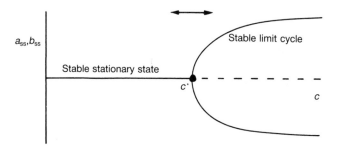

FIG. 13.18.   Variation in the stationary-state stability of species A and B for the Hudson–Rössler model, showing a Hopf bifurcation at c^*.

As a numerical example we will follow Hudson and Rossler and take the following specific parameter values:

$$k_1 p = 0.01 \qquad k_2 q = 0.000\,5 \qquad k_3 = 1.0$$
$$k_4 = 0.11 \qquad K = 0.08 \qquad k_5 = k_{-5} = 0.02. \tag{13.43}$$

Using the usual Hopf analysis we can readily find that the two-variable scheme, with fixed c, has a Hopf bifurcation at $c^* = 1.102$. Now let us consider the behaviour of the full three-variable scheme.

The governing rate equations are

$$\frac{da}{dt} = k_1 p - k_3 ab - k_5 a + k_{-5} c \tag{13.44}$$

$$\frac{db}{dt} = k_2 q + k_3 ab - \frac{k_4 b}{b + K} \tag{13.45}$$

$$\frac{dc}{dt} = k_5 a - k_{-5} c. \tag{13.46}$$

The stationary-state concentrations a_{ss}, b_{ss}, and c_{ss} with the above parameter values are

$$a_{ss} = c_{ss} = 1.1845 \qquad b_{ss} = 8.442 \times 10^{-3}. \tag{13.47}$$

Most significantly, c_{ss} exceeds the Hopf bifurcation value c^* of the two-variable subscheme. Thus, if the subscheme is going to bear any relation to the full model, we may expect that as the concentration of C approaches its stationary state, the concentrations of A and B may begin to oscillate. This is particularly important with the values of the rate constants chosen. These ensure that the A to C interconversion occurs on a rapid timescale compared with the overall evolution of a and b. Thus we can expect the concentrations a and c to be almost 'at equilibrium' with each other, with $k_2 a \approx k_{-2} c$ at all times. If, then, the concentration of A is oscillatory, so will be that of C. In this way, C may vary periodically and, if the amplitude of the oscillation is sufficiently large, will cross the 'bifurcation value' c^* before and after its successive minima. With $c < c^*$, the concentrations of A and B are attracted to the 'stable' pseudo-stationary-state locus of the two-variable scheme, with c increasing in time; as c increases above c^*, there is an oscillatory growth away from the underlying two-variable trajectory, and c begins to fall. This hand-waving description would lead to a motion similar to a trajectory on the surface of a torus in a–b–c space. The computed trajectories are shown in Fig. 13.19. The torus has become strongly distorted, giving an aperiodic response.

This approach to converting two-variable oscillations to three-variable chaos is like the relaxation oscillation analysis and is similarly not quite exact

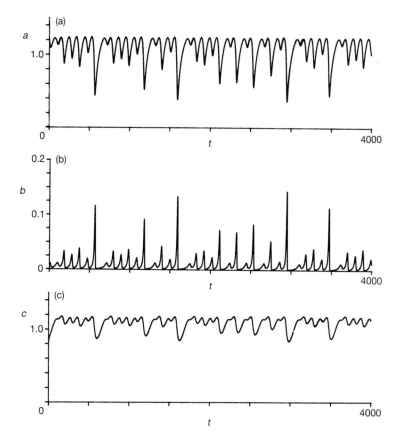

FIG. 13.19.   Relaxation–chaotic oscillations $a(t)$, $b(t)$, and $c(t)$ for the Hudson–Rössler model.

quantitatively. Nevertheless, it successfully illustrates the desired point, namely that chemical mechanisms can provide their own 'internal' forcing. Hudson and Rossler have also been able to extend these arguments to systems with four variables and to predict yet more complex behaviour which they have termed 'hyperchaos'.

13.4.2. Consecutive exothermic reactions in a CSTR

We now turn to an example where full use has been made of the bifurcation analysis based on Floquet multipliers, as described in §5.4.3.

Following Jorgensen and Aris (1983), we take a well-stirred flow reactor fed by a stream of reactant A. This reactant is successively converted to an intermediate B and then to a final product C:

$$A \rightarrow B \qquad \text{rate} = k_1(T)a \qquad (13.48)$$

$$B \rightarrow C \qquad \text{rate} = k_2(T)b. \qquad (13.49)$$

Both steps are simple first-order processes: each is exothermic and each rate constant has an Arrhenius temperature dependence $k_i = A_i \exp(-E_i/RT)$.

Proceeding as in chapter 7, the full form of the three governing equations (two reaction rate equations and an energy balance) are

$$\frac{da}{dt} = \frac{a_0 - a}{t_{\text{res}}} - k_1(T)a \qquad (13.50)$$

$$\frac{db}{dt} = \frac{b_0 - b}{t_{\text{res}}} + k_1(T)a - k_2(T)b \qquad (13.51)$$

$$c_v \sigma \frac{dT}{dt} = \frac{c_v \sigma (T - T_0)}{t_{\text{res}}} + Q_1 k_1(T)a + Q_2 k_2(T)b - \frac{\chi S}{V}(T - T_a) \qquad (13.52)$$

where Q_1 and Q_2 are the exothermicities ($-\Delta H_1$ and $-\Delta H_2$ respectively).

Additional simplifications made for the specific example to be considered are: (i) there is no inflow of B or C, $b_0 = c_0 = 0$; (ii) both reactions have the same activation energy, $E_1 = E_2$; (iii) both reactions have the same exothermicity, $Q_1 = Q_2$; (iv) the inflow temperature is the same as that of thermostatted oven, $T_0 = T_a$; and (v) the temperature dependence of each rate constant can be sufficiently well represented by the exponential approximation (see chapters 4 and 7 for details).

In their analysis, which will form the basis of what follows here, Jorgensen and Aris chose to vary the Newtonian cooling time, keeping the residence time constant during any given experiment. Thus we may use t_{res} as the timescale with which to make the rate equations dimensionless. The resulting forms, with the above simplifications, are

$$\frac{d\alpha}{d\tau} = 1 - \alpha - \frac{\alpha e^{\theta}}{\tau_{\text{ch}}} \qquad (13.53)$$

$$\frac{d\beta}{d\tau} = \frac{\alpha e \theta}{\tau_{\text{ch}}} - \frac{\rho \beta e^{\theta}}{\tau_{\text{ch}}} - \beta \qquad (13.54)$$

$$\frac{d\theta}{d\tau} = \frac{\theta_{\text{ad}} \alpha e^{\theta}}{\tau_{\text{ch}}} + \frac{\theta_{\text{ad}} \rho \beta e^{\theta}}{\tau_{\text{ch}}} - (1 + \tau_N^{-1})\theta. \qquad (13.55)$$

The dimensionless concentrations and temperature excess are given by

$$\alpha = a/a_0 \qquad b = b/a_0 \qquad \theta = E_1(T - T_a)/RT_a^2 \qquad (13.56)$$

and we have the following parameters:

chemical time $\qquad\qquad\qquad \tau_{ch} = 1/k_1(T_a)t_{res}$ $\qquad\qquad$ (13.57)

rate constant ratio $\qquad\qquad\quad \rho = A_2/A_1$ $\qquad\qquad\qquad$ (13.58)

adiabatic temperature rise $\quad\; \theta_{ad} = Q_1 a_0 E_1/\sigma c_v R T_a^2$ $\qquad$ (13.59)

cooling time $\qquad\qquad\qquad\quad \tau_N = t_N/t_{res} = c_v \sigma V/\chi S t_{res}.$ $\qquad$ (13.60)

In what follows we will take $\tau_{ch} = 1.818$, $\theta_{ad} = 17.5$, and $\rho = 0.01$, and vary τ_N across the range

$$0.125 \leqslant \tau_N \leqslant 0.2. \qquad\qquad (13.61)$$

The stationary-state solution is unique in the parameter range of interest.

The upper and lower limits in eqn (13.61) correspond to points of super-critical Hopf bifurcation. The stationary state is unstable in the whole of the region, and near to the limits is surrounded by a stable period-1 limit cycle. We now wish to see how the stability of this oscillatory solution, and any higher-order periodicities which might emerge, varies with τ_N.

The simple pattern of unstable stationary state with a stable period-1 limit cycle is found over almost the whole range of instability. There are just two windows in the parameter range for which more complex responses arise: these are $0.1372 \leqslant \tau_N \leqslant 0.1439$ and $0.1786 \leqslant \tau_N \leqslant 0.1822$, as shown in Fig. 13.20.

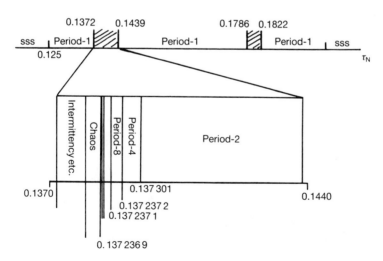

FIG. 13.20.   The variation in the periodicity of the attractor wth the Newtonian cooling time τ_N for the consecutive exothermic reaction model in a CSTR. (Adapted and reprinted with permission from Jorgensen, D. V. and Aris, R. (1983). *Chem. Eng. Sci.*, **38**, 45–53.)

To investigate these windows, we begin with $\tau_N = 0.1440$. Here the system displays a period-1 oscillation, with period $\tau_p = 1.734$. If the heat transfer rate is now increased, so τ_N decreases to $0.143\,897\,7$, the period lengthens to 1.782. At the same time, the critical Floquet multiplier for the period-1 limit cycle approaches -1. There is a period-doubling bifurcation, and for $\tau_N = 0.143\,897\,5$ the stable response is a period-2 limit cycle with a period of 3.562. Figure 13.21 shows the development of the trajectory in the α–θ phase plane through this bifurcation. The period doubling is manifest by a separation of successive circuits near the maximum in the dimensionless temperature rise. This splitting becomes more noticeable and eventually two quite distinct loops exist. The difference between the larger and smaller loops is reflected in the temperature maxima when θ is plotted as a function of time τ.

The new period-2 limit cycle is born with a CFM equal to $+1$. This decreases rapidly as we reduce τ_N further. Eventually, the CFM approaches -1 again: a second period doubling occurs at $\tau_N = 0.137\,307\,1$, where a period-2 with $\tau_p = 3.408$ gives way to a period-4 solution with $\tau_p = 6.816$. The initial splitting in the phase space trajectory again occurs in the vicinity of the maximum temperature rise and the double circuit gives way to four loops as shown in the sequence in Fig. 13.22.

Further period doublings have been located at $\tau_N = 0.137\,237\,2$ (to period-8) and $0.137\,237\,1$ (to period-16) The ranges over which each of these

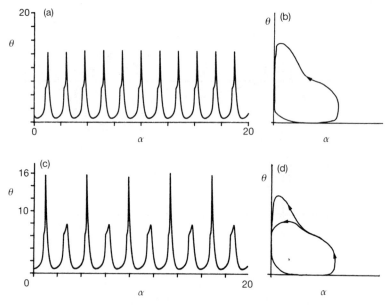

FIG. 13.21.   The period-doubling bifurcation as τ_N is decreased through $0.143\,897\,6$: (a), (b) period-1 with $\tau_N = 0.143\,897\,7$; (c), (d) period-2 pattern for $0.143\,897\,7$.

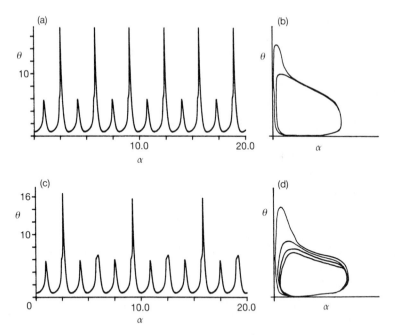

FIG. 13.22.   The second period doubling as τ_N is decreased through 0.137 307 1.

higher-order periodicities holds sway reduce successively, as predicted by the Feigenbaum scenario discussed in §13.1.3. There appears to be a convergence of the cascade as τ_N is reduced to the value 0.137 236 9. Beyond this point the solution is chaotic. Aperiodicity does not survive across the whole range of Newtonian cooling time between its Feigenbaum limit and the return of simple period-1 (at $\tau_N = 0.137\,167$). Rather the character of the solution varies intermittently between chaotic and periodic. A period-doubling cascade based on an initial period-6 oscillation appears at $\tau_N = 0.137\,236\,7$ and converges back to chaos at $\tau_N = 0.137\,210\,5$. Subsequent cascades based on period-5 and then period-4 also occur, with $0.137\,210\,5 < \tau_N < 0.137\,210\,4$ and $0.137\,210\,4 < \tau_N < 0.137\,196\,9$ respectively. Finally, there is a very complex region, even by these standards, in which high-order periodicities and chaos seem to be mixed together.

   The above sequence of exotica occurs, in reverse, in the upper window. It is true that the ranges of parameter values over which much of the 'interesting' patterns occur may be very small, but it is a tribute to the power of the Floquet analysis and even more so to the patience and skill of Jorgensen and Aris that such fine detail can be resolved from such a simple model.

   We may also note that very similar sequences arise for a CSTR in which two non-competitive exothermic reactions proceed, coupled only through the temperature rise and its effects on the two rate constants. The study by Lynch *et al.* (1982) again addressed a set of parameter values for which the stationary-state solution is unique and loses stability via a Hopf bifurcation.

References

Berge, P., Pomeau, Y., and Vidal, C. (1984). *Order within chaos*. Wiley, New York.

Guckenheimer, J. and Holmes, P. (1986). *Nonlinear oscillations, dynamical systems and bifurcations of vector fields*. Springer, New York.

Holden, A. (ed.) (1986). *Chaos*. Manchester University Press.

Hudson, J. L. and Rössler, O. E. (1984). Chaos in simple three- and four-variable chemical systems. In *Modelling of patterns in space and time*, (ed. W. Jäger and J. D. Murray). Springer, Berlin.

Jorgensen, D. V. and Aris, R. (1983). On the dynamics of a stirred tank with consecutive reactions. *Chem. Eng. Sci.*, **38**, 45–53.

Kai, T. and Tomita, K. (1979). Stroboscopic phase portrait of a forced non-linear oscillator. *Prog. Theor. Phys.*, **61**, 54–73.

Kevrekidis, I. G., Schmidt, L. D., and Aris, R. (1986). Some common features of periodically forced reacting systems. *Chem. Eng. Sci.*, **41**, 1263–76.

Kevrekidis, I. G., Schmidt, L.D., and Aris, R. (1986). Forcing an entire bifurcation diagram: case studies in chemical oscillators. *Physica*, **23**, 391–5.

Lynch, D. T., Rogers, T. D., and Wanke, S. E. (1982). Chaos in a continuous stirred tank reactor. *Math. Modelling*, **3**, 103–16.

McKarnin, M. A., Aris, R., and Schmidt, L. D. (1988). Response of nonlinear oscillators to forced oscillations: three chemical reaction case studies. *Chem. Eng. Sci.*, **43**, 2833–43.

May, R. M. (1976). Simple mathematical models with very complicated dynamics: review article. *Nature*, **261**, 459–67.

Ruelle, D. (1980). Les attracteurs étrangés. *La Recherche*, **11**, 132–44.

Schneider, F. W. (1985). Periodic perturbations of chemical oscillators: experiments. *Ann. Rev. Phys. Chem.*, **36**, 347–78.

Thompson, J. M. T. and Stewart, H. B. (1986). *Nonlinear dynamics and chaos*. Wiley, Chichester.

Part 2

Experiments

EXPERIMENTAL SYSTEMS 1:
SOLUTION-PHASE REACTIONS

Although this is a text expounding the value of simplified models of chemical instabilities and oscillations, its techniques would not command conviction if they could not bridge the gap between observation of real systems and quantitative interpretation. In this chapter the Belousov–Zhabotinskii reaction and its currently proposed chemical mechanistic steps are related. All schemes build on the classic work of Field, Körös, and Noyes (FKN) (1972) and the simplified three-variable oregonator condensed scheme is the model. The techniques expounded earlier, especially for dealing with relaxation oscillations and chaos, can be used with much success. After a careful study the reader should be able to:

(1) recognize the validity of chemically guided piecewise approaches to the oscillatory behaviour;

(2) follow the reduction of larger kinetic schemes (nine steps from FKN) to the three-variable condensed representation as the oregonator equations;

(3) use independently based knowledge of rate coefficients for elementary steps to evaluate the quotients appearing in the oregonator and proceed to the logical dimensionless forms with a clear appreciation of where parameters are small (eqns (14.8), (14.9), and (14.10));

(4) exploit techniques suitable to relaxation oscillations using, where suitable, two-variable approximations to the oregonator to evaluate amplitudes and periods;

(5) extend the analysis to open systems and include compound oscillations and bursting.

14.1. The Belousov–Zhabotinskii reaction

In its classic form, the Belousov–Zhabotinskii (B–Z) reaction is the catalyzed oxidation of an organic species by acid bromate ion. The original recipe uses the cerium(III) and (IV) couple as the catalyst, but many other transition metal ions which possess two oxidation states differing by one electron can play the same role: the ferroin–ferriin system with Fe ion complexed by 1–10 phenanthroline is popular for demonstrations as the colour changes between Fe(II) and Fe(III) are easily observed and dramatic (red to blue). Typically, the organic species is malonic acid, $HOOCCH_2COOH$. The reaction may be run in batch or flow systems, with continuous stirring. (In closed systems and the absence of stirring, spatial patterns arise which are also of great interest but which will not be discussed here.) We begin with closed reactors, describing the phenomena observed and then the mechanistic interpretation and mathematical modelling which has been developed for this system.

14.1.1. Oscillatory behaviour in well-stirred closed vessesls

Suitable initial concentrations for a B–Z reaction are $[BrO_3^-]_0 = 6.25 \times 10^{-2}\,M$, $[\text{malonic acid}]_0 = 0.275\,M$, $[Ce(IV)]_0 = 2 \times 10^{-3}\,M$, and $[H^+] = 2\,M$. Under such conditions, the reaction typically shows a pre-oscillatory induction period. At the end of this, the system gives way to oscillations in $[Br^-]$ and $[Ce^{4+}]$. The onset of oscillations usually has the form of a 'hard excitation', with large-amplitude excursions developing and being established immediately. (In the terms of chapter 5, this is reminiscent of a system passing through a subcritical Hopf bifurcation.) After the first few peaks, the output from a Pt–calomel electrode pair becomes very regular, with successive oscillations appearing almost identical in amplitude (typically 130 mV) and period (1 min). This reflects regular oscillations in the concentra-

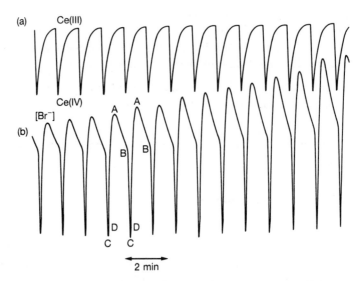

FIG. 14.1. Typical experimental records from (a) platinum electrode and (b) bromide ion electrode for the Belousov–Zhabotinskii reaction in a closed system. In each case the reference electrode is calomel.

tion of the oxidized and reduced form of the catalyst, [Ce(IV)]/[Ce(III)]. An electrode sensitive to bromide ion shows a slight difference as there is a continuous increase in the maximum value obtained, and hence the amplitude, in successive excursions. The largest amplitude from the Br^- electrode is about 100 mV, equivalent to a variation in concentration of about two orders of magnitude. Typical experimental records are shown in Fig. 14.1. The Br^- waveform shows 'relaxation' oscillation character, with periods of relatively smooth change such as the portions A to B and C to D, separated by sharp jumps in concentration such as B to C and D to A.

As the experiment proceeds, there is a gradual lengthening of the oscillatory period, e.g. to 2 min after $1\frac{1}{2}$ h of reaction. This lengthening eventually becomes more pronounced and is accompanied by a decrease in the amplitudes. In the final stages of the oscillatory phase, the pulses can become irregular, but there is an overall decrease to zero amplitude. Oscillation ceases and the ultimate, and very slow, decay to final equilibrium proceeds.

14.1.2. Kinetic mechanism (FKN scheme)

The kinetic mechanism giving rise to oscillations, and to other types of 'non-linear' behaviour, in this reaction is now believed to be well established. There are perhaps some doubts about the details involved in the steps involving the various organic species which participate, but numerical computations with typically 22 reversible reactions have been used relatively successfully to account for qualitative and quantitative observations. A skeleton version of the FKN mechanism is given in Table 14.1.

The important features which lead to oscillatory responses are as follows. The mechanism can be divided into three overall processes. First (*process A*) we have a sequence of three steps which bring about the reduction of bromate to bromine. The reducing agent is bromide ion, initially present perhaps only as an impurity in bromate, but also produced at other stages of the reaction. The reduction proceeds through two-electron transfer steps (R3) and (R2), producing the Br(III) and Br(+ I) species $HBrO_2$ and $HOBr$:

$$BrO_3^- + Br^- + 2H^+ \rightarrow HBrO_2 + HOBr \tag{R3}$$

$$HBrO_2 + Br^- + H^+ \rightarrow 2HOBr. \tag{R2}$$

There then follows a disproportionation between Br(+ I) and Br(− I) in step (R1):

$$HOBr + Br^- + H^+ \rightarrow Br_2 + H_2O. \tag{R1}$$

The overall stoichiometry of process A is obtained by taking (R3) + (R2) + 3(R1), giving

$$BrO_3^- + 5Br^- + 6H^+ \rightarrow 3Br_2 + 3H_2O \tag{A}$$

Table 14.1

FKN mechanism for B–Z reaction

Reaction	Rate constant
(R1) $Br^- + HOBr + H^+ \rightarrow Br_2 + H_2O$	$k_{R1} = 8 \times 10^9 \, M^{-2} s^{-1}$
(R2) $HBrO_2 + Br^- + H^+ \rightarrow 2HOBr$	$k_{R2} = 10^6 \, M^{-2} s^{-1}$
(R3) $BrO_3^- + Br^- + 2H^+ \rightarrow HBrO_2 + HOBr$	$k_{R3} = 2 \, M^{-3} s^{-1}$
(R4) $2HBrO_2 \rightarrow BrO_3^- + HOBr$	$k_{R4} = 2 \times 10^3 \, M^{-1} s^{-1}$
(R5) $BrO_3^- + HBrO_2 + H^+ \rightarrow 2BrO_2\cdot + H_2O$	$k_{R5} = 10 \, M^{-2} s^{-1}$
(R6) $BrO_2\cdot + Ce(III) + H^+ \rightarrow HBrO_2 + Ce(IV)$	$k_{R6} = 6 \times 10^5 \, M^{-2} s^{-1}$
(C1) $CH_2(COOH)_2 \rightleftharpoons (HO)_2C = CHCOOH \text{ (enol)}$	
(C2) $(HO)_2C = CHCOOH + Br_2 \rightarrow BrCH(COOH)_2 + H^+ + Br^-$	$\Big\}$ see text
(C3) $2Ce(IV) + CH_2(COOH)_2 + BrCH(COOH)_2 \rightarrow f\,Br^- + \text{other products}$	

Reactions (R1)–(R6) are regarded as elementary steps, so the individual rates r_i can be obtained, e.g. for (R1) as r_1 as k_{R1} [Br$^-$] [HOBr] [H$^+$] etc. For the empirical rate law appropriate to (C1)–(C3), see text below.

so five bromide ions are consumed per bromate to form three molecules of bromine.

This process serves to reduce the bromide ion concentration but leaves the Ce(IV) concentration substantially unaltered (because of the stoichiometry factor of 5 and the very high ratio of initial bromate to bromide ion concentration, we may also regard [BrO_3^-] as remaining virtually constant during this sequence). Some bromide is returned as Br_2 reacts with the enol form of the malonic acid (MA), giving bromomalonic acid (BrMA):

$$MA \rightarrow enol \tag{C1}$$

$$enol + Br_2 \rightarrow BrMA + H^+ + Br^-. \tag{C2}$$

If all the bromide produced reacts this way we can take (A) + 3(C1) + 3(C2) with a net loss of only two Br$^-$ per cycle through process A.

The second sequence, *process B*, becomes important when the concentration of Br$^-$ becomes low. Then, the species $HBrO_2$, hypobromous acid, can compete as reducing agent for bromate:

$$BrO_3^- + HBrO_2 + H^+ \rightarrow 2BrO_2\cdot + H_2O. \tag{R5}$$

This produces two molecules of the radical BrO_2 (possibly through the prior formation and subsequent dissociation of Br_2O_4). This bromine(IV) species can then be reduced by Ce^{3+} in a single-electron transfer step reproducing $HBrO_2$:

$$BrO_2\cdot + Ce(III) + H^+ \rightarrow Ce(IV) + HBrO_2. \tag{R6}$$

These two steps together constitute an autocatalytic sequence. There is a net increase in hypobromous acid through the cycle and (R5) is the rate-determining step, so the rate of increase of $[HBrO_2]$ depends on the concentration of that species. Taking (R5) + 2(R6) we have

$$BrO_3^- + 3H^+ + 2Ce(III) + HBrO_2 \rightarrow H_2O + 2Ce(IV) + 2HBrO_2.$$

The rate of this process accelerates, and the cerium catalyst is rapidly converted to its oxidized form, as a result of the autocatalysis. There is a sharp colour change (colourless to yellow, or red to blue if ferroin indicator is added to the mixture).

Process A does not cease entirely, but rapidly becomes dwarfed by process B, reflecting the dramatic nature of the exponential growth arising with autocatalysis. Process B is 'switched on' when the bromide ion concentration has fallen far enough for reaction (R5) to compete with (R2). The rates of these steps will be equal when

$$k_{R5}[HBrO_2][H^+][BrO_3^-] = k_{R2}[HBrO_2][H^+][Br^-] \qquad (14.1)$$

i.e. when $[Br^-]$ falls to a 'critical' value $[Br^-]_{cr}$, given by

$$[Br^-]_{cr} = (k_{R5}/k_{R2})[BrO_3^-] \sim 10^{-5}[BrO_3^-]. \qquad (14.2)$$

Thus, for $[Br^-] > [Br^-]_{cr}$, process A dominates, leading to a reduction in the concentration of bromide ion concentration. For $[Br^-] < [Br^-]_{cr}$, the autocatalytic sequence in process B is the more important, leading indirectly to an increase in bromide ion concentration. (In fact, the acceleration in the rate which accompanies the autocatalysis in process B means that this sequence can compete even at bromide ion concentrations above $[Br^-]_{cr}$ as defined above. A more accurate analysis, which we will see later, predicts

$$[Br^-]_{cr} = (1 + 1/\sqrt{2})(k_{R5}/k_{R2})[BrO_3^-] \qquad (14.3)$$

so the system switches from A to B even when the rate of reaction (R2) exceeds that of (R5).)

The 'chain branching' in process B does not continue indefinitely and is limited at high concentrations of $HBrO_2$ by the 'quadratic termination' step

$$2HBrO_2 \rightleftharpoons HOBr + BrO_3^- + H^+. \qquad (R4)$$

The overall stoichiometry of process B is obtained by taking 2(R5) + 4(R6) + (R4):

$$BrO_3^- + 4Ce(III) + 5H^+ \rightarrow 4Ce(IV) + HOBr + 2H_2O. \qquad (B)$$

Thus, so far we have the removal of bromide ions by process A and the autocatalytic oxidation of the metal ion catalyst Ce(III) through process B. The B–Z system can operate as an oscillatory process because the products of these two sequences, BrMA and Ce(IV), can react together regenerating Br^-

and Ce(III) for the reaction. This part of the reaction is not so well under-stood. It is generally accepted that Ce(IV) can oxidize both MA and BrMA, yielding a variety of products. It appears that, whatever the actual chemistry involved, the oxidation of the various organic species can be represented by a single overall (i.e. not elementary) step,

$$2Ce(IV) + BrMA + MA \rightarrow 2Ce(III) + f\,Br^- + \text{other products} \quad (C3)$$

which is, most simply, just the concurrent oxidation of the two organic species. Reactions (C1)–(C3) then constitute *process C*.

The important features here are that this process produces approximately $\frac{1}{2}f\,Br^-$ ions per ceric ion consumed. The empirical rate law appears to be well approximated by

$$r_C = -\,d[Ce(IV)/dt] = k_0[Org][Ce(IV)] \quad (14.4)$$

i.e. it is first order in the oxidized catalyst concentration. Here [Org] repres-ents the total concentration of MA and BrMA. A justification for this form can be proposed because the rate-determining step in the formation of BrMA appears to be the enolization of MA.

The factor f can be regarded as a parameter whose value we have some freedom in choosing. The stoichiometry in (C3) suggests a maximum value of $f = 1$. Higher values might in fact be possible, for instance if BrMA is oxidized more readily than MA or is present in higher concentrations. Values for f less than unity imply that some bromine remains bound to organic species or is removed from the system in some other way.

With the stoichiometry (A) + 3(C), so that all Br_2 produced reacts with the enol, we have

$$3enol + 3MA + 6Ce(IV) + BrO_3^- + 5Br^- + 3H^+$$
$$\rightarrow 3(1 + f)Br^- + 6Ce(III) + 3H_2O. \quad (14.5)$$

Thus the case $f = \frac{2}{3}$ would give no net change in the bromide ion concentra-tion through any given cycle; the case $f = 1$ gives a net increase in $[Br^-]$, similar to that observed experimentally. During this part of the reaction the catalyst is reduced, with an accompanying colour change from yellow to colourless (blue to red in the presence of ferroin). There is no autocatalytic character here so this is a gradual process.

Further mechanistic comments

We have not attempted to give a fully detailed interpretation of the B–Z reaction and the many variations of it which have been developed over the recent years. Such accounts can be found in the book edited by Field and Burger (1985) or elsewhere. One feature which has emerged, and which deserves passing comment, however, touches on the organic reaction subset and, in particular, the regeneration of bromide ion (the reactivation of process A).

The oxidized form of the catalyst can react with both MA and BrMA. We can write these two steps separately:

$$Ce(IV) + MA \rightarrow Ce(III) + products \qquad rate = k_a[Ce(IV)][MA]$$

$$Ce(IV) + BrMA \rightarrow Ce(III) + Br^- + products \quad rate = k_b[Ce(IV)][BrMA]$$

where k_a and k_b are the respective rate constants. The total rate of reduction of Ce(IV) through these steps, r_C, is then given by the sum

$$r_C = k_a[Ce(IV)][MA] + k_b[Ce(IV)][BrMA].$$

Comparing this form with eqn (14.4) we see that the empirical form $k_o[Org]$ requires

$$k_o[Org] = k_a[MA] + k_b[BrMA].$$

The stoichiometric factor f is then

$$f = \frac{2k_b[BrMA]}{k_a[MA] + k_b[BrMA]}.$$

These two forms show that neither the 'rate constant' k_o nor the stoichiometric factor f rate are truly constant during the reaction. The ratio of the two organic species BrMA and MA will change, both during one oscillatory cycle and over the course of the whole experiment. As MA is consumed during the reaction we may expect a gradual increase in f. We will see later that oscillations are only possible if f lies within a certain range (roughly $0.5 < f < 2.4$). Initially [BrMA] is low and hence f may be less than the lower limit. As [BrMA] increases in the pre-oscillatory period, so f may then increase sufficiently for oscillations to develop.

The idea that the stoichiometric factor varies during the reaction and can perhaps be represented as a function of the instantaneous concentrations of some species has become important for modelling some of the complex responses of this system, but we will deal with that aspect later. There is, however, some justified concern about the above interpretation. Typical values for f can be estimated from the known values of k_a and k_b. These suggest that f is rather too low for oscillations under almost all conditions where such behaviour has been observed experimentally.

It is now believed that the Ce(IV) oxidation of BrMA is not the main route by which Br^- is returned to the reaction fold. Instead, HOBr becomes involved in another radical chain process. The oxidized form of the catalyst is supposed to initiate this chain through the reaction

$$RH + Ce(IV) \rightarrow R\cdot + Ce(III) + H^+$$

where RH is some organic species. The chain propagates through $R\cdot$ and $Br\cdot$ radicals through the steps

$$R\cdot + HOBr \rightarrow ROH + Br\cdot$$

$$Br\cdot + RH \rightarrow Br^- + H^+ + R\cdot$$

overall $$RH + HOBr \rightarrow ROH + H^+ + Br^-.$$

This chain sequence increases the number of bromide ions produced per Ce(IV) consumed.

Another reaction which may be important in some variants of the classic B–Z system is the reverse of step (R1), the hydrolysis of bromine:

$$Br_2 + H_2O \rightarrow Br^- + HOBr + H^+. \tag{R-1}$$

This has been invoked for bromide ion regeneration when malonic acid is replaced by an organic species which either cannot be brominated or the bromo-derivative is not oxidized by Ce(IV).

14.1.3. Mathematical model: the oregonator

The qualitative arguments above suggest that the autocatalysis and feed-back mechanisms in the FKN scheme could be capable of accounting for oscillatory behaviour in the B–Z reaction. For quantitative tests, a direct route could proceed by means of numerical integration with the known rate constants and appropriate initial concentrations. More general information can, however, be obtained by the application of the various techniques discussed in earlier chapters. These are most effective if the full mechanism can be reduced to a smaller model, with fewer variables. Two particular types of approximation are used in such a reduction. First, the reactant species are, where appropriate, assumed to be in a large stoichiometric excess so that their concentrations can be held constant and equal to their (known) initial values. This is the pool chemical approximation, which has been justified in chapters 2–5. Secondly, some of the most reactive intermediate species, such as $BrO_2\cdot$ in the present model, may adjust their concentration on such short timescales as to remain in 'pseudo-equilibrium' with the slower concentration variables. Then we can assert that $d[BrO_2\cdot]/dt \approx 0$ etc. at all times, using the stationary-state hypothesis to eliminate these species. Those intermediates and/or reactants which remain after these procedures are the important ones, determining the dynamics of the reaction.

In the oregonator model of the B–Z system, the following reduction of the FKN mechanisms are made. The concentrations of BrO_3^- and of H^+ and the total concentration of organic species [Org] are assumed to be constant. The species $BrO_2\cdot$ is selected as a 'fast' variable which allows the two reactions

(R5) and (R6) to be combined into the autocatalytic sequence discussed above. Following the notation which has become established, we denote the concentrations of the remaining species as follows.:

$$[HBrO_2] = X \qquad [Br^-] = Y \qquad [Ce(IV)] = Z$$

with $[HOBr] = P$, although the latter will not appear actively in the reduced model.

The oregonator scheme is then

$$A + Y \rightarrow X + P \qquad \text{rate} = k_3AY \qquad\qquad (O1)$$

$$X + Y \rightarrow 2P \qquad \text{rate} = k_2XY \qquad\qquad (O2)$$

$$A + X \rightarrow 2X + 2Z \qquad \text{rate} = k_5AX \qquad\qquad (O3)$$

$$2X \rightarrow A + P \qquad \text{rate} = k_4X^2 \qquad\qquad (O4)$$

$$B + Z \rightarrow \tfrac{1}{2}fY \qquad \text{rate} = k_0BZ. \qquad\qquad (O5)$$

Comparing these forms with the FKN mechanism, we see that (O1) is the equivalent of reaction (R3), (O2) that of (R2), and (O4) represents step (R4). The autocatalytic sequence obtained by taking (R5) + 2(R6), with the former as the rate-determining step, is condensed to one reaction, (O3). The part of the mechanism involving organic species, process C, is represented by the single step (O5), with the empirical rate law discussed in the previous section.

The rate constants used for the model here are numbered so as to coincide with the corresponding reactions in the FKN mechanism, but with the appropriate powers of $[H^+]$ multiplied in: thus, $k_3 = k_{R3}[H^+]$ etc.

14.1.4. Analysis of oregonator model

The three reaction rate equations for the concentrations of X, Y, and Z have the form

$$\frac{dX}{dt} = k_3AY - k_2XY + k_5AX - 2k_4X^2 \qquad\qquad (14.6)$$

$$\frac{dY}{dt} = - k_3AY - k_2XY + \tfrac{1}{2}fk_0BZ \qquad\qquad (14.7)$$

$$\frac{dZ}{dt} = 2k_5AX - k_0BZ. \qquad\qquad (14.8)$$

To make these dimensionless, we follow Tyson's suggestions (1979) and take

$$x = \frac{X}{X_0} \qquad y = \frac{Y}{Y_0} \qquad z = \frac{Z}{Z_0} \qquad \tau = \frac{t}{T_0} \qquad\qquad (14.9)$$

where

$$X_0 = k_5 A/2k_4 \qquad\qquad Y_0 = k_5 A/k_2$$
$$Z_0 = (k_5 A)^2/k_4 k_0 B \qquad T_0 = 1/k_0 B.$$

Thus the lower case letters x, y, and z represent the dimensionless concentrations of $HBrO_2$, Br^-, and $Ce(IV)$ respectively. Each concentration is measured on a slightly different scale, as befits the different ranges over which these vary. The chemical timescale chosen is based on the pseudo-first-order rate constant for the regeneration of bromide ion.

The dimensionless rate equations are now

$$\varepsilon \frac{dx}{d\tau} = qy - xy + x(1 - x) \tag{14.10}$$

$$\varepsilon' \frac{dy}{d\tau} = -qy - xy + fz \tag{14.11}$$

$$\frac{dz}{d\tau} = x - z. \tag{14.12}$$

There are three new dimensionless parameters in these equations (ε, ε', and q) defined by

$$\varepsilon = k_0 B/k_5 A \tag{14.13}$$

$$\varepsilon' = \frac{2k_0 k_4 B}{k_2 k_5 A} \tag{14.14}$$

$$q = \frac{2k_3 k_4}{k_2 k_5}. \tag{14.15}$$

The first of these depend on the ratio of the initial concentrations of organic species and bromate ion B/A; q is independent of these concentrations and of $[H^+]$.

If we consider a solution with $[H^+] = 0.8$ M (pH $= 0.097$), the rate constants k_2–k_5 have the following values:

$$k_2 = 8 \times 10^5\,M^{-1}s^{-1} \qquad k_3 = 1.28\,M^{-1}s^{-1}$$
$$k_4 = 2 \times 10^3\,M^{-1}s^{-1} \qquad k_5 = 8.0\,M^{-1}s^{-1}.$$

If we also consider typical initial concentrations $A = [BrO_3^-]_0 = 0.06$ M and $B = [Org]_0 = 0.02$ M, and take $k_0 = 1\,M^{-1}s^{-1}$, the parameters ε, ε', and q are then

$$\varepsilon = 4 \times 10^{-2} \qquad \varepsilon' = 4 \times 10^{-4} \qquad q = 8 \times 10^{-4} \tag{14.16}$$

respectively.

14.1.5. Relaxation oscillations

It is possible to do a full Hopf bifurcation analysis on this three-variable scheme using the formulae derived in chapter 13 etc. This would express the conditions at which oscillatory behaviour begins in terms of the parameters ε, ε', q, and the 'stoichiometric factor' f. These results could then be translated back into dimensional terms by means of the definitions of q etc. However, the form of eqns (14.10)–(14.12) suggests an alternative approach. The two differentials dx/dt and, especially, dy/dt are multiplied by small parameters ε and ε'. From the discussion in chapter 5, we may thus expect any oscillatory responses to have the form of relaxation oscillations.

Let us start with the concentration of bromide, y. Because ε' is particularly small we can expect that the bromide ion concentration will always adjust rapidly to the instantaneous composition of the reacting mixture. Thus y will always remain close to a value such that the right-hand side is equal to zero:

$$y = \frac{fz}{q + x}. \tag{14.17}$$

In this way, the instantaneous bromide ion concentration is fully determined by those of HBrO$_2$ and Ce(IV): we have one less independent variable. The resulting two-variable system is governed by the rate equations

$$\varepsilon \frac{dx}{d\tau} = x(1 - x) + \frac{f(q - x)z}{(q + x)} = g(x, z) \tag{14.18}$$

$$\frac{dz}{d\tau} = x - z \qquad\qquad = h(x, z). \tag{14.19}$$

Now let us consider the dimensionless concentration of HBrO$_2$. Because ε is small, we may again expect that during the course of the reaction x will try to maintain a value with respect to z such that the right-hand side of eqn (14.18) is close to zero. When this cannot be achieved, there will be rapid jumps in x.

This is most easily illustrated by considering the nullclines $g(x, z) = 0$ and $h(x, z) = 0$ in the phase plane, as shown in Fig. 14.2. The condition $h(x, z) = 0$ simply gives the straight line $x = z$, emerging from the origin with unit gradient. Above this line, as drawn in Fig. 14.2, $dz/d\tau$ is negative; below it, $dz/d\tau$ is positive. The condition $g(x, z) = 0$ describes a cubic curve which decreases from infinity at $x = q$, has a minimum and a maximum, and then crosses the axis at $x = 1$.

When the nullclines have a single intersection (stationary state) which lies on the segment of $g(x, z) = 0$ between the minimum and maximum (where z is increasing with x), we expect instability. The resulting oscillation corresponds

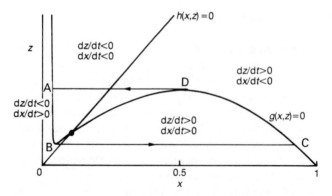

FIG. 14.2.   Schematic representation of the nullclines $g(x, z) = 0$ and $h(x, z) = 0$ for the reduced oregonator model of the Belousov–Zhabotinskii reaction showing the relaxation oscillation cycle ABCD around the unstable stationary-state intersection.

to motion around the loop ABCD, along the $g(x, z) = 0$ manifold between AB and CD, with rapid jumps in x along BC and DA.

14.1.6. Analytical results for amplitude and period

In order to determine the amplitude and period of the relaxation oscillation described above, we need to locate the four poins A–D. The 'jump points', at which the concentration of bromous acid x begins its rapid motion, are simply the turning points in the curve $g(x, z) = 0$. With $q \ll 1$, these are given by

$$\text{B:}\ \ x_B = (1 + \sqrt{2})q \qquad z_B = \frac{1}{f}(1 + \sqrt{2})^2 q \qquad y_B = 1 + \frac{1}{\sqrt{2}} \quad (14.20)$$

$$\text{D:}\ \ x_D = \frac{1}{2} - q \qquad z_D = \frac{1}{4f}(1 + 4q) \qquad y_D = \frac{1}{2}(1 + 4q). \quad (14.21)$$

For A and C, we find

$$\text{A:}\ \ x_A = (1 + 8q)q \qquad z_A = \frac{1}{4f}(1 + 4q) \qquad y_A = \frac{1}{8q} \quad (14.22)$$

$$\text{C:}\ \ x_C = 1 - 6q \qquad z_C = \frac{1}{f}(1 + \sqrt{2})^2 q \qquad y_C = (1 + \sqrt{2})^2 q. \quad (14.23)$$

In these results, the values for the bromide ion y are obtained from eqn (14.17).

First, we can derive the conditions for oscillatory behaviour. The station-ary-state intersection of the two curves will be unstable if x_{ss} lies in the range

$$(1 + \sqrt{2})q = x_B < x_{ss} < x_D = \frac{1}{2} - q. \qquad (14.24)$$

The stationary-state condition for the oregonator equations has $z_{ss} = x_{ss}$ and then either $x_{ss} = 0$ or

$$(1 - x_{ss})(q + x_{ss}) + f(q - x_{ss}) = 0. \qquad (14.25)$$

This quadratic has one positive and one negative root for any given q and $f(q, f > 0)$.

The lower limit of oscillatory behaviour, with $x_{ss} = x_B$, corresponds to a high value for the stoichiometric factor, $f \approx (1 + \sqrt{2})[1 - (1 + \sqrt{2})q]$. With $x_{ss} = x_D$, we have $f \approx \frac{1}{2} + q$. Thus, in the limit of small ε, relaxation oscillations occur over the range

$$0.501 \approx \tfrac{1}{2} + q < f < (1 + \sqrt{2})[1 - (1 + \sqrt{2})q] \approx 2.41. \qquad (14.26)$$

In §14.1.2 it was suggested that f will typically have a value between $\frac{2}{3}$ and 1, so the above analysis predicts that the FKN scheme should then describe oscillatory reaction. Computed concentration histories are shown in Fig. 14.3.

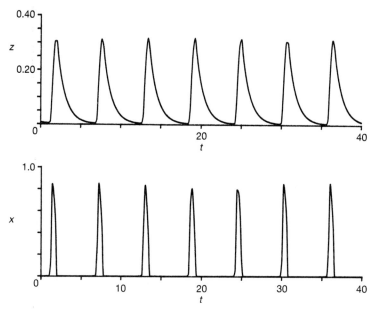

FIG. 14.3. Computed oscillations in the dimensionless concentrations of $HBrO_2(x)$ and oxi-dized catalyst (z) from the oregonator model with $f = 1$, $q = 8 \times 10^{-4}$, and $\varepsilon = 4 \times 10^{-2}$.

If we wish to establish how the oscillatory range depends on the initial concentrations of bromate ions and the organic species, we can perform a Hopf bifurcation analysis on the pair of eqns (14.18) and (14.19). Assuming that q is constant, we then can present the results in the form of an f parameter plane.

The two formulae we need are (i) the stationary-state solution for x, obtained as the positive solution to the quadratic (14.25), and (ii) the condition equivalent to $\mathrm{tr}(\mathbf{J}) = 0$, i.e. $\partial(dx/d\tau)/\partial x + \partial(dz/d\tau)/\partial z = 0$. These give

$$x_{ss} = \tfrac{1}{2}\{1 - f - q + [(1 - f - q)^2 + 4q(1 + f)]^{1/2}\} \qquad (14.27)$$

and

$$\varepsilon = 1 - 2x\left(1 + \frac{fq}{(q + x)^2}\right). \qquad (14.28)$$

Thus, specifying q and f, we may evaluate x_{ss} and then ε for Hopf bifurcation. The resulting locus is shown in Fig. 14.4. In the limit $\varepsilon \to 0$, we recover the relaxation oscillation results of eqn (14.26). As ε increases with the initial ratio of organic species to bromate ion, so the oscillatory range decreases. Oscillations are not possible if $\varepsilon > 0.88$, corresponding to

$$k_0[\mathrm{Org}] > 0.88k_5[\mathrm{BrO_3^-}]. \qquad (14.29)$$

This can be regarded as a condition on the initial concentrations or on the rate constant for the organic oxidation step (bromide regeneration). Clearly, however, the oregonator has a non-zero range of experimental conditions over which oscillatory behaviour is possible.

The next test of the model is its prediction of the 'critical' bromide ion concentration. This is the concentration of Br^- at which the mechanism switches from process A to process B and the autocatalytic increase in $[\mathrm{HBrO_2}]$ (i.e. x) begins. On the phase space nullclines this is represented by

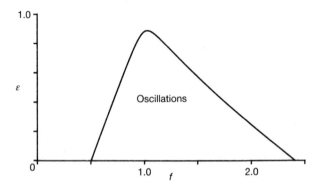

FIG. 14.4.   Locus of Hopf bifurcation points for the two-variable oregonator model of the Belousov–Zhabotinskii reaction, as given by eqns (14.27) and (14.28).

point B, at which $y = y_B = 1 + 1/\sqrt{2} \approx 1.71$. In dimensional terms, from the definitions of y and Y_0, this gives

$$[Br^-]_{cr} \approx 1.71 \, (k_5/k_2)[BrO_3^-] \approx 1.71 \times 10^{-5}[BrO_3^-] \qquad (14.30)$$

which is in outstandingly good agreement with the experimental measurement of $2 \times 10^{-5}[BrO_3^-]$ made by Field et al. (1972).

The oscillatory amplitudes for the three intermediates can be obtained from the maximum and minimum values. For Br^-, $y_{max} = y_A$ and $y_{min} = y_C$; for $HBrO_2$, $x_{max} = x_C$ and $x_{min} = x_A$; for $Ce(IV)$, $z_{max} = z_D$ and $z_{min} = z_B$. In practice, measurements of Br^- and $Ce(IV)$ are most easily made potentiometrically, so the direct reading is of $\log[Br^-]$ etc. In these terms the amplitudes in the bromide and cerium ion concentrations become $\log([Br^-]_{max}/[Br^-]_{min})$ and $\log([Ce(IV)]_{max}/[Ce(IV)]_{min})$ respectively. The latter can be more reliably obtained from experimental records. The formulae above give

$$\log([Ce(IV)]_{max}/[Ce(IV)]_{min}) = \log(z_{max}/z_{min})$$
$$\approx -\log[4(1 + \sqrt{2})^2 q] \approx 1.73 \qquad (14.31)$$

or

$$[Ce(IV)]_{max} \approx 54[Ce(IV)]_{min}. \qquad (14.32)$$

Experimentally, the amplitude is of the same order of magnitude but with $[Ce(IV)]_{max} \approx 10[Ce(IV)]_{min}$.

The maximum bromide ion concentration is predicted to be

$$[Br^-]_{max} \approx (k_5/k_2)[BrO_3^-] \, y_{max} \approx \tfrac{1}{16}(k_5^2/k_3 k_4)[BrO_3^-] \approx \tfrac{1}{640}[BrO_3^-].$$
$$(14.33)$$

We may thus expect a maximum bromide ion concentration of approximately 10^{-4} M.

These formulae also predict that the maximum and critical bromide ion concentrations should be simply related, with

$$[Br^-]_{max} \approx \frac{[Br^-]_{cr}}{8(1 + \sqrt{2})q} \approx 90[Br^-]_{cr}.$$

Experimentally this ratio is found to be smaller, with $[Br^-]_{max} \approx 3[Br^-]_{cr}$. There may be two contributing factors to this discrepancy: first the experimental estimate may be imprecise and, secondly, the implicit assumption that ε is very small may not hold well enough here.

The oscillatory period is comprised mainly of the slow motion from A to B in Fig. 14.2. In dimensionless time, this is given approximately by

$$T_p \approx -\ln\{4[(1 + \sqrt{2})^2 - f]q\}.$$

In order to evaluate this term we must specify a value for the stoichiometric factor f. Taking $f = 1$ for simplicity gives $T_p \approx 4.17$. The period in real time is then obtained from

$$t_p = T_p \times T_0 = k_0[\text{Org}]T_p$$

giving $t_p \approx 210\,\text{s}$. Again we have good order-of-magnitude agreement with the experimental observation, $t_p \approx 110\,\text{s}$.

The various tests, both qualitative and quantitative, described in the previous subsections indicate that the FKN mechanism and its reduced oregonator form serve as a firm basis for the understanding and interpretation of the B–Z reaction in a closed system. Clearly, one would also hope that the same chemical reactions would remain important even when the reaction is operated in a flow reactor. The behaviour shown experimentally in the latter case is of even wider complexity, as we discuss in the next section, but indeed oregonator models can cope very well with these additional aspects.

14.2. The B–Z reaction in well-stirred flow reactors

In a closed vessel, oscillatory behaviour may be a long-lived response, but it is necessarily a strictly transient phenomenon. There is an inevitable approach to equilibrium, with the consequent disappearance of anything other than a monotonic final approach to the thermodynamic state. When the B–Z reaction is run in a CSTR, the constant inflow of fresh reactants allows the system to attain true limit cycle oscillations. These are indefinitely sustainable and, under the simplest conditions, successive excursions have exactly the same amplitude and period.

14.2.1. Experimental observations

At relatively long residence times, the oscillations have very similar wave-forms to those observed over much of the course of a closed vessel experiment. However, the period and amplitude now depend on the reactor residence time as well as the initial (inflow) concentrations of the reactants. At shorter residence times (higher flow rates), the system is maintained further from the chemical equilibrium state. The oscillations can have much smaller amplitude and higher frequency. With further decreases in t_{res}, the system can ultimately undergo a Hopf bifurcation, typically supercritical, so the amplitude tends to zero and a stable stationary state emerges.

The change in waveform just described does not follow a simple smooth variation from large to small amplitude. Instead it occurs through a complex set of bifurcations. For intermediate residence time, the full oscillatory cycle may have differing numbers of both large- and small-amplitude excursions. A typical scenario is that described by Hudson *et al.* (1979). Their experiments used a 25.4 cm³ reactor at 25 °C and for which the inlet concentrations (allowing for dilution effects) were $[MA]_0 = 0.3\,M$, $[BrO_3^-]_0 = 0.14\,M$, $[H^+]_0 = 0.4\,M$, $[Ce(III)]_0 = 10^{-3}\,M$, with $[Br^-]_0$ estimated to be about $3 \times 10^{-6}\,M$.

At the longest residence time, 8.73 min, the reaction shows simple large-amplitude excursion with a period of about 1 min. The corresponding output from a bromide ion electrode is shown in Fig. 14.5(a). This type of response can be denoted as 1^0, indicating that a complete cycle has one large and no small excursions. With $t_{res} = 6.26\,min$, the waveform is as shown in Fig. 14.5(c): each complete oscillation has one large and one small peak, so we denote this 1^1. Although more complex than the previous pattern, this response is still completely periodic. The full period is now approximately 1.33 min and so the change from (a) to (c) does not correspond to a period doubling. In fact, if we look at the behaviour for a residence time in between these two, $t_{res} = 6.76\,min$ say, we find additional complexity, with an alternation between 1^0 and 1^1 types, giving a $1^0 1^1$ or 2^1 pattern as shown in Fig. 14.5(b).

Returning to the 1^1 waveform of Fig. 14.5(c), a further decrease in residence time again sees a change in response, with more small-amplitude character coming in. Figure 14.5(f) shows the regular 1^2 oscillation at $t_{res} = 5.85\,min$; 14.5(h) corresponds to a 1^3 at $t_{res} = 5.50\,min$; and 14.5(j) to the 1^4 at $t_{res} = 5.28\,min$.

Again there are ranges of higher complexity lying between these relatively simple responses. With $t_{res} = 6.18\,min$, the oscillation is comprised of alternate 1^1 and 1^2 excursions (Fig. 14.5(d)). Of particular interest are the waveforms such as that observed with $t_{res} = 5.89\,min$. As shown in Fig. 14.5(e), the response again contains the 1^1 and the 1^2 waveforms interspersed. However, there is now no regular alternation, nor indeed any apparent periodicity. There is no exact repeating sequence and the system exhibits an aperiodic 'mixture' of the two basic patterns. Similarly, at $t_{res} = 5.63\,min$ there appears to be a chaotic pattern built from the 1^2 and 1^3 oscillations (Fig. 14.5(g)). Furthermore, the 1^3 and 1^4 oscillations combine aperiodically at $t_{res} = 5.34\,min$ (Fig. 14.5(i)). In each case, the chaotic pattern exists over a non-zero, although quite narrow, range of residence times.

These aperiodic states are experimentally very stable; some experiments were run for over 28 h and were reproducible from day to day. Other oscillations of the form 1^n, where $n \gg 1$, are also observed, e.g. at $t_{res} = 4.73\,min$ in Fig. 14.5(k). As the number of small oscillations increases, we eventually tend to a 0^1, i.e. no large-amplitude excursions, giving the simple higher-frequency form such as that shown for $t_{res} = 4.69\,min$ in Fig. 14.5(l). The Hopf bifurcation and the consequent disappearance of oscillations occurs at a residence time of about 4.6 min, leaving a stable stationary state.

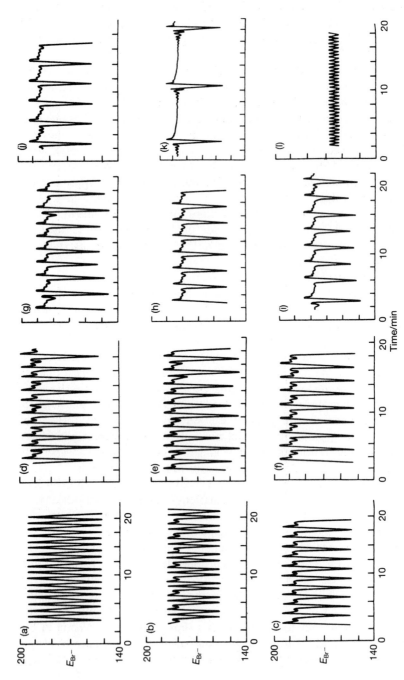

FIG. 14.5. Complex oscillations for the Belousov–Zhabotinskii reaction in a CSTR. The different periodicities and chaotic patterns are described in the text. (Reprinted with permission from Hudson, J. L. et al. (1979). J. Chem. Phys., **71**, 1601–6.)

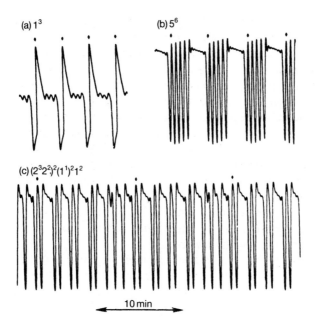

FIG. 14.6. Complex bursting patterns for the Belousov–Zhabotinskii reaction in a CSTR. (Reprinted with permission from Maselko, J. and Swinney, H. L. (1986). *J. Chem. Phys.*, **85**, 6430–41.)

In other experiments, Maselko and Swinney (1986) found waveforms M^N where both M and N differ from unity, e.g. the 5^6 pattern in Fig. 14.6. Such responses can appear as periods of regular (large) oscillations separated by periods of damped oscillatory, or even almost non-oscillatory (quiescent, reaction. This pattern of behaviour has been termed 'bursting'.

14.2.2. Modelling and computations

The first attempts to model the complex patterns described above followed quickly. In 1978 Showalter *et al.* computed the behaviour of a modified oregonator scheme:

$$A + Y \rightleftharpoons X + P \tag{S1}$$

$$X + Y \rightleftharpoons 2P \tag{S2}$$

$$A + X \rightleftharpoons 2W \tag{S3}$$

$$C + W \rightleftharpoons X + Z \tag{S4}$$

$$2X \rightleftharpoons A + P \tag{S5}$$

$$B + Z \rightarrow \tfrac{1}{2} f Y + C. \tag{S6}$$

The major extensions from the original model include the following (i) The autocatalytic process (O3) is replaced by two steps (S3) and (S4) which involve a new species W. The latter can be identified with the radical $BrO_2 \cdot$ in the FKN mechanism. (ii) Reactions (S1)–(S5) are allowed to be reversible, so the product P (HOBr) becomes involved in the kinetics. (iii) The reduced form of the catalyst, Ce(III) or its equivalent, appears explicitly as species C.

The bromide ion regeneration step (S6) is still modelled by a single step, effectively first order in the oxidized catalyst concentration and with the stoichiometric factor f. In their equations, Showalter et al. effectively assumed that the concentration of organic species (B) remains constant, rather than letting B be a variable with its own rate equation. The inflow to the reactor comprises only three species, A (BrO_3^-), Y (Br^-), and C: the rate constants used were not those listed above but an earlier set in which k_2 and k_4 in particular are larger by about three orders of magnitude, and k_0 and f were adjusted to slightly lower values.

The resulting seven-variable scheme was integrated numerically and gave impressive qualitative agreement with many of the experimental waveforms. A typical example is shown in Fig. 14.7 where five small oscillations separate each large excursion. A significant aspect of this work, however, is the absence

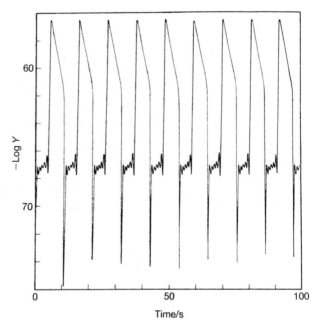

FIG. 14.7. Computed bromide ion concentration showing complex periodic oscillations of 1^5 form from seven-variable reversible oregonator model. (Reprinted with permission from Showalter, K. et al. (1978). J. Chem. Phys., **69**, 2514–24.)

of aperiodic responses. Although the patterns become quite complex, they are all strictly periodic, with no chaotic interspersing of 1^1 and 1^2 for example.

This work has been continued in a similar spirit by Bar-Eli and Noyes (1988). Their kinetic model included one additional reversible step: a reaction between the oxidized catalyst Z and the $BrO_2 \cdot$ radical:

$$Z + W \rightleftharpoons A + C. \tag{S7}$$

More species concentrations were allowed to vary, including [MA] and [H^+] and the reactor had non-zero inflow of MA, H^+, Ce(III), BrO_3^-, and Br^-. More of the experimental complex waveforms were reproduced this way, but again all computed oscillations were strictly periodic: no chaotic patterns were found.

A slightly different approach taken first by Janz et al. (1930) has been designed to retain a model with as few variables as possible. These authors returned to the irreversible oregonator scheme of § 14.1.3 but introduced a new feature. The stoichiometric factor f has been discussed earlier: one may expect that a more accurate representation of the B–Z behaviour could be obtained if the organic chemistry could be modelled in finer detail. For this, Janz et al. argued that at any time during the reaction the instantaneous value for f depends on the mixture composition at that time. To incorporate this, they made f a function of the concentration of the major brominating species HOBr (or P in terms of the reduced model). The particular form taken was

$$f = \frac{FP^2}{K + P^2}. \tag{14.34}$$

The HOBr concentration is now an important variable. It is formed in reactions (O1), (O2), and (O4) and removed by the outflow from the reactor and also by reaction (O5). The bromate ion is also considered as a variable, so the model is described by five rate equations (for A, X, Y, Z, and P).

This adapted scheme has been successfully used to predict and help understand the bursting patterns (the M^N oscillations where M and N are large). Rinzel and Troy (1982) have also been able to reduce the model using a singular perturbation approach similar to the analysis of § 14.1.5. The basic interpretation is that of a stoichiometric factor f whose value oscillates during the reaction in response to the concentration of HOBr. For part of the oscillatory period f is within the region of instability shown in Fig. 14.4, and during this time f is increasing. At other times, the stoichiometric factor has increased sufficiently to pass across the Hopf bifurcation boundary; the system is attracted towards a 'stable' stationary state, but this leads to a reduction in f back into the oscillatory region.

Because of the role played by HOBr, this scheme has become known as the P-feedback oregonator. Recent extensive computations by Showalter et al. (1989) have reproduced and predicted extraordinary complexities for the B–Z reaction, some already observed experimentally by Hudson (1979) or by Maselko and Swinney (1986), and some yet to be found. Again, however, the computed waveforms are always strictly periodic and the bifurcations between different complex responses appear to follow intricate but logical scenarios. Some apparently chaotic behaviour has been observed, but only transiently (although this may last for the equivalent of many hours): all such traces eventually settle on to a periodic pattern.

The interpretation drawn by the series of authors above is that the experimental observations of aperiodicity stem not from the kinetic properties of the B–Z reaction alone. Rather, one can imagine the chemistry trying to settle on to a long-term periodic state, but under some circumstances and over a relatively limited range of experimental conditions, being frustrated in this by features such as imperfect experimental control over e.g. mixing of the incoming reactants or the constancy of the pumping rate during the course of a complete experiment.

An alternative, and apparently more successful, attempt at modelling the aperiodicities in the B–Z system has grown from the collaboration between groups in Bordeaux and Texas. Roux _et al._ (1983) analysed experimental data in terms of a 'delay map'. The experimental data were obtained as a series of

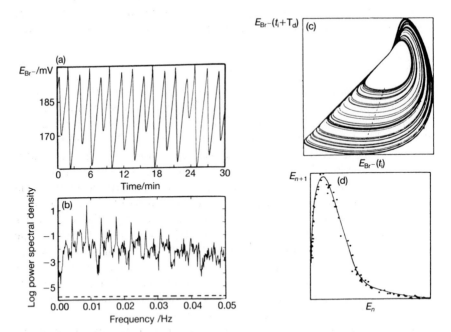

FIG. 14.8. Aperiodic behaviour in the Belousov–Zhabotinskii reactions for the 'Texas' experiments: (a) time series for bromide ion concentration; (b) power spectrum of data in (a); (c) delay map of data in (a); (d) Poincaré section (next return map) for the reconstructed attractor in (c). (Reprinted with permission from Roux, J. C. _et al._ (1983). _Physica_, D **8**, 257–66.)

measurements of bromide ion potential at various equal time increments, a so-called 'time series', as shown in Fig. 14.8(a). The Fourier transform or power spectrum of this series, Fig. 14.8(b), is broad banded rather than comprised of sharp peaks. The delay map shown in Fig. 14.8(c) is then constructed by plotting the potential at time t_i against that at time $t_i + T_d$ for successive t_i, where T_d is a 'delay time'. Roux et al. took $T_d = 8.8$ s. Next, a line is drawn across this trajectory and a 'next return' made by plotting the intersection of the trajectory with this line on one circuit against the intersection on the next circuit. This leads to the curve in Fig. 14.8(d), which shows some resemblance to the one-dimensional cubic map discussed in chapter 13.

This return map, named the Texattractor, can be tested in various ways to confirm that the behaviour is genuinely chaotic and used to work backwards to predict time series from a given starting point. The real challenge is then to obtain similar maps and aperiodic waveforms from a mechanistic model of 'elementary' steps. The particular scheme chosen by the Texas–Bordeaux groups consists of nine reactions, with the six 'inorganic' steps (R1)–(R6) from the FKN set plus three reactions for the organic involvement in bromide regeneration. The latter was taken to involve organic radicals and bromination by HOBr, rather than taking the simple one-step form with a stoichiometric factor f. Thus the reactions are

$$BrO_3^- + Br^- + 2H^+ \rightarrow HBrO_2 + HOBr$$

$$HBrO_2 + Br^- + H^+ \rightarrow 2HOBr$$

$$HOBr + Br^- + H^+ \rightarrow Br_2 + H_2O$$

$$BrO_3^- + HBrO_2 + H^+ \rightleftharpoons 2BrO_2\cdot + H_2O$$

$$2HOBr \rightarrow HOBr + BrO_3^- + H^+$$

$$BrO_2\cdot + Ce(III) + H^+ \rightarrow Ce(IV) + HBrO_2$$

$$HOBr + MA \rightarrow BrMA + H_2O$$

$$BrMA + Ce(IV) \rightarrow Br^- + R\cdot + Ce(III) + H^+$$

$$R\cdot + Ce(IV) \rightarrow Ce(III) + P.$$

Again, the rate constants used were the older set rather than those given in Table 14.1. To simplify the computations, the concentrations of BrO_3^-, MA, H^+, and Ce(III) were held constant, leaving seven variables, namely the concentrations of Br^-, HOBr, $HBrO_2$, $BrO_2\cdot$, Ce(IV), BrMA, and $R\cdot$. Only reaction (R5) is taken to be reversible.

The computations on this model reveal a period-doubling sequence as the flow rate parameter is increased, starting from a simple period-1 (1^0) waveform and ultimately leading to chaos. This chaotic state has periodic windows and eventually gives way to a period-halving sequence back to a 1^1 oscillation. Further period-doubling cascades and regions of chaos lead us through successive 1^i waveforms, with i increasing by 1 at each stage. Not all such features have been found experimentally, but there are similarities between the periodic–chaotic alternations *in vitro* and *in numero*. Some doubt has, however, been cast on both the suitability of the selection of reactions in the model and on the accuracy of the computations.

The *existence* of aperiodic responses in experiments with the B–Z system in a CSTR seems beyond doubt. Whether this chaos does or could arise from purely chemical kinetic features of the mechanism has been the subject of active debate over the past 11 years. Despite strong claims from both sides, it seems unlikely that the issue has yet been resolved. Computations with reduced models will always beg the question of the importance of the parts left out, and there is still a tendency to use the older rate constant 'to aid comparison with previous work' when perhaps it is comparison with experiment which is most relevant. The arguments proceed on an even deeper basis than has been set out in the preceding section, and in an area where both experiment and computations (and particularly their interpretation) are fraught with difficulty. Much has already been learnt from the efforts of all the workers in the field, but it appears that much still remains to come. We will move on to other aspects of these non-linear chemical systems.

14.3. Bistability and the minimal bromate oscillator

The main area of doubt in modelling the B–Z system arises from the reactions involving the various organic species. The role played by this subset of reactions is two-fold: the regeneration of bromide ion and the reduction of the catalyst back to its lower oxidation state. If the reaction is carried out in an open system, these two crucial processes can be replaced by a direct inflow of Br^- and Ce(III). The organic reactions then become redundant, so the malonic acid can be omitted.

The minimal bromate oscillator was constructed following such arguments, using a CSTR fed by BrO_3^-, Br^-, and Ce(III). Because there is much greater certainty about the inorganic chemistry involved in the FKN scheme, this system is well characterized mechanistically. We will discuss the oscillations in this system presently, but begin with the phenomenon of bistability.

14.3.1. Multiple stationary states

A CSTR with inflow of acidified bromate, bromide, and cerous ions may show bistability. For example, with $[H^+] = 1.5$ M, $[BrO_3^-]_0 = 2 \times 10^{-3}$ M, $[Br^-]_0 = 1 \times 10^{-5}$ M, and $[Ce(III)] = 1 \times 10^{-4}$ M the system may settle into either of two stationary states, one with relatively high bromide ion concentration ($[Br^-]_{ss} \approx 3 \times 10^{-6}$ M) and one of low $[Br^-]_{ss}$ ($< 10^{-7}$ M).

As the experimental operating conditions such as inflow concentrations and flow rate are varied, so the two stationary states draw out the upper and lower branches of a hysteresis loop. Various examples, taken from Geiseler (1982), are shown in Fig. 14.9. The following qualitative features emerge: high stationary-state concentrations of bromide are favoured by high flow rates (short residence times), low inflow concentrations of bromate or catalyst ions but high bromide inflow.

Recalling the FKN mechanism in Table 14.1 and the previous discussion, a high stationary-state concentration of Br^- will mean that reactions (R1)–(R3) (process) dominate, so there will be little production of $BrO_2 \cdot$ and hence no significant oxidation of the catalyst. With low $[Br^-]_{ss}$, however, the autocatalytic sequence of reactions (R4)–(R6) will become important.

The bistability results can be presented in a slightly different form. For a sequence of experiments with constant flow rate etc., but varying the inflow concentrations of bromide and bromate, multiple stationary states are only found within a closed cusp-shaped region. Outside this region there is only a single stationary state. Just above the upper limit, the unique state has a high bromide ion concentration, whilst $[Br^-]_{ss}$ is low just below the lower boundary. Note, however, that the system can be taken smoothly from a low to a high bromide ion concentration by varying the inflow concentrations so

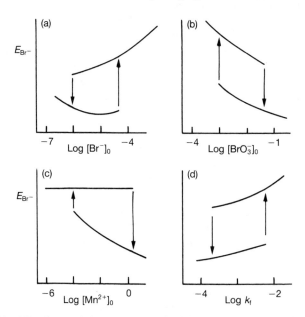

FIG. 14.9. Bistability (hysteresis loops) in the minimal bromate system in a CSTR: (a) varying bromide ion inflow concentration; (b) varying bromate ion inflow concentration; (c) varying catalyst inflow concentration; (d) varying flow rate. (Reprinted with permission from Geiseler, W. (1982). *J. Phys. Chem.*, **86**, 4394–9.)

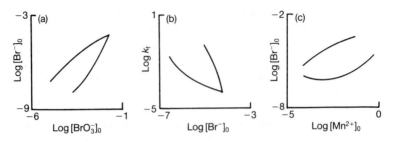

FIG. 14.10. Regions of multiple stationary-state behaviour in various parameter planes for the minimal bromate system in a CSTR: (a) $[Br]_0$–$[BrO_3^-]_0$; (b) flow rate–$[Br^-]_0$; (c) $[Br^-]_0$–$[Mn^{2+}]_0$. (Reprinted with permission from Geiseler, W. (1982). *J. Phys. Chem.*, **86**, 4394–9.)

as to move around the region of bistability instead of crossing it. A typical 'parameter plane' plot is shown in Fig. 14.10. Similar closed regions are found in other parameter planes.

The bistability in this system was first reported by Geiseler and Föllner (1977), and again full-scale computer modelling followed very quickly (Bar-Eli and Noyes 1977, 1978). These numerical studies used the inorganic subset of reactions from the full FKN scheme (a subset first presented by Noyes *et al.* 1971). Good qualitative, and in some respects quantitative, agreement between experiment and computation was obtained despite the use of the older set of rate constants, which differ from those accepted today. Further numerical predictions were later confirmed experimentally by Geiseler and Bar-Eli (1981).

It is relatively simple to adapt the oregonator model for this reduced system. Without the organic species, we may discard the troublesome reaction (O5), retaining just the first four steps in the model. The other modification is the need to introduce flow terms into the rate equations. In the simplest formulation the concentrations of 'major species' such as BrO_3^-, H^+, etc. are still assumed to be constant (and equal to their inflow concentrations). We need only consider two species, $HBrO_2$ (X) and Br^- (Y), and of these only the latter has a non-zero inflow. There is no explicit involvement of the catalyst: the oxidized form (Z in the oregonator) is formed in step (O3) but does not react further if reverse reactions are ignored; the reduced form, which is involved in reaction (R6), does not appear as a variable because the autocatalytic sequence is condensed into one step (O3) with reaction (R5) rate determining.

The rate equations corresponding to (14.6) and (14.7) are now

$$\frac{dX}{dt} = k_3AY - k_2XY + k_5AX - k_4X^2 - k_fX \tag{14.35}$$

$$\frac{dY}{dt} = -k_3AY - k_2XY + k_f(Y_0 - Y) \tag{14.36}$$

where A is the concentration of bromate, k_f is the flow rate, and Y_0 is the inflow concentration of bromide ion, These forms can be made dimensionless by introducing the same x and y from eqn (14.9) to give

$$\frac{dx}{d\tau} = qy - xy + x(1 - x) - kx \tag{14.37}$$

$$\varepsilon \frac{dy}{d\tau} = -qy - xy + \varepsilon k(y_0 - y). \tag{14.38}$$

Here the parameter q is again defined by eqn (14.15) and hence has a value of approximately 8×10^{-4}. In § 14.1.4 the chemical timescale T_0 is that based on the 'organic' reaction step (O5): in the present situation we must employ a different timescale, and we take $T_0 = 1/k_5A$. In this way, ε (which differs from that in the earlier section) and the dimensionless flow rate k are given by

$$\varepsilon = \frac{2k_4}{k_2} \qquad k = \frac{k_f}{k_5A}. \tag{14.39}$$

For stationary states, $dx/d\tau = dy/d\tau = 0$. These conditions then give

$$y_{ss} = \frac{\varepsilon k y_0}{\varepsilon k + q + x_{ss}} \tag{14.40}$$

and x_{ss} as the solution of the cubic equation

$$x_{ss}^3 - [1 - (1 + \varepsilon)k - q]x_{ss}^2 + [\varepsilon k y_0 - (\varepsilon k + q)(1 - k)]x_{ss} - \varepsilon k y_0 q = 0. \tag{14.41}$$

With typical values for q, y_0, and ε (see below), this has either one or three positive real roots for x_{ss}—corresponding to unique or multiple stationary-state solutions—depending on the dimensionless flow rate k.

Typical values for the various parameters in eqn (14.41) are $y_0 \approx 10^3$, $\varepsilon \approx 10^{-3}$, $k \approx 0.1$, and $q \approx 8 \times 10^{-4}$. To leading order then, we can simplify the stationary-state condition to

$$F(x_{ss}, k; \varepsilon, y_0 q) = x_{ss}^3 - (1 - k)x_{ss}^2 + \varepsilon k y_0 x_{ss} - \varepsilon k y_0 q = 0. \tag{14.42}$$

To locate the region of multiple solutions using the methods of chapter 7, we solve the equations $F = F_{x_{ss}} = 0$ simultaneously. The partial derivative form $F_{x_{ss}}$ is given by

$$F_{x_{ss}} = 3x_{ss}^2 - 2(1 - k)x_{ss} + \varepsilon k y_0 = 0. \tag{14.43}$$

This latter condition gives the stationary-state concentration of $HBrO_2$ at the turning points in the x_{ss}–k locus. Working still to leading order, the two roots of (14.43) are

$$x_{ss}^+ \approx \tfrac{2}{3}(1 - k) \qquad \text{and} \qquad x_{ss}^- \approx \tfrac{1}{2}\varepsilon k y_0(1 - k)^{-1}. \tag{14.44}$$

Thus, typically $x_{ss}^+ \approx 0.6$ and $x_{ss}^- \approx 0.05$.

Taking the upper root, x_{ss}^+, fist we find from (14.40) that the corresponding concentration of bromide ion is $y_{ss} \approx \frac{3}{2} \varepsilon k y_0$, which gives

$$[\mathrm{Br}^-]_{ss} \approx 10^{-4} [\mathrm{Br}^-]_0 \qquad (14.45a)$$

whilst for x_{ss}^-, we have $y_{ss} \approx 2(1 - k)$, giving

$$[\mathrm{Br}^-]_{ss} \approx \frac{2k_5}{k_2} [\mathrm{BrO}_3^-]_0 . \qquad (14.45b)$$

Thus, at this second turning point, the stationary-state bromide ion concentration given by eqn (14.45b) is approximately a factor of 10 higher than that given by (14.45a).

Also, by substituting $x_{ss}^{\pm}$ into eqn (14.42), we find the expressions for the experimental conditions governing the turning points in the stationary-state locus. For x_{ss}^+, this gives

$$\varepsilon k y_0 \approx \frac{2}{9} (1 - k)^2 \quad \text{so that} \quad [\mathrm{Br}^-]_0 \approx \frac{k_5^2 [\mathrm{BrO}_3^-]^2}{9 k_4 k_f} \qquad (14.46a)$$

i.e. the critical inflow concentration of bromide ion increases with the square of the inflow concentration of bromate. Turning now to the other root, x_{ss}^-, the corresponding relationship is, to leading order,

$$\varepsilon k y_0 \approx \frac{1}{2} k^2 \quad \text{leading to} \quad [\mathrm{Br}^-]_0 \approx \frac{k_f}{4 k_4} \qquad (14.46b)$$

so here $[\mathrm{Br}^-]_0$ is independent of the bromate inflow.

Equations (14.46a) and (14.46b) give the lower and upper limits on the range of multiplicity, equivalent to the experimental results shown in Fig. 14.10. Working in the $[\mathrm{Br}^-]_0$–$[\mathrm{BrO}_3^-]_0$ parameter plane with k etc. fixed, the lower limit (14.46a) is approximately parabolic. The upper boundary is basically the straight line $[\mathrm{Br}^-]_0 = $ constant. Thus the two limits describe a closed region, merging at a cusp point.

The latter, which represents the disappearance of bistability, can be located by equating $[\mathrm{Br}^-]_0$ from the two expressions. This occurs when $k = 0.4$, so that

$$[\mathrm{BrO}_3^-]_0 \approx 2.5(k_f/k_5) \quad \text{and} \quad [\mathrm{Br}^-]_0 \approx 10^{-3}(k_f/k_5). \quad (14.47)$$

Qualitatively, this very simplified analysis reproduces many of the observed responses and gives some indication of the interrelationship between the various parameters. There are, however, significant quantitative differences between the above predictions and the particular experimental and computational observations of Geiseler and Bar-Eli. For their conditions one of the important assumptions above, namely $k \ll 1$, does not hold; in fact $k \approx 1$. A full analysis of the condition for the loss of multiplicity can be made based on the full stationary-state equation (14.41), again solving $F = F_{x_{ss}} = 0$

simultaneously. Explicit analytical expressions are no longer attainable, but numerical agreement is particularly good.

Before leaving this question of bistability, the multiplicity of stationary states can be illustrated in another useful way — by examining the intersections of the two nullclines specified by $dx/d\tau = 0$ and $dy/d\tau = 0$ respectively. The latter corresponds to the curve given by eqn (14.40) but without the subscripts ss:

$$y \text{ nullcline} \qquad y = \frac{\varepsilon k y_0}{\varepsilon k + q + x} \qquad\qquad (14.48)$$

whilst for the former

$$x \text{ nullcline} \qquad y = \frac{x(1 - k - x)}{x - q}. \qquad\qquad (14.49)$$

Both nullclines are monotonically decreasing functions of x: the y nullcline cuts the y axis at $y \approx y_0$ and asymptotes to $y \approx \varepsilon k y_0$ at large x. The x nullcline is positive for $q < x$ and cuts the x axis at $x = 1 - k$.

The relative positions of the various features of these two curves just described determine the number of intersections. These intersections correspond to stationary states. Figure 14.11 shows the x nullcline and three possible relative orientations for the y nullcline corresponding to different y_0 values. For curves (a) and (c) there is only one intersection, with high and relatively low bromide ion concentration respectively. With an intermediate bromide inflow, curve (b), there are three stationary-state intersections.

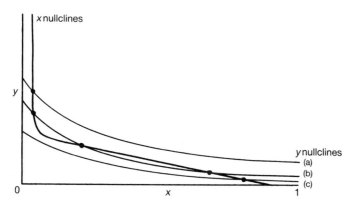

FIG. 14.11.   Representation of bistability in the reduced oregonator model appropriate to the minimal bromate system: for high or low inflow concentrations of bromide ion (y_0), the x and y nullclines have only one intersection; in between, three intersections are possible.

14.4. Oscillations in the minimal bromate system: cross-shaped diagrams

The cusp in the $[Br^-]_0$–$[BrO_3^-]_0$ parameter plane in Fig. 14.11 is a typical feature of many inorganic solution-phase reactions in open systems, especially those which involve an element of autocatalysis. Indeed we have seen a perfect example in the simple cubic autocatalysis of chapters 6–8. In the latter, the appropriate parameter plane is that of residence time t_{res} and inflow concentration of the catalyst b_0, as shown for example in Fig. 6.8. Within the cusp, the system shows bistability. The bistability in these types of reaction arises from the non-linearities in the kinetics. The cusp point represents the classical disappearance of a hysteresis loop, typical of non-linear systems.

The same non-linearity also frequently allows the possibility of Hopf bifurcations and hence sustained oscillations (provided there are at least two independent variables). It has been observed empirically that the boundaries of the region of bistability in the parameter plane appear to continue beyond the cusp point (which is then a crossing point rather than a cusp), as indicated in Fig. 14.12. In the new region thus described, above the cross point, the reaction does not exhibit bistability but shows large-amplitude relaxation oscillations.

This feature, which has become known as the cross-shaped diagram, has had a powerful influence on the recent history of chemical oscillators. Bistable chemical systems are relatively common in CSTRs. The cross-shaped diagram has allowed a relatively systematic route to producing 'designer' oscillations reactions out of bistability. The basic principle is as follows. Given a reaction which exhibits bistability, the search begins for an additional 'feedback'. In the minimal bromate oscillator, this feedback can be identified in terms of the inflow of fresh bromide. At low concentrations of Br^-, in this specific case, there is bistability over a range of $[BrO_3^-]_0$, as indicated previously. However, as $[Br^-]_0$ is increased, this range decreases in

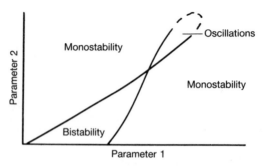

FIG. 14.12.   A typical 'cross-shaped diagram' showing a region of relxation oscillations appearing from the tip of the multistability cusp in the parameter plane.

size. By plotting the boundaries of bistability on the parameter plane, we can readily find the cusp (or crossing) point: beyond that we expect to observe oscillations. Indeed, the upper and lower boundaries directly point out the way towards the oscillatory region.

Following these ideas, in 1981 Bar-Eli predicted the existence of large-amplitude oscillations in the minimal bromate system and was able to produce them in numerical computations with the scheme of Noyes *et al.* (1971). Experimental confirmation appeared rapidly (1982) and independently from Geiseler and from Orban *et al.*

The cross-shaped diagram approach has no rigorous theoretical basis, although De Kepper and Boissonade (1985), in particular, have given it a more respectable footing. It has to some extent confused the relationship between bistability and oscillation, suggesting that the former cause the latter, rather than both equally being consequences of the underlying non-linear kinetics. On the credit side, it has, by virtue of its simplicity and applicability, given rise to by far the largest class of known chemical oscillators, particularly in the inorganic solution phase. Indeed the timescale between the discovery of a new bistable system and the first observations of this type of oscillations in that sytem is becoming so short that one is much less often published without the other. The first oscillator to be 'designed' in this way was the chlorite–iodate–arsenite system, where the chlorite ion acts as the feedback chemical on the iodate–arsenite bistability (indeed a whole family of chlorite oscillators has been discovered in this way). Many new bromate and other oxyhalogen oscillations have been added to the list in the past seven years and, significantly, a growing number of non-halogen systems.

Theoretical analyses have allowed a considerable 'fine structure', beyond the resolution of the early experimental observations, to the cross-shape to be determined. The cubic autocatalysis model with decay again allows these features to be presented particularly clearly. First, the boundaries beyond the cusp ('crossing point') are not simple extensions of the bistability boundaries (which really do meet and end at the cusp point). Rather these are the loci of Hopf bifurcation points and start out from the bistability boundary curves (which are loci of saddle–node bifurcation points) as shown in Fig. 14.13. The starting points are thus degenerate bifurcations of the type considered in chapter 8 where the trace and determinant of the Jacobian matrix vanish simultaneously (double-zero eigenvalues, DZE). The full Hopf bifurcation locus thus starts at one of the DZE points, that on the upper branch say, and runs through the region of bistability. It emerges close to, but not exactly at, the cusp point. The locus then moves off into the region beyond, where the system has only a single stationary state, before looping round and heading back. It eventually finishes at the other DZE point, on the other bistability boundary.

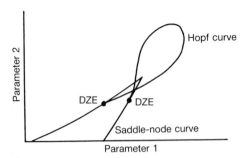

FIG. 14.13. Fine structure in the parameter plane showing the separate existences of the saddle–node and Hopf bifurcation loci, representing the conditions for the onset of multistability and oscillation respectively. Over some conditions multiple stationary states and oscillatory behaviour may coexist.

Thus oscillations and bistability coexist across a non-zero area of the parameter plane. For some experimental conditions this may become a significant and hence easily observable coexistence; for many inorganic reactions, however, it only occurs over a narrow range. Nevertheless, small-amplitude oscillations, such as would be expected close to the Hopf bifurcation points, have been observed even in the minimal bromate system inside the bistable region. One may suspect that as this aspect of detail is more widely appreciated so more examples will be reported.

14.5. The iodate–arsenite reaction

Iodate and iodide ions react together in a redox process to produce iodine I_2. In the presence of a reducing agent such as arsenite, the I_2 is converted back to I^- and the whole process can become autocatalytic. The empirical rate law, either for removal of IO_3^- of production of I^-, has the form

$$- d[IO_3^-]/dt = d[I^-]/dt = (k_q + k_c[I^-])[H^+]^2[I^-][IO_3^-]. \quad (14.50)$$

At constant pH, therefore, we have a mixture of quadratic and cubic autocatalyses, the latter dominating if $k_c[I^-] \gg k_q$.

If this reaction is conducted in a well-stirred closed vessel, it behaves as a *clock reaction*. The initial mixture is colourless and remains so for a given period of time, which depends on the initial concentrations. At the end of this 'induction period', which basically corresponds to the consumption of the arsenite reducing agent, there is then a sharp change in colour and the solution turns brown (or blue if starch is used as an indicator) as a net production of molecular iodine begins.

If the solution is not well stirred, and the reaction is carried out in a test-tube, the first appearance of a colour change occurs at the solution–air interface at the top of the tube. This interface between coloured and colourless solution then propagates down the tube as a travelling wave, of exactly the form discussed in chapter 11, with a constant velocity. Saul and Showalter have shown how the rate law above and the analytical treatment of wave velocity given in §11.1.5 give excellent agreement with experimentally observed wave velocities.

In a CSTR, the arsenite–iodate reaction is bistable, as predicted in chapter 6 for the simple cubic autocatalysis without decay (recall that bistability really refers to three branches of stationary-state solutions, but the middle branch of saddle points is not attainable under normal operating conditions). One of the earliest experimental observations of these multiple stationary states was by Lintz and Weber, who also 'unfolded' the hysteresis loop by allowing a relatively high inflow of iodide relative to iodate, such that $[I^-]_0 > \frac{1}{8} [IO_3^-]_0$.

The analyses of chapter 6 also predict that if the reaction can be modified to allow a simple first-order removal of the autocatalyst I^-, e.g. by reaction, then additional stationary-state patterns such as isolas and mushrooms should be possible. Ganapathisubramanian and Showalter (1984b) achieved such an extra removal, not by reaction but by having an additional inflow to their reactor of pure solvent at a pumping rate which could be varied independently of the inflow of the reactants. In this way, they obtained the extra patterns, as shown in Fig. 14.14. This arrangement does not, however, completely decouple the concentrations of iodide and iodate, in the sense that the solvent flow removes both species at the same rate. Thus, although the more complex stationary-state responses are enabled, the richer dynamical behaviour such as sustained oscillations does not arise.

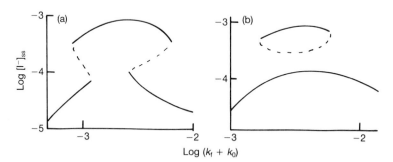

FIG. 14.14. Mushroom and isola patterns in the iodate–arsenite reaction in a CSTR. The patterns are obtained by having the usual inflow of reactants, with flow rate k_f, and a separate inflow of pure solvent, with flow rate k_0. In the experiments, k_f is varied and k_0 held constant. The inflow concentrations are $[IO_3^-]_0 = 1.01 \times 10^{-3}$ M, $[I^-]_0 = 8 \times 10^{-5}$ M, $[AsO_3^{3-}]_0 = 4.98 \times 10^{-3}$ M, pH = 2.23, and $T = 25$ °C. In (a), $k_0 = 4.91 \times 10^{-4}$ s^{-1}, in (b) $k_0 = 6.20 \times 10^{-4}$ s^{-1}. (Reprinted with permission from Ganapathisubramanian, N. and Showalter, K. (1984). *J. Am. Chem. Soc.*, **106**, 816–17.)

Oscillatory reaction can be produced in some modifications of this system, built according to the cross-shaped diagram approach discussed above. The chlorite–iodate–arsenite system was an early triumph of this approach but is rather complicated; the iodate–chlorite system also shows bistability, so this arrangement is coupling two bistable systems, which can even lead to 'tristability' (five branches of stationary states of which two are saddles and three are potentially observable). This system also shows complex oscillations and even aperiodicities, so is more complex than a simple two-variable scheme. Recently, the feedback required for the cross-shaped diagram approach has been achieved with ferrocyanide ion, and relatively simple low-dimension models based on cubic autocatalysis are being developed.

References

Argoul, F., Arneodo, A., Richetti, P., Roux, J. C., and Swinney, H. L. (1987). Chemical chaos: from hints to confirmation. *Acc. Chem. Res.*, **20**, 436–42.

Argoul, F., Arneodo, A., Richetti, P., and Roux, J. C. (1987). From quasiperiodicity to chaos in the Belousov–Zhabotinskii reaction. II *J. Chem. Phys.*, **86**, 3339–56.

Bar-Eli, K. (1981). The behavior of a multistable chemical system near the critical point. In *Nonlinear phenomena in chemical dynamics*, (ed. C. Vidal and A. Pacault), pp. 228–34. Springer-Verlag, Berlin.

Bar-Eli, K. and Geiseler, W. (1983). Oscillations in the bromate-bromide-cerous system. The simplest chemical oscillator. *J. Phys. Chem.*, **87**, 3769–74.

Bar-Eli, K. and Noyes, R. M. (1977). Model calculations describing bistability for the stirred flow oxidation of cerous ions by bromate. *J. Phys. Chem.*, **81**, 1988–90.

Bar-Eli, K. and Noyes, R. M. (1978). Detailed calculations of multiple steady states during oxidation of cerous ions by bromate in a stirred flow reactor. *J. Phys. Chem.*, **82**, 1352–9.

Bar-Eli, K. and Noyes, R. M. (1988). Computations simulating experimental observations of complex bursting patterns in the Belousov-Zhabotinskii system. *J. Chem. Phys.*, **88**, 3636–54.

De Kepper, P. and Boissonade, J. (1985). From bistability to sustained oscillations in homogeneous chemical systems in a flow reactor mode. In *Oscillations and traveling waves in chemical systems*, (ed. R. J. Field and M. Burger), ch. 7, pp. 223–56, Wiley, New York.

Edblom, E. C., Orban, M., and Epstein, I. R. (1986). A new iodate oscillator: the Landolt reaction with ferrocyanide in a CSTR. *J. Am. Chem Soc.*, **108**, 2826–30.

Edblom, E. C., Gyorgyi, L., Orban, M., and Epstein, I. R. (1987). A mechanism for dynamical behaviour in the Landolt reaction with ferrocyanide. *J. Am. Chem. Soc.*, **109**, 4876–80.

Epstein, I. R. and Orban, M. (1985). Halogen-based oscillators in a flow reactor. In *Oscillations and traveling waves in chemical systems*, (ed. R. J. Field and M. Burger), ch. 8, pp. 257–86. Wiley, New York.

Field, R. J. (1985). Experimental and mechanistic considerations of bromate-ion-driven chemical oscillations and traveling waves in closed systems. In *Oscillations and traveling waves in chemical systems*, (ed. R. J. Field and M. Burger), ch. 2, pp. 55–92. Wiley, New York.

Field, R. J. and Burger, M. (eds) (1985). *Oscillations and traveling waves in chemical systems*. Wiley, New York.

Field, R. J., Körös, E., and Noyes, R. M. (1972). Oscillations in chemical systems, part 2. Thorough analysis of temporal oscillations in the Ce-BrO_3^--malonic acid system. *J. Am. Chem. Soc.*, **94**, 8649–64.

Field, R. J. and Noyes, R. M. (1974). Oscillations in chemical systems, part 4. Limit cycle behaviour in a model of a real chemical reaction. *J. Chem. Phys.*, **60**, 1877–84.

Ganapathisubramanian, N. and Showalter, K. (1984a). Washout effects in pumped tank reactors. *J. Am. Chem. Soc.*, **106**, 816–17.

Ganapathisubramanian, N. and Showalter, K. (1984b). Bistability, mushrooms and isolas. *J. Chem. Phys.*, **80**, 4177–84.

Gaspar, V. and Showalter, K. (1987). The oscillatory Landolt reaction. Empirical rate law model and detailed mechanism. *J. Am. Chem. Soc.*, **109**, 4869–76.

Geiseler, W. (1982). Multiplicity, stability and oscillations in the stirred flow oxidation of manganese(II) by acid bromate. *J. Phys. Chem.*, **86**, 4394–99.

Geiseler, W. and Bar-Eli, K. (1981). Bistability of the oxidation of cerous ions by bromate in a stirred flow reactor. *J. Phys. Chem.*, **85**, 908–14.

Geiseler, W. and Föllner, H. H. (1977). Three steady-state situation in an open chemical reaction system. part 1. *Biophys. Chem.*, **6**, 107–15.

Graziani, K. R., Hudson, J. L., and Schmitz, R. A. (1976), The Belousov-Zhabotinskii reaction in a continuous flow reactor *Chem. Eng. J.*, **12**, 9–21.

Györgyi, L. and Field, R. J. (1988). Aperiodicity resulting from external and internal two-cycle coupling in the Belousov–Zhabotinskii reaction. *J. Chem. Phys.*, **92**, 7079–88.

Györgyi, L. and Field, R. J. (1989). Aperiodicity resulting from two-cycle coupling in the Belousov–Zhabotinskii reaction. 2. Modelling of the effect of dead spaces at the input ports of a continuous-flow stirred tank reactor. *J. Phys. Chem.*, **93**, 2865–8.

Hudson, J. L. and Mankin, J. C. (1981). Chaos in the Belousov–Zhabotinskii reaction. *J. Chem. Phys.*, **74**, 6171–7.

Hudson, J. L., Hart, M., and Marinko, D. (1979). An experimental study of multiple peak periodic and non-periodic oscillations in the Belousov–Zhabotinskii reaction. *J. Chem. Phys.*, **71**, 1601–6.

Janz, R. D., Vanecek, D. J., and Field, R. J. (1980). Composite double oscillations in a modified version of the oregonator model of the Belousov–Zhabotinskii reaction. *J. Chem. Phys.*, **73**, 3132–8.

Lintz, H.-G. and Weber, W. (1980). The study of mixing in a continuous stirred tank reactor using an autocatalytic reactor. *Chem. Eng. Sci.*, **35**, 203–8.

Maselko, J. (1980). Experimental studies of chaos-type reactions. The system Mn^{2+}-oxalacetic acid-H_2SO_4-$KBrO_3$. *Chem. Phys. Lett.*, **73**, 194–9.

Maselko, J. and Swinney, H. L. (1986). Complex periodic oscillations and Farey arithmetic in the Belousov–Zhabotinskii reaction. *J. Chem. Phys.*, **85**, 6430–41.

Maselko, J. and Swinney, H. L. (1987). A Farey triangle in the Belousov–Zhabotinskii reaction. *Phys. Lett.*, A **119**, 403–6.

Noyes, R. M., Field, R. J., and Thompson, R. C. (1971). Mechanism of reaction of Br(V) with weak, one-electron reducing agents. *J. Am. Chem. Soc.*, **93**, 7315–16.

Orban, M., De Kepper, P., and Epstein, I. R. (1982). Systematic design of chemical oscillators, part 10. minimal bromate oscillators: bromate-bromide-catalyst. *J. Am. Chem. Soc.*, **104**, 2657–8.

Rinzel, J. and Troy, W. C. (1982). Bursting phenomena in a simplified oregonator flow system model. *J. Chem. Phys.*, **76**, 1775–89.

Roux, J. C., Simoyi, R. H., and Swinney, H. L. (1983). Observation of a strange attractor. *Physica*, D **8**, 257–66.

Rouff, P., Varga, M., and Körös, E. (1988). How bromate oscillators are controlled. *Acc. Chem. Res.*, **21**, 326–32.

Saul, A. and Showalter, K. (1985). Propagating reaction–diffusion fronts. In *Oscillations and traveling waves in chemical systems*, (ed. R. J. Field and M. J. Burger), ch. 11, pp. 419–40, Wiley, New York.

Schmitz, R. A., Graziani, K. R., and Hudson, J. L. (1977). Experimental evidence of chaotic states in the Belousov–Zhabotinskii reaction. *J. Chem. Phys.*, **67**, 3040–4.

Showalter, K., Noyes, R. M., and Bar-Eli, K. (1978). A modified oregonator model exhibiting complicated limit cycle behaviour in a flow system. *J. Chem. Phys.* **69**, 2514–24.

Sørensen, P. G. (1974). Comment. *Faraday Symp. Chem. Soc.*, **9**, 88.

Sørensen, P. G. (1979). Experimental investigation of behaviour and stability properties of attractors corresponding to burst phenomena in the open Belousov–Zhabotinskii reaction. *Ann. NY Acad. Sci.*, **316**, 667–75.

Troy, W. C. (1985). Mathematical analysis of the oregonator model of the Belousov–Zhabotinskii reaction, In *Oscillations and traveling waves in chemical systems*, (ed. R. J. Field and M. Burger), ch. 3, pp. 93–144. Wiley, New York.

Tyson, J. J. (1979). Oscillations, bistability and echo waves in models of the Belousov–Zhabotinskii reaction. *Ann. NY Acad. Sci.*, **316**, 279–95.

Tyson, J. J. (1982). Scaling and reducing the Field-Körös-Noyes mechanism of the Belousov–Zhabotinskii reaction. *J. Phys. Chem.*, **86**, 3006–12.

Tyson, J. J. (1985). A quantitative account of oscillations, bistability and traveling waves in the Belousov–Zhabotinskii reaction. In *Oscillations and traveling waves in chemical systems*, (R. J. Field and M. Burger), ch. 3, pp. 93–144. Wiley, New York.

Winfree, A. T. (1984). The prehistory of the Belousov-Zhabotinskii oscillator. *J. Chem. Educ.*, **61**, 661–3.

EXPERIMENTAL SYSTEMS 2:
GAS-PHASE REACTIONS

Although solution-phase reactions have been more extensively studied, and certainly more widely publicized, a number of spontaneous reactions in the gas phase also provide exemplary oscillators. Here we look at three of the simplest oxidation processes: those of hydrogen, of carbon monoxide, and of acetaldehyde (ethanal). After careful study of this chapter the reader should be able to:

(1) give a mechanistic interpretation for the second explosion limit of the $H_2 + O_2$ reaction in a closed vessel, including the influence of pressure, temperature, and mixture composition;

(2) constuct the ignition limits for the same reaction in a CSTR and understand ignition as an oscillatory process;

(3) appreciate the important role played by the product H_2O as an inhibitor of the $H_2 + O_2$ reaction;

(4) obtain simple analytical expressions for the oscillatory ignition period;

(5) explain the term 'birhythmicity' and understand the effect of residence time on its occurrence;

(6) understand the terms 'cool flame', two-stage ignition, and complete ignition;

(7) give a mechanistic interpretation for these different patterns of reaction;

(8) understand, adapt, and analyse the simple Gray and Yang model for acetaldehyde oxidation.

If the minimal bromate and Belousov—Zhabotinskii reactions are respectively the simplest and most studied solution-phase oscillators, then their gas-phase counterparts must undoubtedly be the oxidations of hydrogen and of acetaldehyde. Like the minimal bromate system, the $H_2 + O_2$ reaction is only oscillatory under continuous flow conditions (both can have clock reaction character in a closed vessel). There are also strong analogies between the acetaldehyde + oxygen and B–Z systems: both show an oscillatory train of excursions in batch, although the influence of reactant consumption, which curtails the oscillations, is much more apparent for the 'cool flames' in the former than for B–Z. In a CSTR, the cool flames can be sustained indefinitely and complex oscillatory responses are also now possible. There is even a simple kinetic model for acetaldehyde oxidation playing the role of the oregonator. This model, known as the Gray–Yang scheme after its inventors, gives a simple representation of the interactions which arise from chemical chain branching (autocatalysis) and the thermal effects through self-heating.

15.1. The hydrogen–oxygen reaction

The behaviour of $H_2 + O_2$ mixtures at reduced pressures (10–100 Torr, i.e. 1.3–13 kN m^{-2}) and ambient temperatures in the range 700–800 K is a relatively familiar part of most undergraduate kinetics courses. For some pressures and temperatures, the reaction between an equimolar mixture, for example, is quite slow, perhaps almost immeasurably so. A repeat of the experiment, with the ambient temperature increased by only a few kelvin or the pressure varied slightly, can show a completely different response with the reactant consumed rapidly (in milliseconds) as the mixture ignites. Figure 15.1 shows a typical ignition diagram of the p–T_a conditions for slow reaction separated from those for ignition by a curved ignition boundary. Conventionally, the three branches of this curve are known as the first, second, and third explosion limits, in order of increasing pressure.

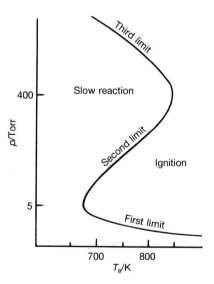

FIG. 15.1.   Typical representation of the pressure–temperature limits for spontaneous ignition in $H_2 + O_2$ mixtures in a closed vessel.

In a closed vessel, ignition is a one-off event: there is complete consumption of the reactants as they are converted to water. During the ignition pulse, high radical concentrations may be achieved and the temperature will increase transiently. This self-heating is a consequence of the reaction but is not believed to be a prime driving force for the onset of the ignition process, so mechanistic interpretations of the p–T_a diagram and modelling are based on isothermal schemes.

In the vicinity of the second explosion limit, qualitative and quantitative modelling requires only five reaction steps:

(0) $H_2 + O_2 \rightarrow 2OH$ initiation

(1) $OH + H_2 \rightarrow H + H_2O$ propagation

(2) $H + O_2 \rightarrow OH + O$ branching

(3) $O + H_2 \rightarrow OH + H$ branching

(4) $H + O_2 + M \rightarrow HO_2 + M$ termination.

Reaction (0) is an initiation process, through which the first radicals in the system are produced. Here we have chosen one possible representation; other choices could have been $H_2 \rightarrow 2H$ or $H_2 + O_2 \rightarrow H + HO_2$, etc. The 'correct' initiation process is exceedingly hard to identify because such reactions are slow and soon swamped by radical production in later steps. Fortunately, the qualitative behaviour is not in the least sensitive to the choice made and the present form is particularly convenient.

15.1.1. Chain branching (autocatalysis)

The group of reactions (1)–(3) form the clockwork of the $H_2 + O_2$ reaction, and indeed make their presence felt one way or another in almost all gas-phase oxidations. They serve to provide the autocatalysis in this scheme, in terms of 'chain branching', increasing the number of 'active species' (radicals) in the system. Some discussion of the autocatalysis in this sub-scheme has been given in chapter 1, but the importance of this part of the mechanism throughout this section is such that it is worth reviewing and commenting further on these aspects.

Consider the fate of a single H atom as it passes through the cycle of these three steps. First the H is transformed by step (2) to two radicals OH and O. The resulting O atom then reacts quickly by step (3), returning an H atom to the system and producing a second OH radical. The system has thus gained two OH radicals. Each of these will be rapidly converted to an H atom by step (1). Here, then, lies the autocatalysis: going through the cycle, one H atom gives rise to three. This constitutes chain branching: in the absence of radical termination, the three can then go on to become nine, then 27, etc. Furthermore, the slowest step (the rate-determining step) is the sequence in reaction (2). Thus the rate r_b at which the H atom concentration is increasing due to this branching sequence is proportional to [H] and is given by

$$r_b = 2k_2[H][O_2]. \qquad (15.1)$$

The factor of 2 just reflects the increase from one to three radicals through the cycle.

15.1.2. Chain termination

If these were the only reactions, the reaction rate would simply increase exponentially, leading to ignition for all experimental conditions. Autocatalysis can, however, be limited by the competition for radicals provided by step (4). This is regarded as a termination step because the 'product' HO_2, although it is formally a radical, is much less reactive than H, OH, and O and in this simple scheme is implicitly allowed to decay to stable products (not radicals). Reaction (4) also involves a 'third body', represented by M. The role of this species is to remove energy from the HO_2 molecule: if this does not occur sufficiently rapidly then the HO_2 will simply redissociate back to H and O_2 on the first vibration of the new H–O bond. The third body is thus any species to which the newly formed HO_2 radical can transfer energy by collision. The species with which the HO_2 is most likely to collide are those present in the highest concentrations, i.e. H_2, O_2, and any added inert gases (and H_2O later in the reaction). The radicals H, OH, and O are always present at concentrations lower by several orders of magnitude and so do not contribute significantly to this stabilization process. One further, and very important, point is that the different species with which HO_2 may collide are not equally efficient at receiving sufficient energy to effect the stabilization process. Thus a collision with H_2 is approximately three times more likely to stabilize the radical (and hence complete the termination step) than a collision with O_2 or N_2. A collision with H_2O is approximately six times more efficient than that with H_2. These different efficiencies mean that the rate of step (4) will vary with the mixture composition. This will account for the variation in the position of the second explosion limit with added inert gas or changes in the initial $H_2:O_2$ ratio and also will be crucial in explaining the oscillatory features in a CSTR.

In modelling terms, the different efficiencies of different species can be incorporated into the rate law by expressing the rate of step (4) as a sum of the individual steps with different third bodies. Thus, if the initial mixture contains H_2, O_2, and N_2 with mole fractions x_{H_2}, x_{O_2}, and x_{N_2} respectively, step (4) can be split into three:

(4$_{H_2}$) $H + O_2 + H_2 \rightarrow HO_2 + H_2$

(4$_{O_2}$) $H + O_2 + O_2 \rightarrow HO_2 + O_2$

(4$_{N_2}$) $H + O_2 + N_2 \rightarrow HO_2 + N_2$.

The total rate of radical termination is the sum of these:

$$r_t = k_{4,\,H_2}[H][O_2][H_2] + k_{4,\,O_2}[H][O_2][O_2] + k_{4,N_2}[H][O_2][N_2]$$

or

$$r_t = k_{4,\,H_2}[H][O_2][M](x_{H_2} + a_{O_2}x_{O_2} + a_{N_2}x_{N_2}) \qquad (15.2)$$

where $[M] = p_{tot}/RT_a$ is the total concentration of third bodies and the coefficients a_{O_2} and a_{N_2} are the 'relative efficiencies' of these species defined by $a_{O_2} = k_{4,O_2}/k_{4,H_2}$, $a_{N_2} = k_{4,N_2}/k_{4,N_2}$. It turns out that O_2 and N_2 are less efficient third bodies than H_2, with $a_{O_2} \approx a_{N_2} \approx 0.3$. The form of eqn (15.2) allows us to account for the influence of mixture composition on the location of the second explosion limit.

15.1.3. Net branching factor and ignition

Using the above information, the full rate equation for the concentration of H can be written in the form

$$\frac{d[H]}{dt} = 2I_0 + 2k_2[O_2][H] - k_{4,H_2}[H][O_2][M](\textstyle\sum a_i x_i). \qquad (15.3)$$

The term $I_0 = 2k_0[H_2][O_2]$ represents the small contribution to the radical pool from the initiation step. Equation (15.3) is frequently reexpressed in a shorter form:

$$\frac{d[H]}{dt} = I_0 + \phi[H]. \qquad (15.4)$$

The new term ϕ is called the 'net branching factor' and reflects the difference between what are effectively the pseudo-first-order rate coefficient for branching and termination:

$$\phi = \{2k_2 - k_{4,H_2}[M](\textstyle\sum a_i x_i)\}[O_2]. \qquad (15.5)$$

Exactly the same form is obtained by performing a stationary-state hypothesis analysis on the concentrations of O and OH and substituting for these species in $d[H]/dt$.

Equation (15.4) is linear in [H] and readily integrated. The resulting form shows that, for $\phi = 0$, the H atom concentration varies exponentially as

$$[H] = \frac{I_0}{\phi}(e^{\phi t} - 1). \tag{15.6}$$

There are two completely different patterns of response. For values of k_2, k_{4,H_2}, etc. such that the net branching factor is negative, H atoms will on average tend to be removed through termination more quickly than they will multiply by branching. With $\phi < 0$, eqn (15.6) predicts that [H] will grow from zero and approach a positive, and generally rather low, 'steady-state' concentration $[H]_{ss} = -I_0/\phi$. If, on the other hand, the experimental conditions are such that the net branching factor is positive, $\phi > 0$, the radical concentration grows exponentially for all time. In this simplest form, [H] tends to infinity because the consumption of the reactants is ignored.

These two responses can be identified with systems exhibiting slow reaction and ignition respectively. The critical case separating these patterns is clearly that with $\phi = 0$. If branching and termination are equally matched there is simply a slow linear increase in the radical concentration, $[H] = I_0 t$. Three concentration histories, typical of $\phi < 0$, $\phi = 0$, and $\phi > 0$ respectively, are shown in Fig. 15.2. In each case, we may expect that the behaviour at long times would be significantly altered by reactant consumption, but the distinction between $\phi < 0$ and $\phi > 0$ at short times will still remain. Returning to eqn (15.5), the term [M] is proportional to the total pressure. Thus we can see that higher pressures correspond to negative values for the net branching factor—slow reaction above the explosion limit—and low pressures, below the limit, to ignition. Similarly, we may expect—and do indeed find from independent kinetic studies—that the reaction rate constant for branching, k_2, will have a much higher temperature coefficient (activation energy) than that for the termination step k_4. In fact, k_4 is found to decrease

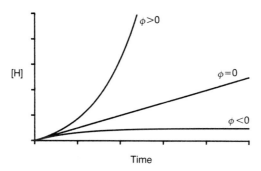

FIG. 15.2. Subcritical, critical, and supercritical evolution of H atom concentration for the $H_2 + O_2$ reaction in the vicinity of the second explosion limit: (a) $\phi < 0$, (b) $\phi = 0$, (c) $\phi > 0$.

slightly with increasing temperature. Thus, increasing the ambient temperature enhances the branching rate relative to termination, predicting ignition at higher T_a and slow reaction at low T_a.

15.1.4. p–T_a ignition limit

The second explosion limit for the $H_2 + O_2$ reaction in a closed vessel is parametrized by the condition $\phi = 0$. This condition can be expressed in the form

$$2k_2 = k_{4,H_2}(p_{tot}/RT_a)(\textstyle\sum a_i x_i). \tag{15.7}$$

In order to predict the p–T_a location of the limit for any given mixture composition, we can rewrite this as

$$p_{tot,\,cr} = 2RT_a(k_2/k_4) \tag{15.8}$$

where the effective termination rate constant $k_4 = k_{4,H_2}(\sum a_i x_i)$. The right-hand side of this equality is an (increasing) function of the ambient temperature. If k_2 and k_4 both have Arrhenius dependences on temperature, a plot of $\ln(p_{tot,\,cr}/T_a)$ versus $1/T_a$ along the limit should be linear, with a gradient proportional to the difference in activation energies, $E_2 - E_4$. We can also see that if the mixture composition is changed, the critical total pressure will vary at constant T_a. Replacement of H_2 by O_2, N_2 or any other species which is a less efficient third body ($a_i < 1$) will decrease the value of k_4 and hence increase the critical pressure. This appears to move the second explosion limit to lower ambient temperatures as illustrated in Fig. 15.3. If, on the other hand, the mole fraction of H_2 is increased at the expense of the less efficient species, or if H_2 is replaced by yet more efficient third bodies such as H_2O, then the critical pressure will be reduced (the limit shifts to higher T_a).

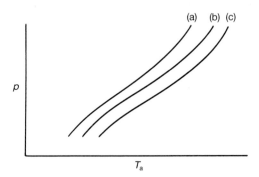

FIG. 15.3. Influence of mixture composition on the p–T_a location of the second explosion limit in a closed vessel: (a) $H_2 + 2O_2$; (b) $H_2 + O_2$; (c) $2H_2 + O_2$. The movement of the limit to higher critical temperatures with increasing hydrogen mole fraction reflects the greater efficiency of H_2 as a third body in the termination step (4).

Table 15.1

Skeleton mechanism for $H_2 + O_2$ reaction

(0)	$H_2 + O_2 \rightarrow 2OH$	$k_0 = 10^8 \exp(-35\,194K/T) \, \text{mol}^{-1}\text{m}^3\text{s}^{-1}$
(1)	$OH + H_2 \rightarrow H_2O + H$	$k_1 = 2.2 \times 10^7 \exp(-2592K/T) \, \text{mol}^{-1}\text{m}^3\text{s}^{-1}$
(2)	$H + O_2 \rightarrow OH + O$	$k_2 = 2.2 \times 10^8 \exp(-8454K/T) \, \text{mol}^{-1}\text{m}^3\text{s}^{-1}$
(3)	$O + H_2 \rightarrow OH + H$	$k_3 = 3.4(T/K)^2 \exp(-4756K/T) \, \text{mol}^{-1}\text{m}^3\text{s}^{-1}$
(4)	$H + O_2 + M \rightarrow HO_2 + M$	$k_{4,H_2} = 5 \times 10^3 \exp(+503K/T) \, \text{mol}^{-2}\text{m}^6\text{s}^{-1}$
		$(a_{O_2} = a_{N_2} = 0.3, \ a_{H_2O} = 6.3)$
(5)	$HO_2 \rightarrow \text{products}$	$k_5 = 50 \, \text{s}^{-1}$
(6)	$HO_2 + H \rightarrow H_2O + O$	$k_6 = 3 \times 10^8 \exp(-2000 \, K/T) \, \text{mol}^{-1}\text{m}^3\text{s}^{-1}$

The discussions above give only a basic introduction to the behaviour of the $H_2 + O_2$ reaction in closed vessels. The simple kinetic scheme is not intended to follow the reaction once the system has crossed the limit and ignition ensued. The slow reaction is particularly well understood for vessels coated with 'aged' boric acid, where it has been used as a tool for evaluating many elementary rate constants. Such a surface preparation virtually excludes reactions at the wall: extra reaction steps must be incorporated into the kinetic scheme to account for radical termination, in particular to model the removal of HO_2 (through conversion to H_2O_2 amongst other fates). Table 15.1 lists some of the reaction steps which will be of relevance to our ensuing discussion, along with the activation energies and pre-exponential factors for the corresponding rate constants.

15.2. The $H_2 + O_2$ reaction in a CSTR

In a closed vessel, ignition is a one-off event. The kinetics of the reaction in the explosion region cannot be easily studied. In a flow system, however, there is a continuous replenishment with fresh reactants and a removal of the products. With residence times of 2–10 s, the $H_2 + O_2$ reaction shows a second explosion limit in a CSTR under very similar p–T_a conditions to those for closed vessels, as shown in Fig. 15.4.

At high pressure, above the limit, or at low ambient temperature, the net branching factor for the system is negative: there is slow reaction. A true stationary state is now achieved, allowing measurement of reactant, product, and intermediate radical concentration. The reaction rate can become sufficiently high for small extents of self-heating to arise—up to 6 K in some cases. If T_a is increased so ϕ becomes positive, the system crosses the limit and ignition ensues. At first the system acts like a closed vessel: the radical concentrations build up and the reaction rate increases. There is a runaway

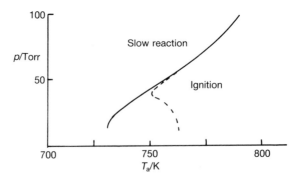

FIG. 15.4. The second p–T_a ignition limit for an equimolar $H_2 + O_2$ mixture in a CSTR with a mean residence time $t_{res} = 8$ s. Above the limit, ignition is at first oscillatory. A second limit (broken curve) separates oscillatory ignition from steady 'flame'.

during which the reactants are consumed and H_2O is produced on a millisecond timescale. The temperature of the reacting gases also increases dramatically during the ignition event. There is virtually complete consumption of H_2 before the rate falls towards zero. After the ignition event, the temperature of the mixture falls back to that of the ambient, mainly by Newtonian heat transfer across the reactor walls. Fresh reactants are pumped into the reactor. There is, however, some delay in any significant onset of reaction, because of the high concentration of H_2O which is a powerful inhibitor of the reaction. The concentration of water vapour decreases exponentially as a result of the outflow, but because of its high efficiency as a third body in reaction (4) $[H_2O]$ must become very small before the net branching factor becomes positive again. Eventually, however, the concentrations of H_2, O_2, and H_2O will be such that a second acceleration of the reaction to ignition follows.

In this way it can readily be understood that ignition can be an oscillatory event in a flow system. Typical experimental records of such behaviour close to the explosion limit are shown in Fig. 15.5. There are, however, many other features—both quantitative and qualitative—observed in this system, whose understanding requires a more rigorous analysis.

15.2.1. Simple oscillatory ignition

For low pressures ($p \lesssim 50$ Torr), and ambient temperatures just above that for the explosion limit, the system exhibits simple oscillatory ignition. Each ignition pulse has the same amplitude, generally measured in terms of the transient temperature rise, the concentration of H_2, O_2, or H_2O, or the intensity of light emission, as shown in Fig. 15.5. The period between

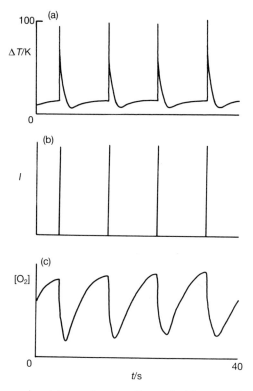

FIG. 15.5. Typical experimental records of oscillatory ignition just above the p–T_a limit: (a) self-heating temperature rise, ΔT; (b) intensity of light emission, I; (c) mass spectrometric record of oxygen concentration ($m/e = 32$). Experimental conditions: equimolar H$_2$ + O$_2$, $t_{res} = 8.5$ s, $p = 60$ Torr, $T_a = 770$ K.

successive pulses is constant for given experimental conditions and generally of similar magnitude to the residence time. This period decreases as the ambient temperature is increased, as shown in curve (a) of Fig. 15.6 which corresponds to a system with $p = 14$ Torr, $t_{res} = 4$ s, and of inflow composition 2H$_2$ + O$_2$.

If the ambient temperature is decreased, back towards the ignition limit, the period becomes very long. In curve (a) of Fig. 15.6, the longest period observed is about 6 residence times. In even more careful experiments with slightly different conditions it has been possible to obtain oscillatory ignitions with a period which is 90 residence times. Such observations require control of the ambient temperature to ± 0.05 K. Presumably, finer control would also reveal even greater lengthening of the period relative to the residence time. Some features of this 'slowing down' close to the limit, and its implications for modelling, will be discussed further in mechanistic terms below.

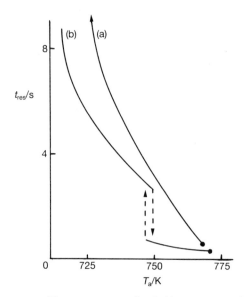

FIG. 15.6.   Dependence on ambient temperature of period between successive ignition pulses for two different residence times: stoichiometric mixture $2H_2 + O_2$ with $p = 14$ Torr. (a) Residence time $\tau_{res} = 4\,s$, showing monotonic decrease in period between onset and disappearance; (b) $t_{res} = 2\,s$, showing birhythmicity.

15.2.2. Steady ignition

At higher ambient temperatures, the period is much shorter than the mean residence time. Here, the ignition pulses are of high frequency and low amplitude: the reactant concentrations do not now build up to values close to the inflow concentration between ignitions, and the concentration of H_2O is always quite high. At high enough T_a, ignition ceases to be oscillatory at all as the amplitude becomes vanishingly small. Note, however, that the period does not decrease to zero. Oscillatory behaviour disappears with a finite period. Reaction now is in the form of a stable steady flame—a stationary state of high reactant consumption. A steady temperature excess is also observed. Typically $\Delta T_{ss} \lesssim 40$ K, but this surprisingly low value is actually the appropriate non-adiabatic 'flame' temperature for the Pyrex-glass CSTRs employed in these studies. Figure 15.4 shows how the limit between oscilla-tory and steady ignition depends on pressure.

If the ambient temperature is decreased again, oscillations regrow, starting at the same T_a at which they ceased on the upward traverse and with the same non-zero period. In the terms of previous chapters, this is typical of a super-critical Hopf bifurcation. The stable stationary state at high T_a is a stable focus (as can be determined experimentally by momentarily perturbing the inflow of one of the reactants). As T_a is decreased, this state loses stability and a stable limit cycle emerges.

15.2.3. Birhythmicity and complex ignitions

Curve (b) in Fig. 15.6 shows how the period between successive ignitions depends on the ambient temperature for a system, with $t_{res} = 14$ Torr, an inflow composition of 2H$_2$ + O$_2$, but at a total pressure of 20 Torr. Again, oscillatory ignition begins on crossing the explosion limit, with long periods relative to the residence time, and the period decreases as the ambient temperature increases. Now, however, the period does not decrease smoothly over the whole range. There are discontinuities in the curve, which consists of two overlapping branches. On raising T_a, the oscillatory waveform can show a sudden jump in character from relative long period and large amplitude to a high-frequency small-amplitude response. Further increasing the ambient temperature causes this high frequency state to give way to a steady flame in the way discussed previously (supercritical Hopf bifurcation). For some (small) range of ambient temperature, the two different oscillatory patterns coexist. Thus we may observe either large-amplitude or small-amplitude excursion at the same reactor conditions, depending on the previous history of the experiment. This bistability and hysteresis between oscillatory states is

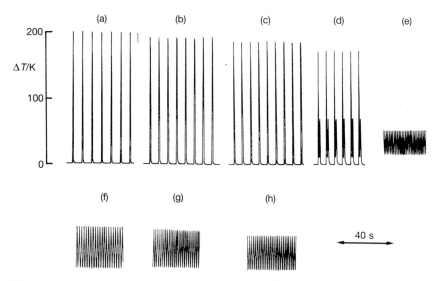

FIG. 15.7.   Typical experimental records (temperature excess ΔT) for a stoichiometric mixture 2H$_2$ + O$_2$ with $t_{res} = 4$ s and $p = 20$ Torr: (a) $T_a = 751$ K; (b) $T_a = 753$ K; (c) $T_a = 756$ K; (d) $T_a = 758$ K; (e) $T_a = 761$ K; (f) $T_a = 751$ K; (g) $T_a = 753$ K; (h) $T_a = 756$ K. Traces (a)–(d) are found as the ambient temperature is increased, (f)–(h) as it is reduced.

similar in principle to that seen in earlier chapters between stationary states. For oscillatory states it is frequently referred to as birhythmicity, and typical experimental traces for the present system are shown in Fig. 15.7. Over some range of conditions, the system has two different stable limit cycles surrounding the unstable stationary state. Furthermore, between and separating these cycles there must be a third unstable limit cycle. The decrease in residence time going from curve (b) to curve (a) in Fig. 15.6 then leads to an unfolding of the hysteresis loop in the locus of period versus T_a. A similar unfolding can be achieved by decreasing the pressure.

Another pattern of reaction which the $H_2 + O_2$ reaction can exhibit under these conditions is that of complex ignitions. Particularly when the system is close to the ends of the birhythmicity hysteresis loop, the full repeating unit in the oscillatory waveform can consist of both large- and small-amplitude excursions. For instance, Fig. 15.7(d) shows what can be referred to as the 1^1 patterns, using the notation applied to the B–Z reaction in the previous chapter. The present reaction has not been as extensively studied under these conditions as have the complex waveforms for the B–Z system, but even so various 1^n patterns with $n \leq 6$ have been characterized, as have additional features such as subcritical Hopf bifurcations into the stable steady flame state.

15.3. Modelling of the $H_2 + O_2$ reaction in a CSTR

For reaction at temperatures below the oscillatory ignition limit, we can expect that the kinetic steps which are important in a CSTR will be the same as those which occur in closed vessels. When we cross the ignition limit, we move into a different regime and can thus expect other steps to play some significant role as well. In this section we discuss first the criterion for the ignition limit itself and then various mechanistic features of the oscillatory ignition events.

15.3.1. Ignition criterion in a CSTR

In §15.1.3, the idea of a net branching factor was introduced. We can use a similar concept for flow reactors. Now, the extra feature for the radicals is an outflow term, e.g. $-k_f[H]$, in the mass-balance equation. Provided the flow rates are relatively small, this is equivalent to allowing an extra termination process, one which is first order in the radical concentration. Thus the net branching factor for a flow system, equivalent to the form of eqn (15.5), becomes

$$\phi_f = \phi - k_f = \{2k_2 - k_{4,H_2}[M](\textstyle\sum a_i x_i)\}[O_2] - k_f. \tag{15.9}$$

The criterion for ignition then is simply $\phi_f = 0$, or

$$2k_2 x_{O_2}\left(\frac{p}{RT_a}\right) = k_{4,H_2} x_{O_2}\left(\frac{p}{RT_a}\right)^2 \sum a_i x_i + k_f. \qquad (15.10)$$

In the experiments described in the previous section the flow rate is such that the k_f term makes only a small correction to the chemical terms. The appropriate root to the quadratic equation (15.10) for the critical pressure can then be expressed as

$$\frac{p_{cr}}{RT_a} = \frac{2k_2}{k_4} - \frac{k_f}{2k_2 x_{O_2}} \qquad (15.11)$$

where $k_4 = k_{4,H_2} \sum a_i x_i$ as before. The first term is just that appropriate to the balance between chain branching and chemical termination. The second term, which involves the relative rates of outflow and branching, may conceivably become important either with very high flows or if the mole fraction of oxygen becomes especially low. In the cases of interest here, however, the first term on the right-hand side is dominant, and so the effects of ambient temperature and mixture composition on the limit in a CSTR are very much the same as in a closed vessel: inert gases act through their influences as third bodies, with differing efficiencies, in the termination step.

15.3.2. Period between oscillatory ignitions

We can use a simple analysis based on a time-dependent net branching factor to predict the period between successive ignition events in a CSTR. To proceed we make a number of assumptions. First, we assume that the actual ignition event is a very fast process, occurring virtually instantaneously and thus making no significant contribution to the period. Secondly, we assume that chemistry does not occur to any extent if the net branching factor ϕ is negative (we will also ignore the difference between ϕ and ϕ_f for simplicity). We will consider a CSTR with an equimolar inflow of H_2 and O_2, so $[H_2]_0 = [O_2]_0 = \frac{1}{2}p_{tot}/RT_a$. For conditions within the oscillatory ignition region, the net branching factor based on the inflow concentration must be positive, i.e.

$$2k_2 - k_{4,H_2}\left(\frac{p}{RT_a}\right)(\tfrac{1}{2} + \tfrac{1}{2}a_{O_2}) > 0. \qquad (15.12)$$

The scenario we consider is as follows. When ϕ becomes positive the chemistry is 'switched on' and ignition rapidly follows. The reactor then contains H_2O and some residual O_2. Because the third-body efficiency of the product is much greater than that of H_2, which it has replaced, the new value for the net branching factor appropriate to this immediate post-ignition composition will be reduced, perhaps to some negative value. In the next stage, H_2O flows out (there is no inflow), so its concentration decreases exponentially, and fresh reactants flow in, so the concentrations of H_2 and O_2 increase exponentially. This replacement of H_2O by less efficient third bodies causes ϕ to increase again, and at some point it passes through zero again leading to another ignition.

Denoting the immediate post-ignition state by $t = 0$, $[H_2O] = [H_2O]_i$ and $[O_2] = [O_2]_i$ then the concentration of H_2O will decrease as

$$[H_2O] = [H_2O]_i[1 - \exp(-t/t_{res})]. \qquad (15.13)$$

Assuming no reaction occurs, the concentrations of H_2 and O_2 are then given by

$$[H_2] = [H_2]_0[1 - \exp(-t/t_{res})] \qquad (15.14)$$

and

$$[O_2] = [O_2]_i + ([O_2]_0 - [O_2]_i)[1 - \exp(-t/t_{res})]. \qquad (15.15)$$

Noting that $[H_2O]_i = [H_2]_0 = \frac{1}{2}(p/RT_a)$ and $[O_2]_i = \frac{1}{2}[O_2]_0$, the time-dependent form for the net branching factor can be written as

$$\phi(t) = 2k_2 - k_{4,H_2}\left(\frac{p}{RT_a}\right)[x_{H_2}(t) + a_{O_2}x_{O_2}(t) + a_{H_2O}x_{H_2O}(t)]$$

$$= 2k_2 - k_{4,H_2}[H_2]_0[1 + a_{O_2} - (1 + \tfrac{1}{2}a_{O_2} - a_{H_2O})\exp(-t/t_{res})]. \qquad (15.16)$$

At $t = 0$, ϕ corresponds to the post-ignition composition and will be negative. As t increases, ϕ increases. The next ignition occurs when ϕ passes through zero, i.e. when

$$2k_2 - k_{4,H_2}[H_2]_0(1 + a_{O_2}) = k_{4,H_2}[H_2]_0(a_{H_2O} - 1 - \tfrac{1}{2}a_{O_2})\exp(-t/t_{res}).$$

Thus the period t_i is given by

$$t_i = t_{res}\ln\left(\frac{a_{H_2O} - 1 - \tfrac{1}{2}a_{O_2}}{(2k_2/k_{4,H_2}[H_2]_0) - (1 + a_{O_2})}\right). \qquad (15.17)$$

This expression has a number of consequences. First, the ignition period is predicted to be a simple multiple of the residence time: t_i is less sensitive to the inflow composition and the rate constants (on which it depends only logarithmically). This result is not too surprising since we have made the assumption that the important processes are those of outflow of the inhibiting product H_2O and inflow of fresh reactants.

Secondly, there will only be an ignition period provided the denominator is positive. This condition is equivalent to requiring $\phi > 0$ for the inflow composition, as specified in eqn (15.12), so that we really are in the ignition region. If T_a is reduced, so that we approach the ignition limit, t_i is predicted to tend to infinity. Such a lengthening of the ignition period has been mentioned above in the experimental studies, and eqn (15.17) has been quite successful in matching the observed results. There is, however, a significant departure from the exact quantitative dependence of t_i on the ambient temperature *very* close to the limit. When T_a exceeds the limit condition by less than about 0.5 K, the observed period becomes much longer than that predicted from the above. Because ϕ never exceeds zero by very much, there is a significant contribution to the period from the chemistry as well as from the flow. Numerical computation with larger schemes and experimental results both show that close to the limit (i.e. for $T_a > T_{a,cr}$ but $T_a - T_{a,cr} \ll T_a$) the induction period lengthens with the form

$$t_i \propto 1/(T_a - T_{a,cr})^{1/2}. \tag{15.18}$$

This dependence, with an induction time inversely proportional to the square root of the 'degree of supercriticality' (i.e. how much T_a exceeds $T_{a,cr}$) is typical of the ignition behaviour in many chemical systems. It has been termed 'slowing down' of oscillatory ignition, in comparison with the slowing down of the decay of transients back to a stationary state discussed in chapter 8.

Returning to eqn (15.17), if the denominator becomes large, say by increasing the ambient temperature so k_2 increases, the induction period can become negative: t_i becomes zero when

$$a_{H_2O} - 1 - \tfrac{1}{2}a_{O_2} = \frac{2k_2}{k_{4,H_2}[H_2]_0} - (1 + a_{O_2})$$

i.e. if

$$2k_2/k_{4,H_2}[H_2]_0 - (a_{H_2O} + \tfrac{1}{2}a_{O_2}) = 0. \tag{15.19}$$

This latter condition is that for the net branching factor for the immediate post-ignition composition to be zero. If the ambient temperature becomes high, then ϕ will still be positive after the ignition, i.e. the ignited state will be maintained by the following gases. Ignition is no longer an oscillatory process: we have a steady flame.

The above analysis can readily be generalized to allow for any inflow composition. The important feature which allows ignition to be oscillatory is the inhibiting effect of the product H_2O through its greatly enhanced efficiency as a third body in the termination step. Thus, we have $a_{H_2O} > 1 + \tfrac{1}{2}a_{O_2}$, so the numerator in (15.17) is positive.

15.3.3. Modelling of complex oscillatory ignition

The simple analysis just presented allows some interpretation of many of the observed experimental features: the ignition process can be oscillatory or, at higher ambient temperatures, can give rise to a steady flame; over much of the oscillatory regime the period between successive ignition events is determined primarily by the flow rate, but close to the limit the chemical timescale can make a significant contribution to the lengthening period. The period is predicted to vary smoothly with the experimental condition and is uniquely

Table 15.2

Baldwin's scheme for $H_2 + O_2$ reaction in the vicinity of the first and second explosion limits

$H_2 + O_2$	$\rightarrow$	$2OH$
$H_2 + O_2$	$\rightarrow$	$HO_2 + H$
$H_2 + OH$	$\rightarrow$	$H_2O + H$
$H + O_2$	$\rightarrow$	$OH + O$
$O + H_2$	$\rightarrow$	$OH + H$
$H + O_2 + M$	$\rightarrow$	$HO_2 + M$
H	$\rightarrow$	inert at wall
O	$\rightarrow$	inert at wall
OH	$\rightarrow$	inert at wall
$HO_2 + HO_2$	$\rightarrow$	$H_2O_2 + O_2$
$HO_2 + H$	$\rightarrow$	$2OH$
$HO_2 + H$	$\rightarrow$	$H_2O + O$
$HO_2 + H$	$\rightarrow$	$H_2 + O_2$
$HO_2 + H_2$	$\rightarrow$	$H_2O_2 + H$
$H_2O_2 + M$	$\rightarrow$	$2OH + M$
$H_2O_2 + H$	$\rightarrow$	$H_2 + HO_2$
$H_2O_2 + H$	$\rightarrow$	$H_2O + OH$
$H_2O_2 + OH$	$\rightarrow$	$H_2O + HO_2$
$H_2O_2 + O$	$\rightarrow$	$OH + HO_2$
$H_2O + O$	$\rightarrow$	$2OH$
$H_2O + H$	$\rightarrow$	$H_2 + OH$
$OH + O$	$\rightarrow$	$O_2 + H$
$OH + H$	$\rightarrow$	$H_2 + O$
$OH + OH$	$\rightarrow$	$H_2O + O$
$H + OH + M$	$\rightarrow$	$H_2O + M$
$H + H + M$	$\rightarrow$	$H_2 + M$
$OH + OH + M$	$\rightarrow$	$H_2O_2 + M$
$O + O + M$	$\rightarrow$	$O_2 + M$
$HO_2 + OH$	$\rightarrow$	$H_2O + O_2$
$H_2O + M$	$\rightarrow$	$H + OH + M$
$H_2 + M$	$\rightarrow$	$2H + M$
$O_2 + M$	$\rightarrow$	$2O + M$

specified. Thus the jumps in period and coexistence associated with birhythmicity do not appear to be accounted for, nor are the more complex waveforms.

In fact, even to obtain the simple single-period oscillations from numerical computations it is necessary to add an extra step to the five-reaction skeleton. This extra step involves two radical species, one of them the relatively stable species HO_2:

(6) $HO_2 + H \rightarrow H_2O + O$

giving the scheme presented in Table 15.1.

If we wish to go further and account for the higher complexities of the oscillations, and to obtain better quantitative matches between experiment and computation, it appears to be necessary to admit that the reaction is non-isothermal. During the ignition event the reaction exothermicity is released in a very short time. The system is almost adiabatic on this timescale, so large extents of self-heating will arise. After the ignition, the temperature will fall back to ambient through Newtonian cooling. The extra coupling and non-linearities introduced by allowing temperature as an extra variable then allow for multiple oscillatory states and for mixed-mode oscillations. Such modelling necessarily requires numerical computation and so has been performed on larger-scale mechanisms, involving perhaps 30 elementary reaction steps. Table 15.2 contains the elementary steps in the Baldwin model (1965, 1974).

15.4. The CO + H$_2$ + O$_2$ reaction

The effect of inert diluents on the p–T_a location of the oscillatory ignition limit in $H_2 + O_2$ systems can be well accounted for in terms of relative efficiencies as third bodies, as discussed above. If, however, the diluent can also become involved chemically, we can expect greater complexity. A particularly interesting case is the addition of carbon monoxide. If we start with an $H_2 + O_2$ mixture and then in a series of subsequent experiments replace increasing proportions of H_2 by CO, the following features emerge. Initial replacement causes the ignition limit to move to lower ambient temperatures: ignition is enhanced. The effect of CO under these conditions can be almost completely explained in terms of its third-body efficiency. CO is less efficient than the H_2 which it has replaced ($a_{CO} \approx 0.4$), so the termination process is slower. The shift in the limit caused by CO is virtually the same as that brought about by replacement of H_2 by N_2.

When x_{H_2} is reduced beyond 0.1, however, the effects of CO and N_2 begin to differ. In particular, for the $H_2 + CO + O_2$ system, the ignition limit now moves back to higher T_a as more H_2 is replaced by CO. Thus we begin to see

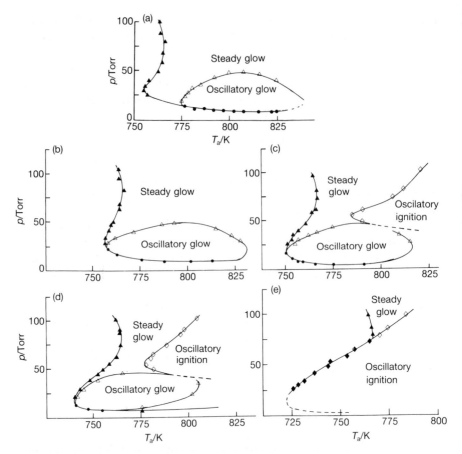

Fig. 15.8. The p–T_a parameter plane for the $CO + O_2$ reaction in a CSTR and the influence of added hydrogen. (a) $CO + O_2$ with no added H_2 showing regions of steady dark reaction, steady glow, and oscillatory glow. (b) $CO + O_2$ with 150 ppm H_2. (c) $CO + O_2 + 1500$ ppm H_2 showing appearance of oscillatory ignition peninsula at the highest ambient temperatures. (d) $CO + O_2 + 7500$ ppm H_2. (e) $CO + O_2 + 10$ per cent H_2, with no oscillatory glow. The reverse sequence (e)–(a) reveals the inhibition of oscillatory ignition which arises as H_2 is replaced by CO.

a chemical effect from CO, which is acting to inhibit ignition. These shifts of the limit with mixture composition are illustrated in Fig. 15.8.

15.4.1. Oscillatory glow

A significant feature which becomes prominent at the lowest mole fractions of H_2 is that as the limit moves to higher temperatures, so it uncovers two regions of different patterns of behaviour: steady glow and oscillatory glow.

Steady glow corresponds to a steady reaction which is accompanied by a continuous emission of chemiluminescence. The emission is a weak, pale-blue glow corresponding to the formation of electronically excited CO$_2$ through the reaction

$$CO + O + M \rightarrow CO_2^* + M.$$

Note that this step also requires a third body to stabilize the recombination product. The glow may be accompanied by a measurable extent of self-heating ($\Delta T \lesssim 6$ K) and of reactant consumption, although at low pressures and ambient temperatures the steady reaction rate may become so small that these become undetectable. Thus, steady glow does not require departures from isothermal operation. The intensity of the emission is markedly increased by increasing x_{H_2}.

Oscillatory glow appears as a non-steady analogue of steady glow. Pulses of chemiluminescence are separated by periods during which the intensity falls to zero. Typical experimental records are shown in Fig. 15.9. This regular periodic emission has been termed the 'lighthouse effect', a term which vividly conveys the observed behaviour. For $x_{H_2} \lesssim 0.001$, there is no detectable

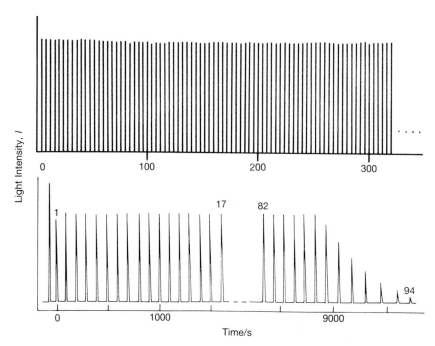

FIG. 15.9.   Two examples of oscillatory glow in the CO + O$_2$ reaction: (a) indefinitely sustained oscillations in a CSTR; (b) a long, but finite, train for a closed system.

self-heating or consumption of CO during the emissions, but, significantly, there appears to be complete consumption of H_2 at each pulse. Again, however, we can look for an isothermal mechanism for this form of oscillation.

The oxidation of CO also shows oscillatory and steady glow in closed vessels. Of course, these are then strictly transient phenomena and must eventually die out as the system approaches its chemical equilibrium state, but they can be long lived, persisting over many hours.

15.4.2. Mechanistic interpretations

The number of additional chemical steps which make a significant contribution to the chemistry in $CO + H_2 + O_2$ under the conditions of interest here is surprisingly small. In fact only two need to be added to the basic $H_2 + O_2$ scheme in order to account for the qualitative features. The reaction

$$(7) \qquad CO + OH \rightarrow CO_2 + H$$

provides the main route to the production of CO_2, but this propagation step has little influence in determining the onset of ignition. A small contribution may come from

$$(8) \qquad CO + HO_2 \rightarrow CO_2 + OH$$

Perhaps the most significant addition is the termination step

$$(9) \qquad CO + O + M \rightarrow CO_2 + M$$

mentioned above as the source of the chemiluminescence. This step removes the active radical O and can compete with reaction (3), which forms part of the branching cycle, when $[CO]/[H_2]$ is large.

The full intricacies and mathematical formulation appropriate to the $CO + O_2$ system are not yet as well developed as for the $H_2 + O_2$ system, but the following tentative explanations appear to be suggested. The main distinction between oscillatory glow and oscillatory ignition lies in the consumption of CO. In both responses there appears to be complete consumption of H_2: for ignition there is also complete consumption of CO, with consequent large temperature excursions and high emission intensities; for oscillatory glow there is very little CO consumption accompanying the H_2 reaction. In the first case we can imagine a full ignition event, with a net branching factor for the whole system, including the extra termination step, being positive at some stage. For oscillatory glow, the $H_2 + O_2$ subsystem appears to have a branched-chain runaway: as the CO/H_2 ratio then increases, the extra termination begins to take control and can inhibit a full ignition. In each case subsequent excursions arise because of the continuous inflow of H_2.

If the above explanation is appropriate for reaction in a CSTR, one might hope that many of the same features will control the oscillatory glow in closed vessels. In fact the experimental conditions in terms of pressure and ambient temperature required are not the same in both systems, so there is scope for additional steps to become important in a closed reactor. In particular, it seems difficult to find a kinetic route to replenish the H_2 concentration between the reaction pulses when there is no inflow. One suggestion is the water–gas shift reaction

$$CO + H_2O \rightleftharpoons CO_2 + H_2$$

which might slowly convert H_2O back to H_2. The reaction in a closed vessel is, however, also very sensitive to any surface pre-treatment or ageing (so successive experiments with apparently identical initial conditions can sometimes give different responses). In all cases it does appear that the $CO + O_2$ reaction proceeds through the $H_2 + O_2$ clockwork, particularly through the branching reaction $H + O_2 \rightarrow OH + O$. Despite strenuous efforts it seems unlikely that anyone has ever been (or will ever be) able to study this reaction without the interference caused by the presence of hydrogen-containing species—if indeed there is any reaction under such a condition.

15.5. Oxidation of acetaldehyde

The reaction between acetaldehyde (CH_3CHO) and O_2 is probably the best characterized of all hydrocarbon oxidation processes. Typical pressures and ambient temperatures of interest are in the range $5 \, kN \, m^{-2} \leq p \leq 25 \, kN \, m^{-2}$ (≈ 40–200 mm Hg) and $450 \, K \leq T_a \leq 650 \, K$. The behaviour exhibited is also influenced by the ratio of CH_3CHO to O_2 and, in a flow system, to the residence time.

In a closed vessel, an oscillatory response known as cool flame oxidation occurs. This mode has some similarities with the lighthouse effect or oscillatory glow of CO oxidation: there is a periodic emission of a pale-blue chemiluminescence. However, with acetaldehyde there is also always a temperature excursion as a result of self-heating. This departure from isothermal reaction is crucial to the cool flame. As we will discuss below, the underlying mechanism for cool flames involves an essential coupling between chemical autocatalysis (through branching and termination steps) *and* thermal effects as the exothermic reaction leads to self-heating. Both parts are not always included in 'explanations' of this system but in fact are equally important. The temperature excursions may be as small as a few kelvin, but usually are of the order of 100–150 K. (Such a temperature rise is only a fraction of the maximum or adiabatic flame temperature, which might be approximately 2000 K, hence the term cool flame.) Each cool flame event leads to the

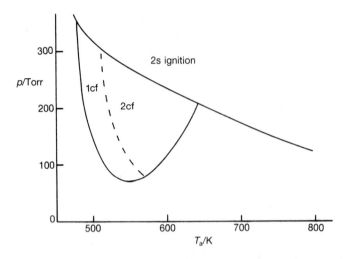

FIG. 15.10. A p–T_a diagram appropriate to the spontaneous oxidation of acetaldehyde in a closed vessel, showing a region of cool flame behaviour and the ignition limit. Ignition is a two-stage process. No long trains of cool flames are possible as reactant consumption soon becomes important.

conversion of perhaps 10–20 per cent of the fuel, so reactant consumption rapidly becomes important in closed systems. Successive cool flames diminish significantly in amplitude, and trains of more than three are uncommon.

The reaction can also show ignition. In a closed vessel this is a one-off event. Figure 15.10 shows a typical p–T_a diagram for this system, with the various limits between slow reaction, ignition, and cool flames. We will not dwell on these results here. The problems of closed reactors associated with reactant consumption are also normally compounded by the absence of stirring, so temperature and concentration gradients can be set up and may be influential in determining the quantitative response. The great steps forward of recent years have built on the introduction of well-stirred flow reactors (CSTRs).

15.5.1. Patterns of reaction in a CSTR

It is convenient to identify five basic, different modes of reaction which are exhibited by acetaldehyde oxidation in a CSTR. These can be termed (I) steady dark reaction, (II) oscillatory ignition, (III) complex ignition, (IV) oscillatory cool flames, and (V) steady glow. Each response occurs within a definite region in the p–T_a diagram; a typical example is shown in Fig. 15.11. We now discuss each of these modes of reaction, and the boundaries between them, in turn.

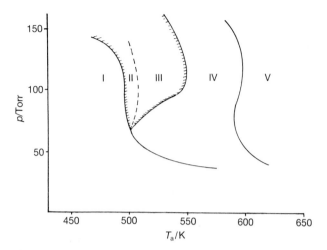

FIG. 15.11.   The p–T_a diagram for acetaldehyde oxidation in a 0.5 dm³ CSTR with a residence time of 3 s, showing five regions: I, steady dark reaction; II, oscillatory (two-stage) ignition; III, complex (multistage) ignition; IV, oscillatory cool flames; V, steady glow.

(a) Steady dark reaction

This response is found at the lowest ambient temperatures. There is no chemiluminescence, but reaction can be followed in terms of measurable stationary-state temperature excesses (self-heating) and by mass spectrometric determination of stationary-state concentration of various reactant and product species. The maximum extents of self-heating are typically about 40 K. The temperature rise, ΔT_{ss}, increases with the ambient temperature, attaining its maximum value at the boundaries between region (I) and regions (II) or (IV). ΔT_{ss} is also a measure of the stationary-state heat release rate R which thus also increases with T_a across this region (R has a positive temperature coefficient).

Typical products of the incomplete combustion in this mode are peracetic acid CH_3CO_3H, methanol CH_3OH, formaldehyde CH_2O, and methane CH_4 as well as CO_2 and H_2O, but no CO is formed.

The stationary state here has stable nodal character: if the inflow is perturbed momentarily and then restored, the return to the stationary state is monotonic (no overshoot).

(b) Oscillatory ignition

In the $H_2 + O_2$ and $CO + O_2$ systems, the one-off ignition in a closed vessel can appear as an oscillatory ignition in a CSTR. In the same way the $CH_3CHO + O_2$ reaction shows a simple oscillatory ignition in a CSTR which appears to be related to the corresponding behaviour in a closed system. The excursions are typically ignition events, with rapid sharp decreases in the concentrations of the reactants and increases in the products (primarily CO and H_2O). There is an intense light emission and a temperature excursion of many hundreds of kelvin (typical records show 400 K, but this must be a lower bound on the actual transient self-heating since the timescale for the ignition event is significantly shorter than the response time for the thermocouple).

The ignition waveform actually shows some important fine structure. The two-stage nature of the ignition in this system is shown in Fig. 15.12; similar detail is observed in the one-off ignition in a closed reactor. The first stage leads to an increase in the gas temperature of up to about 250 K, during which time partial oxidation products such as CH_2O and CH_3OH build up (we will soon see that this is typical of cool flame behaviour). In the present case, however, this initial stage leads on to the rapid acceleration in the reaction and the development of the true ignition stage: the partial oxidation species disappear as CO and H_2O are formed.

Region (II) is entered either from region (I), e.g. by raising the ambient temperature, or from region (III). We will discuss the former here. As T_a is

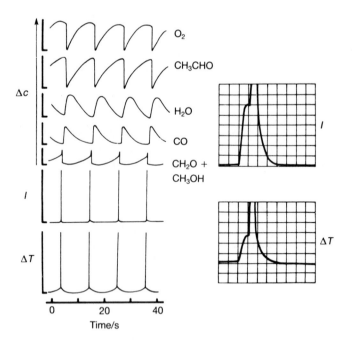

FIG. 15.12. Typical concentration, light emission, and temperature excess records for a two-stage ignition process in a CSTR. The inset oscilloscope records show the two-stage nature of the pulse.

increased across the boundary, there is hard excitation to the ignition pattern, i.e. the system 'jumps' into fully developed large-amplitude ignitions immediately. At short residence times the boundary between regions (I) and (II) may show hysteresis: if T_a is reduced again the 'hard extinction' transition from oscillatory ignition to steady dark reaction occurs at a lower value than the upward transition. At longer residence times, however, this hysteresis disappears and the abrupt change between the two patterns occurs at the same ambient temperature in both directions.

(c) Oscillatory cool flames

It is easier if we discuss region (IV) before region (III). The regular oscillatory pulses of chemiluminescent reaction observed in this region correspond to the cool flames of closed vessels. In a CSTR, a sustained pattern is realized and allows a meaningful kinetic characterization. Some examples of these cool flames are shown in Fig. 15.13. The temperature excursion is always less than 200 K and may become as small as only 10 K at higher ambient temperatures. The chemical species formed during oscillatory cool flame oxidation are CO, H_2O, CH_2O, CH_3OH, and CH_4, with lower yields of CH_3CO_3H, ethane C_2H_6, and hydrogen peroxide H_2O_2. The emission which accompanies these excursions originates from electronically excited formaldehyde CH_2O^*: the emission typically does not fall to zero between the maxima.

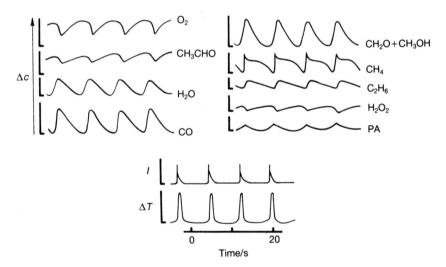

FIG. 15.13. Typical concentration, light emission, and temperature excess records for oscillatory cool flames during acetaldehyde oxidation in a CSTR.

In most reported studies, the oscillatory amplitude of cool flames diminishes smoothly as the ambient temperature is increased. At the same time there is a monotonic increase in the frequency. There have, however, also been observations of birhythmicity in this region, with the coexistence of large- and small-amplitude cool flames over some range of T_a. In all cases, the amplitude of the oscillations tends to zero at the highest ambient temperatures in this region; a supercritical Hopf bifurcation occurs at the boundary between regions (IV) and (V) and a stable, chemiluminescent, stationary state emerges.

(d) Complex ignition

In this region, the waveform contains both ignition and cool flame features. Some examples are shown in Fig. 15.14. A complete period consists of a number, n, of cool flame excursions and a two-stage ignition pulse. Thus we can talk of an $n + 2$ stage ignition. (In region (II) we could say that there $n = 0$.) Returning to region (III), then close to the boundary with region (II) the ignition excursions are interspersed by just one cool flame ($n = 1$ or three-stage ignition). As T_a is increased more cool flames appear between the

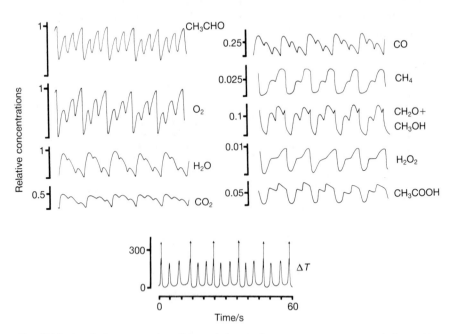

FIG. 15.14.   Typical concentration, light emission, and temperature excess records for multi-stage ignition during acetaldehyde oxidation in a CSTR.

ignitions. At present the record is a seven-stage ignition ($n = 5$) observed close to the boundary with region (IV).

(e) Steady glow

As mentioned above, region (V) lies at the highest temperatures. As it is entered from region (IV), by raising T_a, the oscillatory glow gives way to steady glow via a Hopf bifurcation. The chemiluminescence is still from excited CH_2O, and other partially oxidized products such as CH_3OH are formed as well as CO_2, CO, and H_2O. The stationary-state temperature rise can be up to 80 K, close to the boundary with region (IV). However, in contrast to the steady dark reaction mode, ΔT_{ss} decreases with increasing T_a in this region: the heat release rate has a negative temperature coefficient or n.t.c. (It is important to distinguish between heat release rate and reaction rate here; the n.t.c. arises primarily because the exothermicity of the reaction changes, decreasing as the ambient temperature increases, across this region.)

The stationary state is a stable focus: small perturbations decay by means of a damped oscillatory return.

15.5.2. Kinetic mechanism

The most successful numerical computations for the acetaldehyde + oxygen system involve 60 different elementary steps and allow for non-isothermal operation of the CSTR. Many of the steps in this large scheme are included to give a good quantitative match during the full ignition process. If we restrict our ambitions to a qualitative interpretation of the cool flame behaviour and the onset of ignition, a smaller scheme can be abstracted. Table 15.3 lists 25 steps for this purpose.

The first four reactions provide an initiation process. In particular, it leads to the production of methyl radicals CH_3. The important steps for the cool flame clockwork are (5) to (15), and of these the heart is the pair or reactions (6) and (7). The latter in fact constitute the forward and reverse steps of the addition process

$$CH_3 + O_2 \rightleftharpoons CH_3O_2. \tag{6,7}$$

It is this equilibrium, and its temperature dependence, which holds the key to the observed behaviour. At low reactant temperatures, the equilibrium lies to the right, in favour of CH_3O_2. This methylperoxy species is then further oxidized through reaction (11):

$$CH_3O_2 + CH_3CHO \rightarrow CH_3O_2H + CH_3CO. \tag{11}$$

This produces a CH_3CO to continue the chain, through step (5), and the species CH_3O_2H. The latter decomposes in a subsequent step to produce two

Table 15.3
Reduced kinetic model for acetaldehyde oxidation

(1)	$CH_3CHO + O_2$	$\rightarrow$	$CH_3CO + HO_2$
(2)	$CH_3CO + O_2$	$\rightarrow$	CH_3CO_3
(3)	$CH_3CO_3 + CH_3CHO$	$\rightarrow$	$CH_3CO_3H + CH_3CO$
(4)	CH_3CO_3H	$\rightarrow$	$CH_3 + CO_2 + OH$
(5)	$CH_3CO + M$	$\rightarrow$	$CH_3 + CO + M$
(6)	$CH_3 + O_2$	$\rightarrow$	CH_3O_2
(7)	CH_3O_2	$\rightarrow$	$CH_3 + O_2$
(8)	$CH_3 + CH_3$	$\rightarrow$	C_2H_6
(9)	$CH_3O_2 + CH_3O_2$	$\rightarrow$	$2CH_3O + O_2$
(10)	$CH_3O + CH_3O$	$\rightarrow$	$CH_3OH + CH_2O$
(11)	$CH_3O_2 + CH_3CHO$	$\rightarrow$	$CH_3O_2H + CH_3CO$
(12)	CH_3O_2H	$\rightarrow$	$CH_3O + OH$
(13)	$OH + CH_3CHO$	$\rightarrow$	$CH_3CO + H_2O$
(14)	$CH_3O + O_2$	$\rightarrow$	$CH_2O + HO_2$
(15)	$OH + CH_2O$	$\rightarrow$	$HCO + H_2O$
(16)	$HCO + M$	$\rightarrow$	$H + CO + M$
(17)	$HCO + O_2$	$\rightarrow$	$CO + HO_2$
(18)	$OH + CO$	$\rightarrow$	$CO_2 + H$
(19)	$H + O_2$	$\rightarrow$	$OH + O$
(20)	$H + O_2 + M$	$\rightarrow$	$HO_2 + M$
(21)	$O + CH_2O$	$\rightarrow$	$HCO + OH$
(22)	$HO_2 + CH_2O$	$\rightarrow$	$HCO + H_2O_2$
(23)	$HO_2 + HO_2$	$\rightarrow$	$H_2O_2 + O_2$
(24)	$H_2O_2 + M$	$\rightarrow$	$2OH + M$
(25)	$OH + H_2O_2$	$\rightarrow$	$HO_2 + H_2O$

radicals:

$$CH_3O_2H \rightarrow CH_3O + OH. \tag{12}$$

This process is known as 'degenerate branching', to distinguish it from a direct branching step such as $H + O_2 \rightarrow OH + O$. Because of the branching, the reaction rate increases. The methoxy radical reacts with O_2 through step (14), but the formaldehyde molecule produced is in an excited electronic state, allowing for the chemiluminescent emission observed experimentally. The overall reaction is exothermic, so the accelerating reaction leads to self-heating and higher reactant temperatures.

At higher temperatures, however, the equilibrium formed by steps (6) and (7) shifts to the left, in favour of the dissociated methyl and O_2. At high T, the reaction pathway involves simple methyl radical chemistry, principally the termination through the recombination reaction (8). The total radical concentration falls and, hence, so does the reaction rate. Finally, in the cycle, the temperature of the gas will fall back towards T_a either by Newtonian heat loss or by the outflow of heated gas. The equilibrium then moves back to the right, and the next cool flame can be initiated.

It is important that the above arguments involve both the chemical non-linearity of degenerate branching *and* the thermal feedback effects on the equilibrium steps. Neither a purely kinetic nor a purely thermal mechanism can explain the full complexity of acetaldehyde cool flames.

The final group of reactions, (16) to (25), becomes important as the system moves towards a full ignition. A linking role is played by an increasing formaldehyde concentration, through step (15), which produces the formyl radical HCO. This, in turn, leads to CO and then H atoms. Once hydrogen atoms have been produced, the key branching reaction (19) can occur. This scenario requires the ignition to be at least a two-stage event, with control in the first stage with reactions (5) to (14) and in the ignition phase with steps (16) to (25). If the cool flame stage does not produce a sufficiently high CH_2O concentration, the ignition steps do not come into play: one or more additional cool flame stages may be needed before the ignition stage is initiated, or indeed the latter may not occur at all.

15.5.3. Other hydrocarbon oxidations

Cool flame oscillations are characteristic of the spontaneous oxidation reactions of a great many hydrocarbon fuels. One particularly important effect of this mode of reaction is that of engine knock which occurs during the heating of petrol–air mixtures in the compression stage of the internal combustion cycle. If we denote a general hydrocarbon as RH, we may then hope to prepare a generalized version of Table 15.3.

An initiation process involving $RH + O_2$ can produce alkyl radicals of the form R, and again there is an important equilibrium of the form

$$R + O_2 = RO_2.$$

A degenerate branching sequence can then be proposed on similar principles to reactions (11) and (12) above, i.e.

$$RO_2 + RH \rightarrow RO_2H + R$$

$$RO_2H \rightarrow RO + OH.$$

In general, however, this is not a satisfactory explanation. With acetaldehyde, the –CHO group presents a particularly labile hydrogen atom. The energy barrier for the intermolecular H abstraction in step (11) is relatively low, approximately 44 kJ mol^{-1}. For other hydrocarbons, however, the relevant bond may be a primary, secondary, or tertiary C–H: the H abstraction then has a significantly higher activation energy, $E > 60 \, kJ \, mol^{-1}$, and so is too slow to account for observed cool flame behaviour.

As a more general route, intramolecular H atom abstraction appears to offer branching potential. In particular, abstraction from a β C–H bond permits the reaction to proceed through a favoured six-atom ring. Thus an alkyl radical may go through a sequence of the form

$$\overset{\displaystyle \cdot}{-CH}-CH_2-CH_2- \; + \; O_2 \quad \rightarrow \quad \overset{\displaystyle O-O\cdot}{\underset{\displaystyle |}{-CH}}-CH_2-CH_2-$$

$$\overset{\displaystyle O-O\cdot \quad H}{\underset{\displaystyle | \qquad \; |}{-CH}}-CH_2-CH- \quad \rightarrow \quad \overset{\displaystyle O_2H}{\underset{\displaystyle |}{-CH}}-CH_2-\overset{\displaystyle \cdot}{CH}-$$

$$\overset{\displaystyle O_2H}{\underset{\displaystyle |}{-CH}}-CH_2-\overset{\displaystyle \cdot}{CH}- \; + \; O_2 \quad \rightarrow \quad \overset{\displaystyle O_2H}{\underset{\displaystyle |}{-CH}}-CH_2-\underset{\displaystyle \underset{\displaystyle O_2}{|}}{CH}-$$

$$\overset{\displaystyle O_2H}{\underset{\displaystyle \underset{\displaystyle \underset{\displaystyle H \quad \cdot O-O}{|}}{|}}{-C}}-CH_2-CH- \quad \rightarrow \quad \overset{\displaystyle O_2H}{\underset{\displaystyle \underset{\displaystyle O_2H}{|}}{-\overset{\displaystyle \cdot}{C}}}-CH_2-CH-$$

$$\overset{\displaystyle O_2H}{\underset{\displaystyle \underset{\displaystyle \underset{\displaystyle O_2H}{|}}{|}}{-\overset{\displaystyle \cdot}{C}}}-CH_2-CH- \quad \rightarrow \quad -CO + CH_2-CH + 2OH.$$

This sequence of O_2 additions and intramolecular H atom abstractions finally leads to fragmentation of the fuel into smaller molecular species as well as providing degenerate branching. The detailed structure of the original hydrocarbon is also important as the rates of different steps will depend crucially on whether the particular H atom being abstracted is primary, secondary, or tertiary. Thus, there will be different kinetic implications for branched- or straight-chain hydrocarbons.

The alternative non-branching fate for the R radical is the formation of the conjugate alkene, P say, by

$$R + O_2 \rightarrow P + HO_2.$$

15.5.4. Gray and Yang model

Many of the qualitative features described above can be accounted for in terms of a simple model proposed by Gray and Yang in 1969 (Gray, 1969; Yang and Gray, 1969). This scheme involves chain-branching and thermal feedback and can be written in terms of the following steps:

(1) initiation $A \rightarrow X$ rate $= k_i a$

(2) branching $X \rightarrow 2X$ rate $= k_b x$

(3) termination $X \rightarrow S_1$ rate $= k_{t1} x$

(4) termination $X \rightarrow S_2$ rate $= k_{t2} x$.

Thus, there is an initiation process in which the fuel A produces a radical intermediate X. Step (2) is a branching reaction which competes with two termination steps (3) and (4). Conservation of mass actually requires at least one other species, perhaps O_2, to participate in the branching process, but this is not made explicit in the usual form of this model.

If the consumption of the reactant A is ignored (a pool chemical approximation, in the spirit of chapters 2 to 5), we need only consider one concentration—that of X. However, a second important variable arises crucially in this model. The various steps are taken to be exothermic and the reactor is assigned a finite heat transfer coefficient. Self-heating can occur and the reacting gas temperature may vary. The thermal feedback acts through the temperature dependence of the reaction rate constants, which have an Arrhenius form $k_i(T) = A_i \exp(-E_i/RT)$ etc.

The mass- and energy-balance equations for this scheme appropriate to a closed vessel can be written as

$$\frac{dx}{dt} = k_i a + \phi x \tag{15.20}$$

$$C\frac{dT}{dt} = q_i k_i a + \theta X - \chi(T - T_a). \tag{15.21}$$

Here ϕ is the net branching factor and θ a quantity related to the exothermicities q of the three steps (2)–(4):

$$\phi = k_b - k_{t1} - k_{t2} \qquad \theta = q_b k_b + q_{t1} k_{t1} + q_{t2} k_{t2}.$$

These are both functions of temperature through the rate constants. The quantities C and χ are the effective heat capacity and heat transfer coefficient respectively.

In the simplest form of this model, termination step (3) is taken to be a surface reaction with no activation energy and no exothermicity, i.e.

Table 15.4

Details of Gray and Yang Scheme for thermokinetic oscillations

(1) $A \rightarrow X$	$k_i = A_i N_0^2 \exp(-E_i/RT)$	$A_i = 1.6 \times 10^{10} \, \text{mol}^{-1} \, \text{cm}^3 \, \text{s}^{-1}$ $E_i = 24 \, \text{kcal mol}^{-1}$ $q_i = 0$
(2) $X \rightarrow 2X$	$k_b = A_b N_0 \exp(-E_b/RT)$	$A_b = 1.38 \times 10^8 \, \text{s}^{-1}$ $E_b = 7 \, \text{kcal mol}^{-1}$ $q_b = 4 \, \text{kcal mol}^{-1}$
(3) $X \rightarrow S_1$	$k_{t1} = A_{t1} N_0^{1/2}/d$	$A_{t1} = 3.3 \times 10^3 \, \text{mol}^{1/2} \, \text{cm}^{-1/2} \, \text{s}^{-1}$ $E_{t1} = 0$ $q_{t1} = 0$
(4) $X \rightarrow S_2$	$k_{t2} = A_{t2} N_0 \exp(-E_{t2}/RT)$	$A_{t2} = 7.8 \times 10^{10} \, \text{s}^{-1}$ $E_{t2} = 16 \, \text{kcal mol}^{-1}$ $q_{t2} = 20 \, \text{kcal mol}^{-1}$
(5) $A + X \rightarrow 2X$	$k_{b2} = A_{b2} \exp(-E_{b2}/RT)$	$A_{b2} = 3.7 \times 10^{12} \, \text{mol}^{-1} \, \text{cm}^3 \, \text{s}^{-1}$ $E_{b2} = 25 \, \text{kcal mol}^{-1}$ $q_{b2} = 92 \, \text{kcal mol}^{-1}$
$C = 11 N_0 \, \text{cal cm}^{-3} \, \text{K}^{-1}$		$\chi = 1.89 \times 10^{-4} \, \text{cal cm}^{-3} \, \text{K}^{-1} \, \text{s}^{-1}$

$E_{t1} = q_{t1} = 0$, and k_{t1} is taken to be proportional to the square root of the total pressure and to the inverse of the vessel diameter. The initiation step can also be considered to be virtually thermoneutral, so $q_i \approx 0$.

If $N_0 = p_0/RT_a$ represents the total initial concentration of reactants, the functional forms and values for the thermokinetic parameters suggested by Gray and Yang are given in Table 15.4 (the extra step (5) will be discussed later).

The choice here of the inequality $E_{t2} > E_b > E_{t1}$ is particularly significant as we will see below.

(a) Stationary states

The stationary-state radical concentration and temperature are given by solutions of the equations

$$x_{ss} = -k_i a/\phi(T_{ss}) \tag{15.22}$$

$$q_i k_i a + \theta(T_{ss}) x_{ss} - \chi(T_{ss} - T_a) = 0. \tag{15.23}$$

The form of eqn (15.22) shows that T_{ss} must be such that $\phi(T_{ss})$ is negative. These relationships are most easily used parametrically, i.e. given a particular value of T_{ss} the appropriate ambient temperature and radical concentrations can then be calculated.

For given T_a etc., eqns (15.22) and (15.23) may have multiple solutions, corresponding to multiple stationary states. Thus we may expect jumps between different solutions as the ambient temperature is varied. Also, we have a pair of highly non-linear rate equations, so oscillatory solutions arising from Hopf bifurcation may also be possible.

(b) Thermal diagram

A convenient way of representing the behaviour of this model is to use a thermal diagram—an equivalent construction to the flow diagrams used in chapter 1 and elsewhere. Eliminating the radical concentration x_{ss} between eqns (15.22) and (15.23), the stationary-state condition can be written in the form

$$q_i k_i a - \frac{k_i a \theta(T_{ss})}{\phi(T_{ss})} = \chi(T_{ss} - T_a). \qquad (15.24)$$

$$\underbrace{}_{R_i} \quad \underbrace{\phantom{\frac{k_i a \theta(T_{ss})}{\phi(T_{ss})}}}_{R} \quad \underbrace{\phantom{\chi(T_{ss} - T_a)}}_{L}$$

If we plot the left-hand side $(R_i + R)$ of this equation as a function of the reacting gas temperature T, we obtain a locus of the form shown in Fig. 15.15. This sum represents the total rate of heat release at the stationary state. There is an initial exponential rise in $R_i + R$ as T increases, but eventually the locus shows a maximum, followed by a minimum. Thus, for the branch of the curve between the extrema, this model exhibits a region of *negative temperature coefficient* (n.t.c.) in which the rate of heat release decreases with increasing temperature. This arises because the high activation energy (termination step (4)) becomes dominant over branching at these temperatures. There is, however, an ultimate increase in R; even without net branching the reaction is exothermic and thermal runaway can occur.

The right-hand side of eqn (15.24) describes a straight line with a gradient given by the heat transfer coefficient χ and an intercept corresponding to the ambient temperature T_a. Stationary-state solutions are located by the intersections of R and L. Again we will see that the conditions for which R and L become tangential are of great importance.

If the ambient temperature is low or if the heat transfer conditions are such that L is steep, the first intersection of the heat generation and loss lines corresponds to only small extents of self-heating and to a low radical concentration. Such a situation holds for $T_{a,1}$ in Fig. 15.15: the intersection marked I is typical of a slow reaction state. The second intersection, marked II, is not a stable stationary-state solution: it is a saddle point. Other intersections between R and L are possible, and these will make their presence felt soon.

If the ambient temperature is increased to $T_{a,2}$, the heat release and heat loss rate R and L become tangential: the lowest two intersections merge at a saddle–node bifurcation Beyond this, the system may jump to a higher

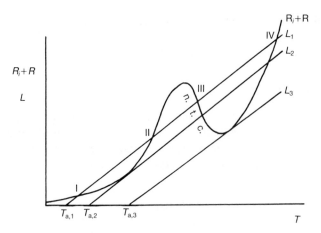

FIG. 15.15. Variation of the heat generation rate $R_i + R$ and the heat transfer rate L for the Gray and Yang model for hydrocarbon oxidation with three different ambient temperatures. The third intersection III lies on a region of negative temperature coefficient (n.t.c.).

intersection such as III. This lies on, or close to, the branch of R corresponding to the n.t.c. There is much more self-heating and a higher radical concentration in this state. We will discuss the stability of this solution below. If the ambient temperature is raised even further, e.g. to $T_{a,3}$, intersections III and IV merge. There are no stationary-state solutions within the range of Fig. 15.15 (although one corresponding to an ignited state with an unrealistically high temperature excess exists).

(c) Stability of stationary states

A typical variation in local stability of the stationary-state intersections along the different branches of the heat release rate curve R is indicated in Fig. 15.16. For the lowest branch I, the solutions are stable nodes. Those corresponding to intersections II and IV are saddle points. It is the third intersection, in the vicinity of the n.t.c., that is of most interest. The solutions lowest along this (decreasing) branch, i.e. those with highest temperatures, are generally stable, either nodes or foci. As we move up this branch, to lower gas temperatures, however, there may be a Hopf bifurcation. Analysis reveals that the bifurcation is supercritical for the parameter values in Table 15.4; a stable limit cycle emerges and grows as we move towards the maximum.

This course of events, with oscillatory behaviour and an n.t.c., parallels the experimental observations of cool flames. If we continue to move up this third branch, the limit cycle associated with the oscillations grows and may ultimately give rise to the formation of a homoclinic orbit. In fact, there are two possibilities here: the homoclinic orbit can be formed with either of the

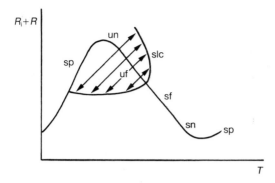

FIG. 15.16.   Typical variation in local stability of the third stationary-state intersection along the branch of negative temperature coefficient. There is a supercritical Hopf bifurcation as we move up the branch from which a stable limit cycle, corresponding to cool flame oscillations, emerges.

two saddles, II or IV. If the homoclinic orbit is formed with the lower saddle, the oscillatory response dies and the system jumps back to the lowest stationary state I. On the other hand, if the limit cycle collides with the saddle IV, the oscillation gives way to an excursion to high gas temperatures, typical of an ignition. The present model cannot, however, treat this latter state at all well: reactant consumption must clearly be vitally important during the ignition.

(d) Influence of total pressure

The pressure of the reacting gases is an important parameter. It helps determine the relative positions of many of the features discussed above. For instance, an important question is that concerning the first tangency (merging of intersections I and II) and the Hopf bifurcation point on the third branch. If the continuation of the tangent heat loss line L cuts the third branch above the Hopf point, intersection III to which the system is moving will be unstable. We then have the possibility of the jump taking us on to a stable limit cycle: there is a jump to cool flames. On the other hand, if the tangent loss line cuts the third branch below the Hopf point, we can expect only damped oscillations at most (but see below).

A further feature is that the third branch may even become completely hidden, as shown in Fig. 15.17. Once the tangency at the first and second intersections has been passed the system may then jump straight into a full ignition.

Finally, the total pressure can influence which of the two branches of saddles the cool flame limit cycle forms its homoclinic orbit with.

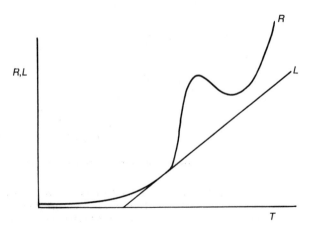

FIG. 15.17.   The thermal diagram for the Gray and Yang model at higher pressures for which
the region of negative temperature coefficient is not accessible through tangency.

(e) Open systems

The above discussion of the Gray and Yang model is in the spirit of
a closed vessel experiment. Some difficulty arises in interpreting the results
when the inevitable effects of reactant consumption are admitted. In fact, as
observed experimentally, such effects are particularly important in this system: no long trains of almost undamped oscillations are found here. Thus it is
impossible to make any meaningful distinction between the limit cycle behaviour and the damped oscillatory approach to a stable focus on the third
branch in the model. When we go to open systems, however, explicit differences emerge. Also, we can imagine varying the ambient temperature both up
and down during a series of experiments to explore the possible hysteresis
effects.

The mass- and energy-balance equations are not altered significantly if
reactant consumption is still ignored: the outflow of radicals appears as an
extra termination step in the net branching factor and the net outflow of
energy with the heated gases increases the gradient of the loss line L. Thus
R and L have qualitatively the same form as those in Figs 15.15 to 15.17. The
fully ignited state, intersection V, will still not be modelled well. Nevertheless,
we can see the origin of the transitions between the slow reaction state and
the steady glow (when intersection III is a stable focus) or cool flames (when
III is an unstable focus surrounded by a stable limit cycle). Reducing the
ambient temperature in either of these cases will then traverse the region of
n.t.c. The system will not return to the lowest branch before exhibiting
hysteresis.

Qualitatively, we can also explain the onset of ignition as a transition from cool flames as the ambient temperature is descreased: the cool flame limit cycle forms a homoclinic orbit with the saddle intersection IV. If the upper state V is also unstable there may be a large-amplitude limit cycle to settle on to corresponding to oscillatory ignition. On decreasing T_a further, this latter limit cycle may form a homoclinic orbit with the saddle point II, giving an extinction back to the slow reaction state I.

(f) Complex oscillations and ignition

The simple Gray and Yang scheme does not account for the ignition process because reactant consumption is ignored. It also cannot account for complex ignitions: the model as discussed so far has two variables, so only simple period-I limit cycles can occur. Wang and Mou (1985) have gone some way towards improving the model in both these respects. They have explicitly included fuel consumption (but not that of O_2). This raises the model to a three-variable scheme. Also, they have introduced an additional step to the mechanism:

(5) high-temperature branching $A + X \rightarrow 2X$ rate $= k_{b2}ax$.

This step involves a direct branching between the fuel and the radical. It is exothermic and has a higher activation energy than the other steps.

The reaction rate and energy-balance equations given by Wang and Mou can be written as

$$\frac{da}{dt} = k_f(a_0 - a) - k_i a - k_{b2}ax \tag{15.25}$$

$$\frac{dx}{dt} = -k_f x + k_i a + \phi x + k_{b2}ax \tag{15.26}$$

$$C\frac{dT}{dt} = q_i k_i a + \theta x + q_{b2}k_{b2}ax - \chi(T - T_a). \tag{15.27}$$

(Apparently these equations have no heat transfer term associated with the outflow. If the flow rate is not going to be varied during a given experiment, however, this extra effect can be included by redefining χ.)

Using the data in Table 15.4, a p–T_a ignition diagram of the correct qualitative form can be predicted. Wang and Mou obtained an improvement in the fit with experiment by varying the rate constants slightly, and it should be noted that the Gray and Yang set are not derived from known rate data. Most significantly, this model predicts complex oscillatory waveforms, of the general form 1^n, with successive large excursions separated by a number, n, of small cool flames. These responses exist over a relatively narrow range of parameter values, e.g. 586–590. 1 K for a total pressure of 560 mm Hg. The complex ignition region is bounded at low ambient temperature by simple oscillatory ignition and by the cool flame region at high T_a, as observed experimentally.

A close examination of the behaviour in the region of complex oscillations reveals a wealth of fine structure. If we traverse the region by increasing the ambient temperature, so entering from the region of simple (two-stage) ignition, then the waveform develops through the sequence $1^1, 1^3, 1^5, \ldots$, i.e. 1^n with n odd and increasing. After a region of apparent irregularity, a different sequence with $1^8, 1^6, 1^4$, and 1^2 (n even and decreasing as T_a increases) emerges. Finally, there is a small range of T_a for which the oscillation has a high periodicity of about 1^{12}, found close to the boundary with the cool flame region IV.

These latter details have not been confirmed experimentally. Gray et al. (1981) observed simply that n increases through the sequence 1,2,3,4 as T_a is increased. Thus, there may be some discrepancy between what is after all a very simplified model and the real acetaldehyde kinetics, but clearly the model does capture much of the essence of hydrocarbon oxidations.

15.6. Summary

The class of chemical oscillators is frequently supposed to consist exclusively of inorganic solution-phase systems such as the B–Z reaction and its derivatives. Many classifications of the latter can be found. In this chapter we have presented three examples of another set of chemical oscillators—those which take place in the gas phase. They are clearly chemical systems and are also exemplary as oscillators. In fact, almost all of the non-linear behaviour found with solution-phase systems has been similarly observed, and modelled, in the gas phase (there is perhaps so far no observation equivalent to the pattern formation of the B–Z in in unstirred media). For the three reactions presented here, the chemical mechanisms are certainly as well established and understood as the B–Z system, perhaps even more so.

If we have helped here to set the record straight concerning gas-phase systems, we should also at least mention other oscillatory manifestations. Chapter 12 describes the theory of oscillations and multistability in heterogeneous catalysis: the prime example here again involves carbon monoxide oxidation, generally over platinum or palladium in a variety of experimental realizations. In heterogeneous systems, furthermore, the electrodissolution of metal electrodes, such as iron into acidic solution, has provided a rather unexpected source of periodic and aperiodic behaviour. Various techniques, including those described in this book and others, have been brought to bear on these systems. It would take more space (and energy) than we have here to discuss these situations in the depth they deserve, so we leave that for another day, and another book!

References

Baldwin, R. R., Jackson, D., Walker, R. W., and Webster, S. J. (1965). Use of the $H_2 + O_2$ reaction in evaluating velocity constants. *10th Int. Symp. on. Combustion.*, pp. 423–33. Combustion Institute, Pittsburg, PA.

Baldwin, R. R., Fuller, M. E., Hillman, J. S., Jackson, D., and Walker, R. W. (1974). Second limit of hydrogen + oxygen mixtures: the reaction $H + HO_2$. *J. Chem. Soc. Faraday Trans. I*, **70**, 635–41.

Baulch, D. L., Griffiths, J. F., Pappin, A. J., and Sykes, A. F. (1988). Stationary-state and oscillatory combustion of hydrogen in a well-stirred flow reactor. *Combust. Flame*, **73**, 163–85.

Baulch, D. L., Griffiths, J. F., Pappin, A. J., and Sykes, A. F. (1988). Third-body interactions in the oscillatory oxidation of hydrogen in a well stirred flow reactor. *J. Chem. Soc. Faraday Trans. I*, **84**, 1575–86.

Bond, J. R., Gray, P., Griffiths, J. F., and Scott, S. K. (1982). Oscillations, glow and ignition in carbon monoxide oxidation II. Oscillations in the gas-phase reaction in a closed system. *Proc. R. Soc.*, **A381**, 293–314.

Chinnick, K., Gibson, C., Griffiths, J. F., and Kordylewski, W. (1986). Isothermal interpretations of oscillatory ignition during hydrogen oxidation in an open system. *Proc. R. Soc.*, **A405**, 117–28.

Dixon-Lewis, G., and Williams, D. J. (1977). In *Comprehensive chemical kinetics*, (ed. C. H. Bamford and C. F. H. Tipper), Vol. 17. Elsevier, Amsterdam.

Gibson, C., Gray, P., Griffiths, J. F., and Hasko, S. M. (1984). Spontaneous ignition of hydrocarbons and related fuels: a fundamental study of thermokinetic interactions. *20th. Int. Symp. on Combustion*, pp. 101–9. Combustion Institute, Pittsburgh, PA.

Gonda, I., and Gray, B. F. (1983). The unified thermal and chain branching model of hydrocarbon oxidation in a well stirred continuous flow reactor. *Proc. R. Soc.*, **A389**, 133–52.

Gray, B. F. (1969). Unified theory of explosions, cool flames and two-stage ignitions, part 1. *Trans. Faraday Soc.*, **65**, 1603–13.

Gray, B. F. and Jones, J. C. (1984). The heat release rates and cool flames of acetaldehyde oxidation in a continuously stirred tank reactor. *Combust. Flame*, **57**, 3–14.

Gray, P., and Scott, S. K. (1985). Isothermal oscillations and relaxation ignitions in gas-phase reactions: the oxidations of carbon monoxide and hydrogen. In *Oscillations and traveling waves in chemical systems*. (ed. R. J. Field and M. Burger), ch. 14, pp. 493–528. Wiley, New York.

Gray, P., Griffiths, J. F., Hasko, S. M., and Lignola, P.-G. (1981). Oscillatory ignitions and cool flames accompanying the non-isothermal oxidation of acetaldehyde in a well stirred flow reactor. *Proc. R. Soc.*, **A374**, 313–39.

Gray, P., Griffiths, J. F., Hasko, S. M., and Lignola, P.-G. (1981). Novel, multiple stage ignitions in the spontaneous combustion of acetaldehyde. *Combust. Flame*, **43**, 175–86.

Gray, P., Griffiths, J. F., and Scott, S. K. (1984). Branched-chain reactions in open systems: theory of the oscillatory ignition limit for the hydrogen + oxygen reaction in a continuous-flow stirred-tank reactor. *Proc. R. Soc.*, **A394**, 243–58.

Gray, P., Griffiths, J. F., and Scott, S. K. (1985). Oscillations, glow and ignition in carbon monoxide oxidation in an open system, I. Experimental studies of the ignition diagram and the effects of added hydrogen. *Proc. R. Soc.*, **A397**, 21–44.

Gray, P., Griffiths, J. F., and Scott, S. K. (1985). Oscillations, glow and ignition in carbon monoxide oxidation in an open system, II. Theory of the oscillatory ignition limit in the c.s.t.r. *Proc. R. Soc.*, **A402**, 187–204.

Gray, P., Griffiths, J. F., Pappin, A., and Scott, S. K. (1987). The interpretation of oscillatory ignition during hydrogen oxidation in an open system. In *Complex chemical reaction systems*, (ed. J. Warnatz and W. Jäger), pp. 150–9. Springer, Berlin.

Griffiths, J. F. (1985). Thermokinetic interactions in simple gaseous reactions. *Ann. Rev. Phys. Chem.*, **36**, 77–104.

Griffiths, J. F. (1985). Thermokinetic oscillations in homogeneous gas-phase oxidations. In *Oscillations and traveling waves in chemical systems*, (ed. R. J. Field and M. Burger), ch. 15, pp. 529–64. Wiley, New York.

Griffiths, J. F. (1986). The fundamentals of spontaneous ignition of gaseous hydrocarbons and related organic compounds. *Adv. Chem. Phys.*, **64**, 203–303.

Griffiths, J. F., and Scott, S. K. (1987). Thermokinetic interactions: fundamentals of spontaneous ignition and cool flames. *Prog. Energy Combust. Sci.*, **13**, 161–97.

Griffiths, J. F., and Sykes, A. F. (1989). Numerical studies of a thermokinetic model for oscillatory cool flame and complex ignition phenomena in ethanal oxidation under well-stirred flowing conditions. *Proc. R. Soc.*, **A422**, 289–310.

Harding, R. H., Sevcikova, H., and Ross, J. (1988). Complex oscillations in the combustion of acetaldehyde. *J. Chem. Phys.*, **89**, 4737–42.

Lignola, P.-G. and Reverchon, E. (1987). Cool flames. *Prog. Energy Combust. Sci.*, **13**, 75–96.

Pugh, S. A., Schell, M., and Ross, J. (1986). Effects of periodic perturbations on the oscillatory combustion of acetaldehyde. *J. Chem. Phys.*, **85**, 868–78.

Pugh, S. A., DeKock, B., and Ross, J. (1986). Effects of two periodic perturbations on the oscillatory combustion of acetaldehyde. *J. Chem. Phys.*, **85**, 879–86.

Pugh, S. A., Kim, H.-R., and Ross, J. (1987). Measurements of [OH] and [CH₃CHO] oscillations and phase relations in the combustion of CH₃CHO. *J. Chem. Phys.*, **86**, 776–83.

Wang, X.-J. and Mou, C. Y. (1985). A thermokinetic model of complex oscillations in gaseous hydrocarbon oxidation. *J. Chem. Phys.*, **83**, 4554–61.

Yang, C. H. (1969). Two-stage ignition and self-excited thermokinetic oscillation in hydrocarbon oxidation. *J. Phys. Chem.*, **73**, 3407–13.

Yang, C. H. and Gray, B. F. (1969). On the slow oxidation of hydrocarbon and cool flames. *J. Phys. Chem.*, **73**, 3395–406.

Yang, C. H. and Gray, B. F. (1969). Unified theory of explosions, cool flames and two stage ignitions, part 2. *Trans. Faraday Soc.*, **65**, 1614–22.

Index